CHEMISTRY RESEARCH AND APPLICATIONS

IONIC LIQUIDS

SYNTHESIS, CHARACTERIZATION AND APPLICATIONS

CHEMISTRY RESEARCH AND APPLICATIONS

Additional books in this series can be found on Nova's website
under the Series tab.

Additional e-books in this series can be found on Nova's website
under the e-book tab.

CHEMISTRY RESEARCH AND APPLICATIONS

IONIC LIQUIDS

SYNTHESIS, CHARACTERIZATION AND APPLICATIONS

ANSON BROOKS
EDITOR

nova publishers
New York

For permission to use material from this book please contact us:
Telephone 631-231-7269; Fax 631-231-8175
Web Site: http://www.novapublishers.com

Library of Congress Cataloging-in-Publication Data

Ionic liquids (Nova Science Publishers)
Ionic liquids : synthesis, characterization and applications / [edited by] Anson Brooks.
pages cm. -- (Chemistry research and applications)
Includes bibliographical references and index.
ISBN: 978-1-63117-716-3 (hardcover)
1. Ionic solutions. I. Brooks, Anson, editor of compilation. II. Title.
QD541.I55 2014
541'.372--dc23

2014011753

Published by Nova Science Publishers, Inc. † New York

CONTENTS

PREFACE

Research into ionic liquids (ILs) has been a very rapidly growing discipline in recent years. ILs have attracted very considerable attention because of their unique properties, which may be useful in new processes and technologies. They have already been studied for a variety of applications such as alternative solvents in organic synthesis and catalysis, reaction media for biocatalysis and biotransformation, in the separation sciences, electrolytes in batteries and solar cells, alternative lubricants, and as media for tissue preservation. ILs have excellent solvation properties, are thermally, chemically and electrochemically stable, and their vapour pressure is negligible. Nevertheless, the same properties that make them attractive replacements for other volatile industrial compounds may render them hazardous to ecosystems. This book discusses several topics that include the pharmaceutical aspect of ILs; the application of ILs to lignin extraction and depolymerization; recent developments in ionic liquid toxicity assessments; the role temperature, irradiation and oxidation has on the stability of ILs; and analytical methods for determining the constituents of ILs.

Chapter 1 - In the last decades, ionic liquids (ILs) progressed from chemical curiosities to interesting biological compounds apprehending the attention of researchers of distinct areas, from chemistry to pharmacology.

Initially, ILs were explored as materials for diverse applications due to the possibility of synthesize compounds with targeted chemical properties combined with selected physical properties. More recently, the emergence of ILs with biological activity revolutionized the scientific focus of these compounds and opened interesting perspectives regarding their pharmaceutical application. From the pharmaceutical point of view, an IL approach, in the design of novel active pharmaceutical ingredients (APIs), appears to be appropriate as it enables the chemical manipulation of the compounds with specific objectives related with the manufacturing process, the stability of the formulation, bioavailability and eventual adverse effects. Furthermore, even though crystallinity confers advantages during isolation, processing and storage of the drug, it is known that solid forms of APIs often suffer from low solubility and polymorphic conversion which can influence negatively the bioavailability of the drug and ultimately its therapeutic effect.

On a distinct perspective, the possibility of engineering the properties of ILs by manipulating anion-cation combinations, in association with their solvent properties and in some cases water-miscibility, are considered promising characteristics regarding the applicability of ILs as solvents or carriers of pharmaceutical drugs.

In this chapter it is intended to expose the pharmaceutical potential of ILs through the discussion of their utilization either as APIs or solvents/carriers of pharmaceutical drugs. The discussion will be centered on the benefits of the IL approach for the development of novel drug candidates considering not only physico-chemical aspects but also the pharmaceutical profile of the developed active pharmaceutical ingredients with IL properties. Considering the utilization of ILs as solvents of drugs or as part of drug delivery systems, it is anticipated that the discussion will be focused on the efficiency and toxicity of the systems and on their influence on the pharmacokinetics and pharmacodynamics of the vehiculated drugs. It is also planned to debate the motivations for the pharmaceutical usage of ILs as well as their peculiar properties that launched them in this context. Currently, the pharmaceutical utilization of ILs is one of the most relevant applications of these solvents, with impact on the safety and effectiveness of the involved processes and with benefits in terms of pharmaceutical formulations and pharmacological activity. It is the author's belief that this chapter is adequate for the book in project and can greatly enhance its acceptance and interest by readers in distinct research areas.

Chapter 2 - Today, in the chemical industry, solvents are used in large quantities. Therefore, solvents establish a significant part of the environmental performance of a process and also impact on safety and health issues. In order to get greener processes especially in the case of new biorefineries it is essential the use of green solvents. Ionic liquids as alternative solvents and catalyst are investigated in the area of biomass deconstruction to obtain valuable chemicals, in order to minimize the impact of the use of volatile organic solvents in the environment.

Until now, the ionic liquids were generally applied in the cellulose solubility and its conversion into a biofuel among others applications. However, lignin is also a key target to take into a count as a renewable source of chemicals and several studies are gaining in importance in the last years. The lignin being the most abundant residue in the paper industries, due to its structure can be developed as aromatics source.

With this purpose, the traditional delignification processes like alkali and organosolv were compared with ionosolv in order to obtain superior quality lignin, suitable to be depolymerised. It was found that ionosolv lignin has greater purity compared to organosolv and alkali lignins, avoiding the attached hemicelluloses and sulphate salts in case of alkali lignins. The recyclability of the ionic liquid was also studied to assess the economic cost of the process. In addition, they were applied as solvent and catalyst in order to depolymerize lignin to valuable products, as vanillin and syringaldehide.

Chapter 3 - Room temperature Ionic Liquids (ILs) have received high interest in academia within the last decades due to their use as catalysts or co-reactants in various organic and inorganic syntheses as well as in polymer and material science. Based on the inert nature of the Ionic Liquids, in most of the reactions ILs undergo no measurable conversion and, hence, become attractive even for industry. In especially, systems that require solely catalytic IL amounts are potential candidates for further investigations on the industrial sector.

Subject of this chapter is to show the use of Ionic Liquids as catalysts for the recently reported imine base / isocyanate (IBI) mediated radical polymerization of methyl methacrylate (MMA). Investigations on this quaternary reactive system allowed both an enhancement of the polymerization properties of the IBI mixtures and the gain of deeper insights to the reaction mechanism of this novel radical polymerization type. Typical IBI mixtures initiate the radical polymerization at slightly elevated temperatures or by adding high molar amounts of the imine base and the iscyanate compared to the monomer equivalent; the addition of Ionic Liquid traces to the polymerization mixture gives poly-merization of the monomer even at room temperature for all IBI mixtures and an enormous increase in the brutto reaction rate.

Compared to the non-IL containing reaction mixture, the average molecular weight of PMMA isolated from the catalyzed polymerization is significantly increased.

There are no experimental observations that the Ionic Liquids are bonded to the resulting polymer chains and ILs can be obtained from the reaction mixture almost quantitatively, thus their interaction as catalyst at the IBI polymerization is indicated. However, the IL component is measured to be part of the IBI brutto polymerization rate law which underlines its significance on the whole polymerization process.

Mechanistic studies on the IL catalyzed IBI polymerization have solidified the understanding of the initial step of this reaction. To monitor the interaction of the Ionic Liquid with MMA and the individual IBI components, band shifts at the attenuated total reflection Fourier transformation middle infrared spectrums (ATR FT MIR) are investigated and comparison with the appropriate Kamlet-Taft Parameters is done that reveal the activation of isocyanate and MMA sites by the IL as a pre-step to the IBI radical formation.

Chapter 4 - Ionic liquids (ILs) have demonstrated an enormous potential in the electrodeposition of various materials including metals, alloys and oxides, providing opportunities to overcome the limitations of water-based solvents by extending the working electro-chemical window and increasing current efficiencies. Also of relevance is the capability to tailor the redox chemistry and to control the nucleation characteristics, promoting nano-sized crystalline structures in the deposited layers. In addition, reduced health and environmental impacts are possible due to the low volatility of ILs, while maintaining suitable properties such as high chemical and thermal stability, paramount to a variety of industries to which electrodeposition is essential and that face stringent environmental regulations.

In this work electrochemical aspects of electrodeposition of gold using ILs are examined. Recent research has suggested a number of energy related applications for gold in batteries, fuel cells and hydrogen fuel processing, which include not only the need for coatings on less noble substrates but also the incorporation of gold as an electroactive deposit. The use of ionic liquids to engineer / develop new gold structures is a valid trend to investigate.

Gold deposition from a solution of 1-butyl-1-methylpyrrolidinium dicyanamide (BMP-DCA) containing $HAuCl_4$ and water is reported. The study includes the electrochemical characterization of the solvent, emphasizing admissible water contents and conditions for the deposition of thin films under laboratory atmospheric conditions. The electrochemistry of the reduction process itself is characterized by cyclic voltammetry as irreversible and under diffusion control. Diffusion coefficients of 1.15×10^{-7} cm^2 s^{-1} (293 K) and 2.60×10^{-7} cm^2 s^{-1} (333 K) for the Au^{3+} ionic species are estimated.

Electrodeposition conditions for gold nanocrystalline layers on copper, nickel and glassy carbon substrates are discussed and its relation with microstructure is established.

This is followed by the performance characterization of selected deposits in a phosphate buffer solution (pH 7.3) where well-defined voltammetric peaks associated with the oxygen chemisorption and reduction is found in all the substrates. Finally, the electrochemical behavior and stability of the gold deposited surfaces are fully addressed.

Chapter 5 - For Zn-anode rechargeable batteries, there are a number of shortcomings associated with using traditional KOH aqueous electrolytes. These include drying-out of the electrolyte due to water evaporation and dendrite formation at the anode, which severely impair battery performance (e.g., cycle life and capacity) and limit their application. It is, therefore, critical to either modify conventional KOH aqueous electrolytes or explore alternative electrolytes to eliminate these bottlenecks to the development of a feasible Zn-anode rechargeable battery system. Room temperature ionic liquids (RTILs) in recent years have been increasingly recognized as potential electrolytes or electrolyte components for rechargeable batteries. Compared with alkaline electrolytes, a simple electrolyte system composed of an RTIL as the sole component faces the challenge of improving its low conductivity before it can be practically applied in a battery. In this chapter, water and/or dimethyl sulfoxide modified RTILs (composed of pyrrolidinium or imidazolium cations and bis(trifluoromethanesulfonyl)imide or dicyanamide anions), with improved conductivity and Zn redox kinetics, are investigated as potential electrolyte systems for Zn-anode rechargeable batteries. In addition, with the capability of modifying metal deposit morphology, RTILs are found to be beneficial for depressing Zn dendrite formation in KOH aqueous electrolytes. An electrolyte, composed of 9M KOH + 5 wt% ZnO with a hydrophilic RTIL (0.5 wt% 1-ethyl-3-methylimidazolium dicyanamide), appears to be a promising electrolyte system for Zn-anode rechargeable batteries.

Chapter 6 - Due to the global challenge of the face of depleting fossil carbon resources and growing concerns about environmental issues, the increasing focus has paid on the conversion of the renewable lignocelluloses biomass into biomaterials, chemicals and energy to replace fossil-based products. Recently, ionic liquids (ILs) as desirable green solvents and reaction media have attracted increased interest in the lignocellulose biomass conversion because of their non-volatile, non-flammable, non-corrosive and thermally stable properties. In this chapter, the pretreatment and fractionation of lignocellulosic biomass with ILs as green solvents are summarized up, which are promising and challenging processes in a lignocellulosic biomass biorefining industry. The fractioned main components such as cellulose, hemicelluloses and lignin can efficiently converted into biopolymers and biomaterials with functional properties in ILs as non-derivatizing solvents. Main components derivatives with high DS and high yields could be synthesized by esterification, etherification, oxidation and grafting copolymerization. Moreover, another important focus is discussed on the synthesis of chemicals with low M_w from lignocellulosic biomass in ILs media. The important platform chemicals such as furfural, 5-hydroxymethyl furfural, levulinic acid, 5-ethoxy-methyl-2-furfural, and lignin-depolymerized compounds can be obtained from main components of lignocellulosic biomass using ILs as green reaction media or high efficient catalyst. The difficulties encountered in the process of lignocelluosic biomass conversion using ILs are critically discussed. ILs display the great potential for application on a commercial scale due to their broad structural diversity, high dissolution ability, great thermal stability and recyclability.

Chapter 7 - In respond to the growing need for sustainable development, the efficient utilization of biomass resources is becoming increasingly important. Lignocellulose biomass mainly composed of cellulose, hemicellulose and lignin is an abundant but highly underutilized renewable resource. The limited types of solvents that readily dissolve ligocellulose, the lack of an environmental-friendly method to extract lignocellulose components, and the difficulty in efficiently modifying lignocellulose are the major reasons that limit the further application of lignocellulose biomass. Since ionic liquid had emerged as the functional solvent and reaction media of biomass, the dissolution, fraction and modification of lignocellulose biomass in ionic liquid have been intensively studied in the past decade. To date, hundreds of papers have been published on these topics. ILs have provided a new platform for the "green" and efficient conversion and utilization of lignocellulose resources. In this chapter, the applications of ILs in the dissolution, fraction and chemically modification of lignocellulose biomass will be systematically introduced. The scientific mechanisms behind the dissolution and fraction of lignocellulose will be explained. The latest progress in applications of ILs in modification of lignocellulose will be introduced to the readers.

Chapter 8 - The on-going development and use of ionic liquids (ILs) require better understanding of their effects on the environment. A considerable amount of data on the toxicity of ILs towards the aquatic and terrestrial environments and also to different trophic levels, including primary producers (e.g., bacteria, algae and plants), primary consumers (e.g., invertebrates) and secondary consumers (e.g., fish) has been published over the past decade. Toxicity has been evaluated on the basis of test systems at different levels of biological complexity, from molecules and cells to organisms. A wide variety of IL species with different head groups, alkyl and functionalized side chains and anions has been investigated to establish not only their overall toxic effect but also the influence of structural elements on their biological activity. This chapter reviews the achievements and current status of environmental risk assessment of ILs. The problem how ecotoxicological experiments that consider diverse mixtures of organisms, abiotic conditions and additional stressors verify environmentally "safe" concentrations of ILs is discussed.

Chapter 9 - ILs show unique properties, which are useful for their application in industry. They are very promising solvents and reagents, however it has to be kept in mind that in practice ILs should be stable under application conditions. After used the ionic liquids should be recovered and reused. Therefore the knowledge of their resistant to temperature, irradiation, oxidizing and reducing agents and activity lose during application is necessary. The attention must also be paid to the purity of ILs used to chemical, photochemical and electrochemical processes, due to their lower stability in presence of small amount of water and mineral acids.

In the review biomedical, industrial and scientific applications of ILs were presented. The influence of operating conditions such as: temperature, the type of irradiation, the redox potential, electrical current and purity of IL on its stability during application are discussed. Additionally the effect of chemical structure of ILs on the degradation pathway were considered.

Sooner or later ILs applied in industry will be waste, and efficient degradation process should be used to their disposal. Most of ILs are poorly biodegradable and do not inert to water organisms. Advanced Oxidation Processes appear to be promising in this respect. The progress in wastewater treatment and disposal of used ILs in oxidative conditions, were also discussed.

Chapter 10 - Room-temperature ionic liquids (RTILs) salts with melting points below 100 °C have been intensively studied in recent years. Unique physiochemical properties and broad application possibilities of ionic liquids make their industrial use a matter of the nearest future. Therefore, analytical methods are required to determine their purity and identification and quantitative determination of their components in various matrices. The primary aim of this review is to outline current developments in chromatographic and electrophoretic methods applied to the analysis of routinely used ionic liquid cations and anions. For qualitative and quantitative determination of ionic liquid components the highly efficient reversed-phase liquid chromatography, ion chromatography and ion-pair chromatography methods are most recently used. An alkyl-bonded stationary phases, phenyl-bonded stationary phases or on silica packing functionalized with mixed polar and apolar functions stationary phases are applied. The usefulness of hydrophilic interaction chromatography as an interesting alternative to the classical RP-HPLC as well as IPC, especially for short-chain, polar cations separation also is discussed. Capillary electro-phoresis, and recently also the isotachophoresis are also being developed.

Chapter 11 - In this communication, separation of aliphatic and aromatic hydrocarbons from artificial gasoline by ionic liquids, [EMpy][ESO_4] or [EMim][ESO_4], has been investigated theoretically using the Wilson, NRTL, UNIQUAC, UNIFAC and ASOG activity models. For this purpose, phase equilibria of ternary systems containing pure hexane, heptane, octane, cyclohexane, cyclooctane, or methylcyclohexane as artificial gasoline and pure toluene, benzene, xylenes or ethyl benzene [BTEX] as aromatic compounds has been predicted. In each system, binary interaction parameters between components or subgroups have been obtained and reported. Model predictions were compared with the experimental phase equilibrium data. The results show that NRTL activity model is more capable in prediction of the phase equilibrium data with the root mean square deviation (rmsd) of 0.26 % for the studied systems.

In: Ionic Liquids: Synthesis, Characterization and Applications ISBN: 978-1-63117-716-3
Editor: Anson Brooks

Chapter 1

IONIC LIQUIDS: A PHARMACEUTICAL PERSPECTIVE

Paula C. A. G. Pinto and M. Lúcia M. F. S. Saraiva
REQUIMTE, Department of Chemistry, Faculty of Pharmacy,
University of Porto, Portugal

ABSTRACT

In the last decades, ionic liquids (ILs) progressed from chemical curiosities to interesting biological compounds apprehending the attention of researchers of distinct areas, from chemistry to pharmacology.

Initially, ILs were explored as materials for diverse applications due to the possibility of synthesize compounds with targeted chemical properties combined with selected physical properties. More recently, the emergence of ILs with biological activity revolutionized the scientific focus of these compounds and opened interesting perspectives regarding their pharmaceutical application. From the pharmaceutical point of view, an IL approach, in the design of novel active pharmaceutical ingredients (APIs), appears to be appropriate as it enables the chemical manipulation of the compounds with specific objectives related with the manufacturing process, the stability of the formulation, bioavailability and eventual adverse effects. Furthermore, even though crystallinity confers advantages during isolation, processing and storage of the drug, it is known that solid forms of APIs often suffer from low solubility and polymorphic conversion which can influence negatively the bioavailability of the drug and ultimately its therapeutic effect.

On a distinct perspective, the possibility of engineering the properties of ILs by manipulating anion-cation combinations, in association with their solvent properties and in some cases water-miscibility, are considered promising characteristics regarding the applicability of ILs as solvents or carriers of pharmaceutical drugs.

In this chapter it is intended to expose the pharmaceutical potential of ILs through the discussion of their utilization either as APIs or solvents/carriers of pharmaceutical drugs. The discussion will be centered on the benefits of the IL approach for the development of novel drug candidates considering not only physico-chemical aspects but also the pharmaceutical profile of the developed active pharmaceutical ingredients with IL properties. Considering the utilization of ILs as solvents of drugs or as part of drug delivery systems, it is anticipated that the discussion will be focused on the efficiency and toxicity of the systems and on their influence on the pharmacokinetics and pharmacodynamics of the vehiculated drugs. It is also planned to debate the motivations for the pharmaceutical usage of ILs as well as their peculiar properties that launched them in this context. Currently, the pharmaceutical utilization of ILs is one of the most relevant applications of these solvents, with impact on the safety and effectiveness of the involved processes and with benefits in terms of pharmaceutical formulations and pharmacological activity. It is our belief that this chapter is adequate for the book in project and can greatly enhance its acceptance and interest by readers in distinct research areas.

ABBREVIATIONS

Ionic Liquid(s)	IL(s)
Research and Development	R&D
GlaxoSmithKlein	GSK
Active pharmaceutical ingredient(s)	API(s)
Ethylammonium nitrate	$EtNH_3 [NO_3]$
1-butylpyridinium chloride-aluminum chloride	bmpy $[Cl]-AlCl_3$
Tetrafluoroborate	BF_4
Hexafluorophosphate	PF_6
Generally recognized as safe	GRAS
Volatile organic solvent(s)	VOCs
Food and Drug Administration	FDA
1-butyl-3-methylimidazolium tetrafluoroborate	bmim $[BF_4]$
NSAID(s)	non-steroidal anti-inflammatory drug(s)
Ring closing metathesis	RCM
Triethylammonium acetate	TEAA
Support ionic liquid phase	SILP
1-butyl-3-methylimidazolium bis(trifluoromethylsulfonyl)imide	bmim $[NTf_2]$
n-propyl-2,3-dimethylimidazolium bis(trifluoromethylsulfonyl)amide	dmim $[[NTf_2]$
DPEN	(R,R)-1,2 diphenylethylenediamine
1-hexyl-3-methylimidazolium tetrafluoroborate	hmim $[BF_4]$
1-ethyl-3-methylimidazolium tetrafluoroborate	emim $[BF_4]$
1-butyl-2,3-dimethylimidazolium hexafluorophosphate	dbmim $[PF_6]$
1-ethyl-3-methylimidazolium chloride-aluminum chloride	emim $[Cl]-AlCl_3$
1-butyl-3-methylimidazolium chloride-iron chloride	bmim $[Cl]-FeCl_3$
1-butyl-3-methylimidazolium chloride-aluminum chloride	bmim $[Cl]-AlCl_3$
1-butyl-3-methylimidazolium chloride-zinc chloride	bmim $[Cl]-ZnCl_2$
1-ethylpyridinium tetrafluoroborate-zinc chloride	etpy $[BF_4]-ZnCl_2$
1-butyl-3-methylimidazolium hexafluorophosphate	bmim $[PF_6]$
1-ethyl-3-methylimidazolium bis(trifluoromethylsulfonyl)imide	emim $[NTf_2]$
methyltrioctylammonium bis(trifluoromethylsulfonyl)imide	moma $[NTf_2]$

1-butyl-1-methylpyrrolidinium bis(trifluoromethylsulfonyl)imide	bmpyr [NTf_2]
1-hexyl-1-methylpyrrolidinium bis(trifluoromethylsulfonyl)imide	hmpyr [NTf_2]
methyl *tert*-butyl ether	MTBE
1-methylimidazolium acetate	mim [OAc]
1-methoxyethyl-3-methylimidazolium methanesulfonate	moemim [Ms]
1-methoxyethyl-3-methylimidazolium trifluoroacetate	moemim [TFa]
1-butyl-3-methylimidazolium trifluoroacetate	bmim [TFa]
1-butyl-3-methylimidazolium methanesulfonate	bmim [Ms]
1-butyl-3-methylimidazolium hydroxide	bmim [OH]
dimethylimidazolium dimethylphosphate	dmim [$(CH_3O)_2PO_2$]
1-hexyl-3-methylimidazolium acetate	hmim [OAc]
1-ethyl-3-methylimidazolium acetate	emim [OAc]
1-methylimidazolium hydrogen sulfate	hmim [SO_4]
1-butylimidazolium tetrafluoroborate	hbim [BF_4]
1-hexyl-3-methylimidazolium hexafluorophosphate	hmim [PF_6]
1-octyl-3methylimidazolium hexafluorophosphate	omim [PF_6]
1-ethyl-3-methylimidazolium methanesulfonate	emim [Ms]
1-ethyl-3-methylimidazolium trifluoromethanesulfonate	emim [TfMs]
dimethylimidazolium dimethylphosphate	dmim [$(CH_3O)_2PO_2$]
1-hexyl-3-methylimidazolium chloride	hmim [Cl]
1-tetradecyl-3-methylimidazolium bromide	tmim [Br]
1-hexyl-3-methylimidazolium bromide	hmim [Br]
Ionic liquid(s) with pharmaceutical activity	IL-API(s)

1. Introduction

The pharmaceutical industry faces nowadays unprecedented challenges and amendments. Drug development became a question of millions of dollars and the emergence of generic drugs led to dramatic decreases on the sales of huge market successes [1]. With this, the investment on research and development (R&D) tends to decelerate since the risks associated with drug discovery seem to not be in accordance with the current economic demands. It is known that 95% of the experimental drugs that reach clinical trials fail to be both effective and safe. Before this, 40% of the novel drug candidates never reach this stage due to poor pharmaceutical properties such as reduced solubility or diminished permeation of the blood-brain barrier [2].

Additionally, the demands regarding the environmental impact of drug discovery are on the frontline of the guidelines of Green Chemistry [3].

In this scenario, the pharmaceutical industry has been struggling to keep the high standards of R&D whit the demands of preserving economically viability and integrate environmental sustainability. Novel approaches and management models have been implemented and at the same time the academic community started to invest on innovative pathways for the development and synthesis of novel drug candidates. Additionally, there has been also investment on the reformulation of old synthetic and production processes to make them more adequate to the current economic and environmental demands. Some pharmaceutical companies are nowadays sensitive to this matter and are aware of the environmental consciousness of the general population. Companies like Pfizer and GlaxoSmithKlein (GSK) created internal guidelines (in accordance with the Green Chemistry principles) that aim, among others, the reduction of greenhouse emissions, the minimization of the impact of R&D (from drug discovery to usage and disposal) and the sustainable utilization of water. On the chemical perspective, one of the most obvious and intuitive approach is the search for greener alternative solvents in research, development, and manufacturing of novel active pharmaceutical ingredients (APIs) [4].

In this context, ionic liquids (ILs) emerged as a new class of compounds with properties that empower their pharmaceutical application in a variety of strands. In April 2000, industry and academia joined efforts and ideas on a NATO Advanced Research Workshop on the topic of "Green Industrial Applications of Ionic Liquids" opening interesting perspectives regarding the future utilization of these compounds.

Even though the industrial applications in this field are still scarce, the peculiar properties of ILs envision immense and successful possibilities illustrated by an increasing number of literature examples [5]. ILs are being explored as solvents or reagents in a variety of pharmaceutical processes and as reaction media for common synthetic routes of both known and novel drugs [6]. Additionally, ILs can be part of distinct drug delivery systems for drugs with reduced bioavailability [7-10].

Ultimately, the biological activity of ILs can be explored to form active pharmaceutical ingredients (APIs) with tunable physico-chemical and biological properties [11-14]. Still, changes at the industrial level demand a strict and rigid protocol in order to guarantee the safety and viability of the developed drugs.

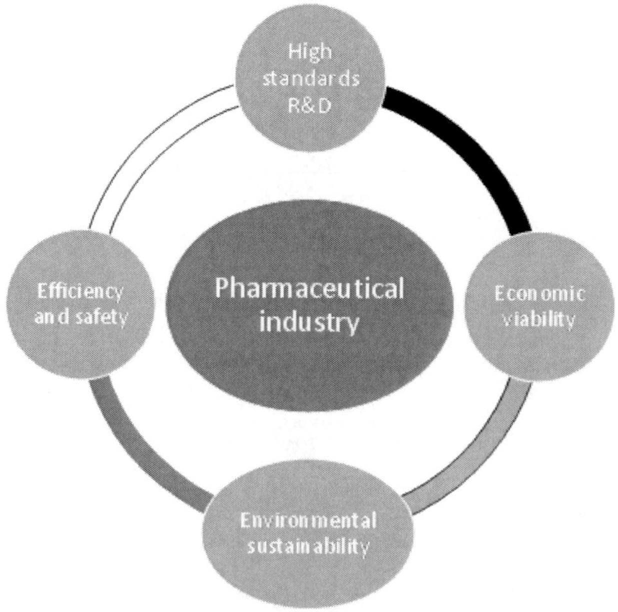

Figure 1. Current demands of the pharmaceutical industry.

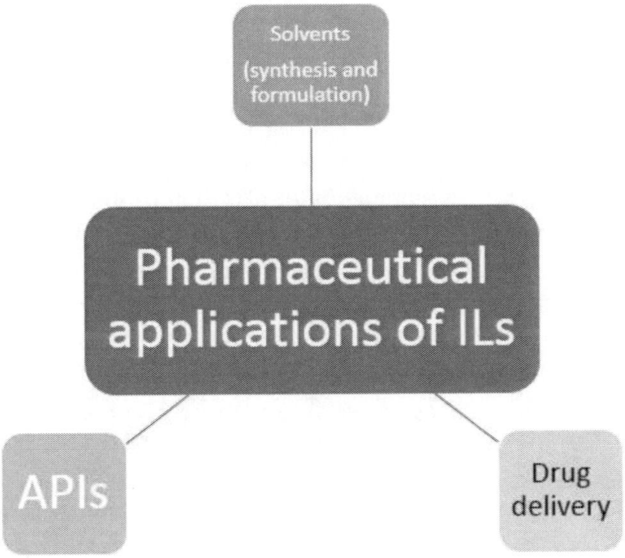

Figure 2. Pharmaceutical applications of ILs.

So, aspects like pharmaceutical profile, safety and human toxicity, environmental impact and biodegradability must be deeply addressed and explored to promote the acceptance of ILs by pharmaceutical industry.

The following discussion will be focused on the evolution of the IL concept to justify the pharmaceutical potential of these solvents. The main features and advantages of the abovementioned pharmaceutical possibilities will be emphasized and where possible, literature and industrial examples will be explored and debated.

2. Historical Perspective of ILs

Most documents focused of the history of ILs go back to the 19th century, to the "red oil" obtained from Friedel-Crafts reactions and to 1914 where the first IL of history, ethylammonium nitrate ($EtNH_3$ $[NO_3]$), was synthesized [15]. However, ILs as we know them today can find their origin in the 1960´s when U.S Air Force Academy started a project based on the replacement of LiCl-KCl molten salt electrolyte in thermal batteries [6]. By that time, the involved researchers managed to synthesize and characterize a totally ionic, non-aqueous solvent, 1-butylpyridinium chloride-aluminum chloride (bmpy [Cl]-$AlCl_3$), an alkyl-chloroaluminate that triggered the interest of chemists on ILs, especially those on the electrochemistry field. This was the first compound fitting the definition of IL that has been accepted for almost a century since Waldens´ studies in 1914: a salt which melts at or below 100ºC.

This IL constitutes the first element of the so called first generation of ILs that was mainly composed of dialkylimidazolium and alkylpyridinium cation derivatives combined with chloroaluminate and other metal halides anions. These compounds were largely studied by Osteryoung group [16, 17] and other researchers for their tunable and unique physical properties such as density, viscosity and thermal stability, among others. However, the majority of these compounds are liquid only on a very narrow composition range and demand particular conditions of handling since chloroaluminates are water and oxygen sensitive. Additionally, the typical cations are very prone to reduction. Thus, these ILs are not suitable to be applied in biotransformations and not surprisingly the investigations on this field advanced to the synthesis of air and water stable compounds with wider liquid range. This was achieved initially through the combination of alkylimidazolium cations (less predisposed to reduction than the pyridinium ones) with water–stable anions like tetrafluoroborate (BF_4), hexafluorophosphate (PF_6) and acetate, among many others. These ILs gathered wide attention from the scientific community not only because of their low reactivity with water but also because of their large electrochemical window. These properties enabled the synthesis of novel ILs with tunable chemical properties combined with selected physical properties and that could be used as materials for diverse applications. These compounds constitute the second generation of ILs and are known for their lower melting points, negligible vapor pressure and moderate polarity. The boost on ILs research occurred between 1999 and 2000 after the emergence of this second generation, with an exponential increase of applications in a variety of fields from biocatalysis to nanotechnology. The major drawback of these ILs is still their toxicity that even though can be modulated is inherent to the majority of the compounds and can be similar to that of chlorinated and aromatic solvents.

As expected, the investigations followed the route to non-toxicity. This conducted to the utilization of stable, safe and biodegradable ions sometimes with biological activity and derived from materials generally recognized as safe (GRAS). The third generation of ILs is then composed of ILs with tunable chemical and physical properties and with reduced toxicity, like for instance ILs based on amino acids. Considering these particularities it is not unexpected that this new generation of ILs broadened the applicability of these compounds and triggered the investigation of their biotechnological and pharmaceutical potential. In this historical perspective it is important to highlight that even though the biological and pharmaceutical properties of ILs were explored before, namely through the investigation of antimicrobial, antifungal and anti-cancer activities of imidazolium, ammonium and phosphonium ILs [18-21], only at this stage this property was explored on purpose and recurring to selected materials.

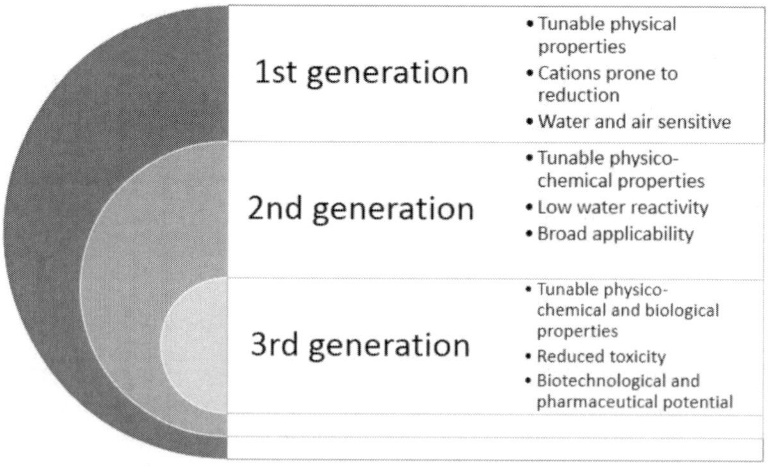

Figure 3. Historical perspective of the evolution of ILs according to their properties and applications.

3. Applicability of ILs in the Pharmaceutical Field

As stated before, the peculiar properties of ILs make them suitable for distinct pharmaceutical applications and only their toxicity has hindered major developments in earlier years. Still, there are several situations in which toxicity can be modulated and others where toxicity can be a desirable property. With the forthcoming of the 3rd generation of ILs novel possibilities arose.

This chapter will focus on the major possible applications of ILs in the pharmaceutical perspective: reaction media for the synthesis of APIs, as vehicles of drugs (solvents and drug delivery) and as APIs themselves.

3.1. Synthesis of Pharmaceutical Drugs in IL Media

In the pharmaceutical industry the use of volatile organic solvents (VOCs) is common and most of the times these solvents are an important part of the whole process from synthesis of APIs (or intermediates) to manufacturing operations like granulation and extraction [22, 23].

Considering that solvent use is about 80-90% of mass utilization in a typical pharmaceutical industry, the use of VOCs becomes a serious concern either from the human or environmental perspective as the toxicity of the production process is largely dependent on the nature of the selected solvents [24]. Moreover, the final formulations will unavoidably contain residual amounts of solvent. Even though there are strict rules and residual solvents limits improved by Food and Drug Administration (FDA) and other organizations and Pharmacopoeias, the risks and adverse effects of VOCs on human health are well known [25, 26]. It is then desirable that novel approaches could be implemented, as it has been already stated on the introduction section.

The current concerns of both academia and industry resulted already on several attempts to substitute VOCs by ILs in drug´s synthetic procedures [27, 28]. This approach is based on the demonstrated ability of ILs to drastically change the rate and selectivity of organic chemical reactions [6]. Currently, it is demonstrated that behind the environmental and safety issues, ILs can have also impact on the reactivity of the species involved in organic synthesis due to a fine balance of entropic and enthalpic contributions that involve the interactions inside the IL and between the ions and dissolved species. In fact, ILs can interact with solutes through dipolar and dispersion forces and act as strong hydrogen-bond acceptors [6]. The most important properties that set ILs apart from other solvents include:

- existing as molten salts, in many cases even below room temperature
- highly polar
- negligable vapor pressure
- nonflammable
- wide liquid range (typically > 300 °C grade)
- electrically conducting
- noteworthy dissolution properties (of both organic and inorganic compounds)
- high thermostability.

Table 1. Synthesized compounds with pharmaceutical interest through the implementation of organic reaction in IL

Synthesized Compound	Biological activity	Reaction	Ionic Liquid	Ref
Isoxazoline derivatives	Antimicrobial	Diels-Alder	bmim [BF_4]	[32]
Pyrrole-fused polycyclic heterocycles (alkaloyds)	Alkaloids	Diels-Alder	bmim [BF_4]	[33]
Pyrano/Thiopyrano fused polycyclic heterocycles	Bactericide	Diels-Alder	TEAA	[34]
Regioselective arylated products	Pharmaceutical intermediates	Palladium catalyzed	bmim [PF_6] bmim [BF_4]	[35]
Arylboronates	Pharmaceutical intermediates Anti-tumoral	Palladium catalyzed	bmim [PF_6] bmim [BF_4] hmim [BF_4] emim [BF_4]	[30]
(S)-Naproxen	NSAID	RCM	bmim [BF_4]	[36]
Hydrogenated aromatic ketones	Pharmaceutical intermediates	RCM	Imidazolium ILs	[37]
Pravadoline	NSAID	Friedel-Crafts	Imidazolium (PF_6 and chloroaluminate) catalysts	[38]
Substituted indoles	Pharmaceutical intermediates	Friedel-Crafts	Imidazolium chloroaluminate catalysts	[39]
Benzophenone derivatives	Pharmaceutical intermediates Sunscreens components	Friedel-Crafts	Imidazolium chloroaluminate catalysts	[40]
Benzophenone derivatives	Pharmaceutical intermediates Sunscreens components	Friedel-Crafts	Imidazolium chloroindate(III) catalysts	[41]
Coenzyme Q10	Antioxidant	Friedel-Crafts	Pyridinium chloroaluminate catalysts	[42]
Kinetic resolution of ibuprofen	NSAID	Biocatalysis (*Candida rugosa* lipase)	bmim [PF_6] bmim [BF_4] bmim [$MeSO_4$] mmim [$MeSO_4$] bmim [$OctylSO_4$] $(C_6H_{13})_3C_{14}H_{29}PN_3$ $(C_4H_9)_3CH_3PTOs$	[43]
Kinetic resolution of ibuprofen	NSAID	Biocatalysis (lipases from*Aspergillus niger AC-54, Candida antarctica, Candida rugosa, Rhizomucor miehei* and *Thermomyces lanuginose*	bmim [PF_6] bmim [BF_4]	[44]
Caffeic acid phenethyl ester analogues	Anti-tumoral	Biocatalysis (*Candida antarctica* lipase)	bmim [NTf_2]	[45]
(R)-1-(4-chlorophenyl)ethanol	Pharmaceutical intermediate	Whole cell biocatalysis (*Lactobacillus kefir*)	bmim [PF_6] bmim [NTf_2] moma [NTf_2]	[46]
(R)-1-(4-chlorophenyl)ethanol	Pharmaceutical intermediate	*Escherichia coli Saccharomyces cerevisiae*	bmim [PF_6] bmim [NTf_2] moma [NTf_2] hmim [PF_6] hmim [NTf_2] bmpyr [NTf_2]	[47, 48]

Synthesized Compound	Biological activity	Reaction	Ionic Liquid	Ref
			hmpyr [NTf_2]	
Chiral alcohols	Pharmaceutical intermediate		[PF_6], [NTf_2] and [FAP] based hydrophobic ILs	[49]
(S)-3-chloro-1-phenyl-1-propanol	Pharmaceutical intermediate	*E. coli*	bmim [PF_6] bmim [NTf_2] emim [BF_4] mmim [$MeSO_4$]	[50]
Nucleosides derivatives	Pharmaceutical intermediates		momim [Ms] momim [TFa] bmim [TFa] bmim [Ms]	[51, 52] [53, 54]
2,3-Disubstituted-1,3-thiazolinin-4-one derivatives	Antiparasitic	Three component condensation (aldehyde, amine and mercaptoacetic acid)	bmim [BF_4] bmim [PF_6]	[55]
2-Aryl-4,5-diphenyl imidazoles, pyrazole derivatives, oximes	Pharmaceutical intermediates	Sonochemical organic synthesis	bmim [OH] emim [OAc] hmim [SO_4]	[56-58]
Quinoline derivatives		Sonochemical organic synthesis	hbmim [BF_4]	[59]

The possibility of developing tailor-made ILs for particular purposes offers exciting opportunities in the organic chemistry field enabling the creation of tasks specific compounds. Moreover, ILs can often form biphasic systems with classic VOCs which allows the simple isolation of reaction products by extraction of the ionic liquid layer. Their ionic nature enables immobilisation and recycling of many transition metal catalysts.

Many organic reactions have been studied in IL media sometimes with remarkable results [6, 29]. In a small number of cases the registered differences were small, not meeting anticipated high expectations or justifying extra expense. Herein we intend to focus on chemical reactions that can have significant utility at the pharmaceutical level and that in fundamental literature studies are not always immediately linked to that type of application. The selected reactions were: Dies-Alder, transition metal catalysed (including Heck reaction and ring-closing metathesis), Friedel-Crafts reactions not forgetting enzyme catalysed synthesis. Some examples of successful attempts to perform synthesis of APIs in ILs will be presented and discussed.

The abovementioned reactions have been extensively studied in IL media [6, 27, 29-31]. Considering that nowadays drug synthesis can be based on the concept of convergent synthesis, according to which several fragments are prepared in parallel paths and combined at the end of the process, the informations collected from those reactions if not totally adequate to a whole synthesis process can be important in one of those parallel paths.

The most important achievements in this field are compiled in Table 1.

3.1.1. Diels-Alder Reaction

The Diels-Alders reaction is an organic chemical reaction, more specifically, a cycloaddition, between a conjugated diene and a substituted alkene to form a substituted cyclohexene system [60]. This reaction enables the synthesis of complex polycyclic molecules with adequate control of stereochemistry. Even though this type of reaction accounts for only 5 to 11 % of C-C bond-forming reactions performed under Good Manufacturing Practice in the pharmaceutical industry [61], it is largely explored by scientists in academic drug development and thus deserves a special attention on this chapter. Among the most important drugs synthesized recurring to Diels-Alder cycloaddition reaction, it is important to highlight several types of vitamins, prostaglandins and steroids [60]. Generically, the Diels-Alder reaction applied to organic synthesis exhibits higher reaction rates in water than in organic solvents, but the need to work with moisture sensitive substrate hinders the industrial implementation of the reaction in aqueous media [27]. This explains the extensive investigation of this reaction in ILs as alternative reaction media [6, 27, 28]. From the available studies it is important to highlight that the majority of the tested ILs incorporate Lewis acidic anions and conduct to a general increase of both reactivity and selectivity of the studied reactions, in less drastic conditions [62-65]. Besides the advanced knowledge on Diels-Alder reaction in IL media with interest for drug development, it is also possible to find recent concrete examples of the synthesis of pharmaceutical drugs or intermediates in ILs [32-34].

The synthesis of isoxazoline derivatives can be performed in 1-butyl-3-methylimidazolium tetrafluoroborate (bmim [BF_4]) with enhanced reaction rates, improved yields, and high selectivity. The IL can be recycled and reused without loss of reactivity or selectivity. The synthesized isoxazoline derivatives are precursors of peptides with activity against gram positive and gram negative organisms and can constitute the basis for the development of new broad spectrum antimicrobial agent [32]. The same IL was used for the synthesis of pyrrole-fused polycyclic heterocycles through intramolecular 1,3-dipolar cycloaddition with higher rates and yields than in conventional organic solvents [33]. Along with the reutilization of the IL, the reaction products can also be readily separated by extraction. The synthesized compounds are of great interest since they are chemically related with well-known alkaloids such as myrmicarin and polyavolensin offering promising perspectives regarding the synthetic approach to natural products. Triethylammonium acetate (TEAA) was applied as greener solvent on a domino reaction based on the Diels Alder approach for the synthesis of pyrano and thiopyrano fused heterocycles thiopyrano fused heterocycles with bactericide activity [34]. Higher yields and shorter reaction times were attained.

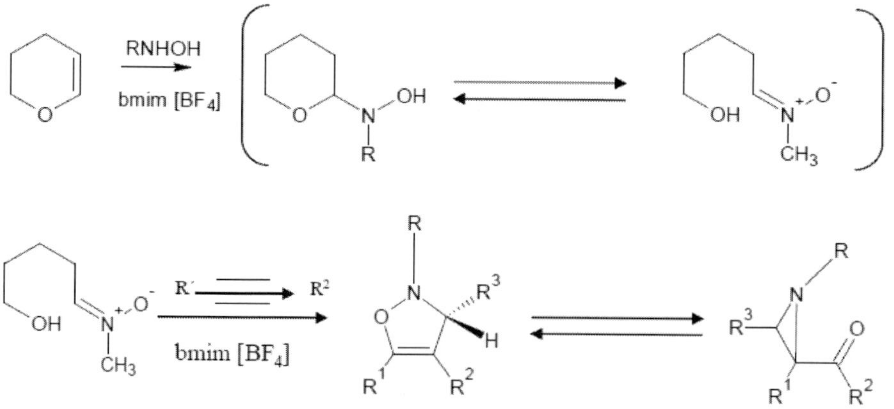

$R = CH_3/C_6H_5/C_6H_{11}/C_6H_5CH_2/4\text{-}HOC_6H_4/2\text{-}HOC_6H_4$
$R^1 = Ph$; $R2 = COOCH_3$
$R^1 = R2 = COOCH_3$
$R^1 = R2 = COOH$
$R^3 = \text{-}(CH_2)_4OH$

Scheme 1. Synthesis of isoxazoline derivatives by Diels-Alder reaction in the presence of bmim [BF_4] [32].

3.1.2. Transition Metal Catalysed Reactions

Transition metal catalysed reactions have been widely used for decades in the pharmaceutical industry for the synthesis of drugs or intermediates. Specifically, palladium catalyzed reactions, including Heck reaction, represented, in 2005, about 60% of the C-C bond forming reactions in medicinal chemistry [61]. This is partially explained by the fact that the implementation of convergent synthesis demands final coupling steps compatible with a wide variety of functional groups. The palladium technology is in accordance with these demands and at the same time avoids the excessive use of protective groups. However, in the pharmaceutical field, this technique poses problems associated with the contamination of the synthesized APIs with palladium, demanding specific removal strategies. This problem can be circumvented recurring to heterogeneous catalysis or removal strategies like nanofiltration or scavenging methods [66].

In the group of transition metal catalysed organic reaction, the Heck reaction and ring closing metathesis (RCM) will be highlighted due to their importance and available information. The Heck reaction is a classic example of a palladium catalyzed carbon-carbon bond formation process recognized as a good method to attach olefins to (hetero)aromatic compounds [67, 68]. In normal reaction conditions, the Heck reaction is not totally regioselective and the palladium catalysts are not stable, leading to high catalyst consumption and difficulties in the implementation of the reaction. In this context, ILs were applied as solvents for the Heck reaction for the first time in 1996. There are several aspects that justify the massive application of ILs in metal catalysed organic reactions. Firstly, besides the environmental and safety issues, the use of a non-volatile solvent simplifies the work set-up for distillation and avoids the problems related to the formation of azeotropic mixtures, typical of VOCs. Additionally, metal catalysts are either soluble or can be immobilized in ILs. In the latter case, the resulting solution can be immobilized into a solid support resulting on a support ionic liquid phase (SILP) catalysts that combines the advantages of both homogeneous and heterogeneous catalysis making this approach very sensitive process technologies.

Many researchers explored this strategy sometimes with slightly different approaches [68] providing important information for future developments in the pharmaceutical field.

Also on the pharmaceutical field, palladium catalyzed reactions have been already implemented in IL media with positive results. The arylation of allylic alcohols by aryl bromides was comparatively studied in imidazolium ILs (1-butyl-3-methylimidazolium - bmim [PF_6], bmim [BF_4] and 1-butyl-3-methylimidazolium bis(trifluoromethyl-sulfonyl)imide bmim [NTf_2]) and molecular solvents [35]. The regioselectivity of the reaction is highly affected by the nature of the solvent, with ILs being the best option for the formation of branched arylated products with the best β/α ratios reported till that time. Amongst the tested ILs, bmim [PF_6] is the most favorable compound in terms of catalyst activity. This reaction conducts to the formation of substituted allylic alcohols with regiocontrol that constitute useful intermediates for pharmaceutical synthesis. The utilization of ILs overcomes deficient regioselectivity, the major problem of this kind of arylation. The synthesis of arylboronates by palladium catalyzed cross-coupling can be performed in alkylimidazolium ILs incorporating BF_4 or PF_6 anions [30]. Catalyst decomposition is however observed with ILs incorporating the PF_6 anion as well as with emim [BF_4]. From the tested ILs, bmim [BF_4] and 1-hexyl-3-methylimidazolium tetrafluoroborate (hmim [BF_4]) are the most appropriate solvents, with increased catalyst stability in the case of bmim [BF_4]. Furthermore, with this IL the reaction time is shorter than in conventional molecular solvents and the catalyst can be recycled from the IL. The synthesized boronate derivatives are important building blocks in organic synthesis and can present biological activity with applicability as anti-tumoral drugs [69]. Still in the field of transition metal catalyzed reactions, RCM is a ruthenium (Ru) catalyzed transformation that conducts to the formation of carbon–carbon double bonds [70]. This reaction belongs to the broad group of olefin metathesis and has gained popularity in the last decade since it enables the synthesis of functionalized heterocycles with applicability in several fields [71]. RCM has been widely explored for the synthesis of pharmaceutical compounds or intermediates such as azepine derivatives [72, 73], gambieric acid along with analogues [74] and acyclonucleoside phosphonates [75], among others. Considering the ability of ILs to stabilize metallic catalyst and the abovementioned advantages of transition metal catalyzed reaction in ILs, Ru mediated RCM soon started to be explored in this reaction media [6, 27, 76, 77]. Even though the initial applications of RCM in ILs revealed a beneficial effect of the tested compounds on the catalytic efficiency, they suffered from reduced stability of the catalyst, a behavior that was previously observed in molecular solvents [78]. The key step on the development of this strategy was the modification of the catalyst through functionalization with an imidazolium group [79, 80]. The obtained catalyst showed to be far more stable than the conventional ones with the additional possibility of being stored for several months. On a distinct perspective there are also reports on the combined use of ILs and VOCs [81] and on the study of RCM in imidazolium ILs under microwave irradiation [82].

RCM in IL media has also been explored during the synthesis of compounds with pharmaceutical interest. (S)-Naproxen can be quantitatively obtained from the asymmetric hydrogenation of 2-arylacrylic acids catalyzed by a Ru catalyst immobilized in bmim [BF_4] [36]. The enantioselectivities are similar or higher than those obtained with the homogeneous reaction. The catalyst can be recovered and reused without loss of catalytic activity or selectivity. A similar approach was adopted by Ngo and co-workers who managed to perform the asymmetric hydrogenation of aromatic ketones. In this case, polar bisphosphonic acid-derived Ru catalysts were immobilized in imidazolium ILs [37] for the production of pharmaceutical intermediates free of metal contaminants. The catalytic performance is highly affected by the nature of the IL with n-propyl-2,3-dimethylimidazolium bis (trifluoromethylsulfonyl)amide (dmim [NTf_2]) providing the best results in terms of catalyst efficiency, when compared to bmim [BF_4] and bmim [PF_6]. As before, both IL and immobilized catalyst can be recycled and reused without leaching of Ru into the organic compounds.

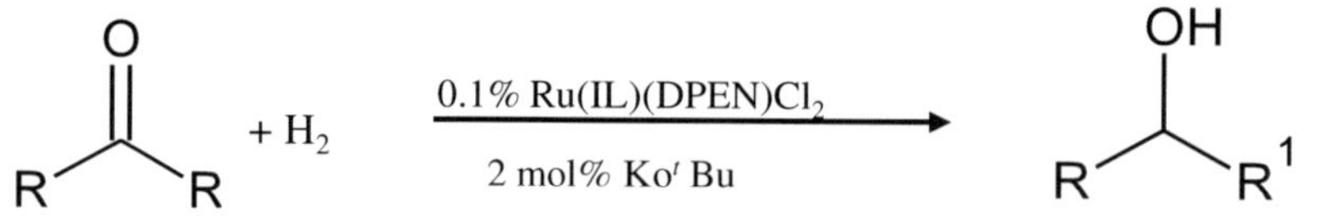

IL = bmim [BF_4]; bmim [PF_6]; dmim [[NTf_2]

DPEN = (R, R)-1,2-diphenylethylenediamine

Scheme 2. Asymmetric hydrogenation of chiral ketones catalyzed by Ru-IL catalysts [37].

3.1.3. Friedel-Crafts Reaction

The Friedel-Crafts acylation reaction represents a synthetic process of great interest to organic chemists of academia and industry being involved in the production of several pharmaceuticals. Since its first description in 1887, this reaction has been increasingly recognized as the method of choice to perform the alkylation of arenes and heteroarenes and form C-C and C-heteroatom bonds [83]. Even though the Friedel-Crafts reaction is traditionally catalyzed by Lewis acids, it is also possible to accelerate the reactions recurring to strong Brønsted-acids including sulfuric acid, hydrofluoric acid or super acids. More environmentally benign options involving alkylation by means of alcohols and styrenes were implemented recently as an alternative to alkyl chlorides. The environmental aspects are worsen by the large amounts of VOCs generally utilized in these reactions [84]. In this context, ILs soon started to be explored in this kind of reaction with interesting results [85]. The first application of an IL in a Friedel-Crafts reaction was accomplished in 1986 and involved the effective utilization of a mixture of 1,3-dialkylimidazolium chloride and aluminum chloride as catalyst [85]. The success of the application is explained by the strong Lewis acidity of acidic chloroaluminate ILs. Since then, it was found that the ILs provide also better alkylate quality than those obtained with conventional catalysts. There is a vast list of applications of IL-catalysts in diverse Friedel-Crafts reactions in a variety of fields, all of them confirming the adequacy of these solvents to the performed reactions [28, 83, 84, 86-88].

Due to the importance of the Friedel-Crafts reaction for the pharmaceutical industry it is not surprising that this reaction was studied in IL media for the production of compounds with pharmaceutical interest. The first report on the synthesis of a pharmaceutical drug was presented by Earle and co-workers [38]. Based on their past experience with Friedel-Crafts reaction of indoles, the researchers managed to synthesize pravadoline, through a combination of Friedel-Crafts reaction and a nucleophilic displacement reaction. In the presence of bmim [PF_6], the alkylation of 2-methylindole with 1-(N-morpholino)-2-chloroethane occurs at room temperature with a yield of 95%. The change of the IL to 1-butyl-2,3-dimethylimidazolium hexafluorophosphate (dbmim [PF_6]) leads to an increase of yield to 99%.

It must be also highlighted that the production of acidic aqueous wastes is decreased due to the fact that the reaction, performed entirely in IL, does not require a Lewis acid catalyst reducing the acidity of the reaction with consequent reduction of wastes and by-products leading also to increased yields. Moreover, in the adopted scheme the product is easily separated and the solvent is recycled.

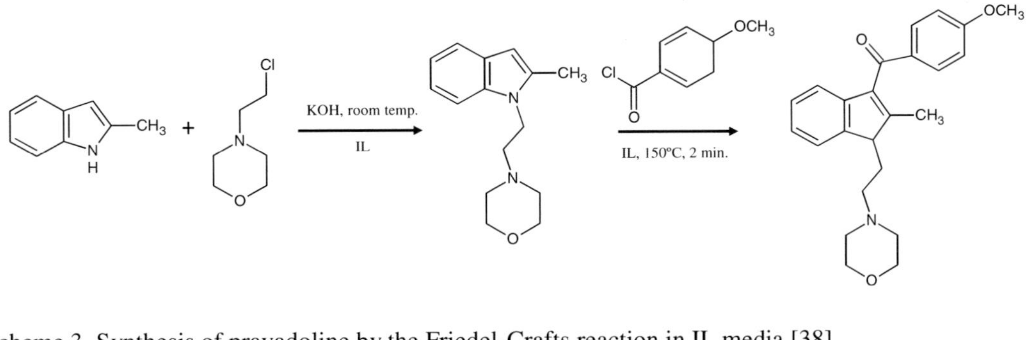

Scheme 3. Synthesis of pravadoline by the Friedel-Crafts reaction in IL media [38].

This reaction was also tested in 1-ethyl-3-methylimidazolium chloride-aluminum chloride (emim [Cl]-$AlCl_3$), a chloroaluminate IL, with results similar to those obtained with a conventional Friedel-Crafts approach and yields between 40 and 89%, with the drawback of producing large amounts of acidic waste.

This latter approach can also be explored for the production of substituted indoles by Friedel-Crafts acylation of indoles at room temperature [39]. The obtained products are indoles substituted at different positions with diverse functional groups that can be used as pharmacophore blocks for the synthesis of several molecules with pharmaceutical interest. Chloroaluminate ILs integrated also a broader study of Friedel-Crafts acylation for the synthesis of benzophenone derivatives recurring to distinct IL catalyst [40]. Amongst the tested ILs, 1-butyl-3-methylimidazolium chloride-iron chloride (bmim [Cl]-$FeCl_3$), exhibited much higher catalytic activity than that registered for 1-butyl-3-methylimidazolium chloride-aluminum chloride (bmim [Cl]-$AlCl_3$), 1-butyl-3-methylimidazolium chloride-zinc chloride (bmim [Cl]-$ZnCl_2$) and benzene. With this catalyst, reaction yields up to 97% were obtained with shorter reaction times. As before, it was possible to easily isolate the reaction products and the ILs could be recycled and reused with both economic and environmental advantages. As an alternative to the use of imidazolium based ILs, Chen and co-workers explored the potential of pyridinium ILs combined with $AlCl_3$, $ZnCl_2$, $FeCl_3$, $SnCl_2$, $SnCl_4$ and CuCl for the synthesis of an intermediate of coenzyme Q_{10} by Friedel-Crafts alkylation under microwave irradiation [42]. Generally, the catalytic activity of the ILs enhanced with the increase of their Lewis acidity with 1-ethylpyridinium tetrafluoroborate-zinc chloride (etpy [BF_4]-$ZnCl_2$) showing the best catalytic activity, with a yield of 89% after 150 seconds of reaction. The method exhibits the same economical and safety advantages described for the abovementioned approaches. In an attempt to provide alternative pathways for this important reaction, Earle and co-workers proposed chloroindate(III) ILs as catalysts for Friedel-Crafts acylations for the production of benzophenone derivatives [41]. The study involved the comparison of the synthesis performance in several chloroindate ILs and 1,2-dichloroethane. It was demonstrated that the best yield was obtained when indium(III) chloride was combined with bmim [NTf_2] either as a solution or as a binary IL (bmim [NTf_2]-$InCl_3$), with reaction yields between 75 and 96%. No catalytic activity was observed whit bmim [OTf] or bmim [BF_4]. The importance and significance of the obtained results is only fully understood considering that the aromatic compounds used for the synthesis are usually non-reactive. This issue is overcome with this approach.

3.1.4. Biocatalysis

Even though nowadays stereoselectivity can be achieved through metal transition catalysis, biocatalysis still remains an important tool for the pharmaceutical industry on the search for stereoisomer drugs instead of racemates [89, 90]. Besides the issue of selectivity, some of the most important features of enzymes, as industrial catalysts, are related with their aptitude to accept a wide range of substrates, even artificial ones, and with their ability to operate in non-aqueous environments [6]. The use of enzymes in synthetic processes does not require the use of protective groups, even in complex molecules, with simplification of the process and reduction of the associated costs. Several industrial processes apply nowadays enzyme-catalyzed reactions for the large-scale production of APIs. As examples, Schering-Plough synthesizes an azole antifungal agent in hundred-kilogram quantities and Bristol-Myers Squibb pharmaceutical research group published a number of plant scale chemoenzymatic synthesis performed in organic media. For years, the use of water as solvent in enzymatic reactions limited the field of application of enzymes in biocatalysis and the productivity of some processes, particularly those involving hydrophobic substrates [91]. As a consequence, the applicability of catalysis in non-aqueous solvents was implemented and discussed and as a result, new reaction media were suggested [92]. The possibility of working with hydrophobic substances, the decrease in microbial contamination and reduction of side reactions are the main advantages related with the development of procedures in non-aqueous media. Organic solvents helped to reduce some problems such as the insolubility of hydrophobic compounds and propagation of radicals in aqueous solutions offering benefits related with enzyme stability and selectivity [93, 94]. Due to the well-known drawbacks of VOCs, and reported inactivation phenomena, ILs emerged as an alternative for biocatalysis in the last decade. The applicability of ILs in biocatalysis was explored intensively through the study of an uncountable number of enzymes in huge variety of ILs with structural variability. The results of the majority of these studies are compiled and discussed in several reviews [95-97].

The research in this field indicates that biocatalytic reactions in ILs exhibit higher selectivity and faster rates with enhanced enzyme stability [98], making this association very promising for organic synthesis [90]. The relationship between ILs and enzymes has been discussed by many authors and the main explanations were pointed out. The polarity of ILs seems to be one of the most important factors affecting enzyme stability and selectivity. Indeed, polar solvents increase the solubility of polar substrates conducting to faster and more selective reactions [99]. Other studies reveal that the viscosity of ILs can also have a strong impact on enzyme´s activity [98]. High viscosity may produce a decrease in the rate of diffusion of the species involved in the process with reduction of the interaction of the enzyme with its substrate which can influence negatively the yield of reaction.

The hydrogen bond capacity of ILs has been also considered very important for protein stability and maintenance of its native structure [100]. The classic enzyme thermal unfolding is prevented by the entrapment of enzymes, with small amounts of water, in the hydrogen bond network of ILs [101]. The suppression of side reactions due to the decrease of water in the reaction media has also been associated with the increased yield of enzymatic reactions in ILs [102].

Even though the investigations in the field of biocatalysis in ILs began by the replacement of organic solvents, nowadays there are studies involving biphasic solvents in which water miscible IL are used in aqueous systems. Regarding biphasic systems, the possibility of enzyme reutilization has already been studied with success [102]. There is also the possibility of work in three phase systems in which ILs, organic solvents and water can be associated to modulate the solubility of substrates and enzyme activity [103].

The first study describing an enzymatic reaction in ILs involved the use of a protease, thermolysin, for the synthesis of Z-aspartame, with similar reaction rates but higher enzyme stability than those obtained with traditional organic solvents [104]. After this, many works involving the use of proteases and lipases, among many others, in ILs have been developed, confirming the advantages of these solvents in biocatalysis [6, 105].

In this context, ILs were soon considered interesting alternatives to VOCs on pharmaceutical synthesis and other bioprocess involving pharmaceutical drugs [106]. The kinetic resolution of ibuprofen catalyzed by *Candida rugosa* lipase in IL media is an excellent example of the successful combination of biocatalysis and ILs [43]. Chirality is a key issue for ibuprofen due to the differences registered between S-(+)-ibuprofen and R-(-)-ibuprofen, with the latter being 100-160 times less active than the first one. For this study, the authors selected seven ILs with structural variability and compared the results with those obtained with isooctane. Bmim [PF_6] improved both the kinetic resolution and the enantioselectivity with high enzyme stability. Studies of the time course conversion and enzyme stability as a function of time in both isooctane and bmim [PF_6] confirmed that this IL is a valid alternative for the kinetic resolution of ibuprofen. Later on, several types of lipases (from *Aspergillus niger AC-54, Candida antarctica, Candida rugosa, Rhizomucor miehei* and *Thermomyces lanuginose*) were studied to perform the resolution of (RS)-ibuprofen in two-phase systems containing bmim [PF_6] and bmim [BF_4] [44]. It was noticed that *C. rugosa* lipase conducted to the best results in terms of conversion degree, enantiomeric excess of remaining acid and *E*-value in a two phase system composed of bmim [PF_6] and isooctane. The influence of the IL was more pronounced on the *E*-value being less significant regarding the conversion degree. A similar tendency was observed with *A. niger* lipase. Acceptable results were also attained with *C. antarctica* lipase in terms of esterification activity and enantioselectivity. This enzyme can also be used for the synthesis of nucleoside drugs esters resorting to an IL as additive [107]. The lipase catalyzed transterification of ribavirin was performed in mixtures of acetone and increasing concentrations of imidazolium ILs (bmim [BF_4], bmim [PF_6] and emim [BF_4]). The best results in terms of transterification extent were obtained with bmim [BF_4] 10%, with an increase of the reaction rate of about 3.5 times when compared with the assay in the absence of IL. These observations were correlated to the increase of solubility of the nucleoside analogue due to the presence of the IL and to enzyme conformational changes. In the presence of bmim [BF_4] the nucleoside synthesis was carried out with excellent regioselectivity, fast reaction rate and high yield.

Resorting also to *C. antarctica* lipase, Kurata and co-workers performed the synthesis of caffeic acid phenethyl ester analogues, with antiproliferative effect on tumor cells, in bmim ILs with variable anions [45]. The transesterification reaction of methyl caffeate with various alcohols to produce the abovementioned analogues is favored in the presence of bmim [NTf_2]. Considering the registered changes in activity, it was concluded that the activity of enzyme is anion dependent. The optimized synthetic system resulted on the production of 2-cyclohexylethyl caffeate and 3-cyclohexylpropyl caffeate with conversion yields of 97.6 and 93.8%, respectively. Both compounds exhibit antiproliferative activity higher than caffeic acid and comparable to 5-fluorouracil.

In the field of biocatalysis, it is also important to highlight whole cell catalyzed processes which can present major advantages over the use of isolated enzymes [108]. In biotransformations involving more than one enzyme, the use of whole-cell biocatalysts (recombinant or not) is almost mandatory as it avoids costly and time consuming enzyme purification.

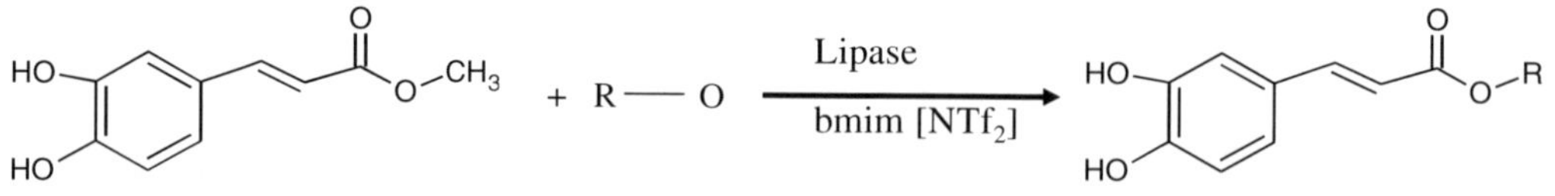

Scheme 4. Synthesis of caffeic acid phenethyl ester in IL media [45].

In addition, the use of cofactors can be avoided and the process can benefit from the intracellular regeneration of cofactor when the membrane of the whole-cell biocatalyst remains stable during the catalytic route. This approach is generally very adequate to asymmetric synthesis due to low water solubility of substrates or products and inhibitory effects of the reactants on the biocatalyst. Considering the issue of membrane integrity, the selection of the solvent in whole cell biocatalysis is of the most importance and can enhance or highlight the mentioned advantages of this catalytic methodology. In this context, it has been demonstrated that ILs can be successfully employed in whole cell biocatalysis in replacement of VOCs [46-49, 108]. Most of the referred applications of whole cell biocatalysis in IL media have pharmaceutical interest. This approach has been mainly applied to the asymmetric synthesis of chiral alcohols [46-50] which are frequently employed as pharmaceutical intermediates [109]. Pfruender and co-workers demonstrated for the first time that a cellular cofactor regeneration system can be active in the presence of water-immiscible imidazolium and ammonium ILs (bmim [PF_6], bmim [NTf_2] and methyltrioctylammonium–NTf_2, moma [NTf_2]) [46]. This important demonstration is a result of the study of the *Lactobacillus kefir* catalyzed asymmetric reduction of chloroacetone to (R)-1-(4-chlorophenyl)ethanol in ILs, used as substrate reservoir and extracting agent. The results of the synthetic process were compared with those obtained with commonly used VOCs, like n-octanol and methyl tert-butyl ether (MTBE). In the presence of bmim [NTf_2] the chemical yield of the synthesis of the chiral alcohol can be doubled and the purity of the product is excellent. With the other tested ILs the results are very acceptable comparing with MTBE. It was also evidenced that the tested ILs do not affect *L. kefir* membrane integrity and can even enhance it, probably due to changes in morphology. Similar results were obtained in a study of the same process catalyzed by *Escherichia coli and Saccharomyces cerevisiae* [47, 48] confirming that the selected ILs can be applied on the whole cell biocatalyzed production of pharmaceutical chiral alcohols. A set of other commercially available ILs proved to be also good options for the asymmetric reduction of ketones using recombinant *E. coli* as biocatalyst [49, 110]. As before, the tested compounds do not affect negatively the intracellular regeneration of NADH nor cell membrane integrity. More recently, Choi and co-workers confirmed the adequacy of imidazolium ILs, with special focus on bmim [NTf_2], for the production of chiral alcohols through the *E. coli* catalyzed synthesis of (S)-3-chloro-1-phenyl-1-propanol, a chiral precursor of anti-depressant drugs [50].

3.1.5. Other Applications

Nucleoside chemistry is one of the most important areas in pharmaceutical research providing compounds with wide application in cancer and viral chemotherapy. The problems associated with the insolubility of nucleoside derivatives have hindered the development of novel strategies and ILs soon started to be explored as reaction media for the solubilization and synthesis of these compounds [51-55, 111]. Uzagare et al., performed a deep study of the solubility of 2´-deoxyribonucleosides in several imidazolium ILs [51]. The IL 1-methoxyethyl-3-methyl imidazolium methanesulfonate (moemim [Ms]) proved to be a better option for the solubilization of the nucleosides than any other IL or VOC. The acylation and peracylation of the 2´-deoxyribonucleosides can be performed in this IL with good isolation of the reaction product with yields between 89 and 95%. The same IL was used for the selective benzoylation of nucleosides using benzoyl cyanide [52]. Highly efficient benzoylations in mild reaction conditions (25ºC) are achieved with selective benzoylation of sugar hydroxyl groups over amine groups and preferential production of O-benzoylated derivatives. The major drawback of this strategy is the formation of the highly toxic HCN, as reaction by-product. As an alternative, the authors studied the process in the presence of 1-methoxyethyl-3-methylimidazolium trifluoroacetate (moemim [TFa]) which exhibits the ability to dissolve both deoxyribo- and ribonucleosides [111]. This IL is a good alternative for the selective benzoylation of nucleosides with high yields, under ambient conditions. These results were of the most importance for the development of three nucleoside based anti-viral drugs namely stavudine, brivudine and trifluridine in three ILs, moemim [Ms], moemim [TFa] and bmim [TFa] [53]. In all cases, the selected ILs showed better solvent properties than the commonly used VOCs like *N,N*-dimethylacetamide. Moreover, the amount of required solvent is reduced due to the high solubility of the synthesized compounds in the ILs. Higher reaction rates are attained with this strategy. The same ILs, with the addition of bmim [Ms], were also applied to the synthesis of antiviral 5-halouracil nucleosides which can be also used as building blocks for the synthesis of other antiviral drugs [54]. As in the abovementioned studies, ILs proved to be good reaction media for this synthesis and in this case could be recovered up to 4 cycles without loss of yield.

Two imidazolium ILs were also selected as reaction media for the synthesis of 2,3-disubstituted-1,3-thiazolinin-4-one derivatives with antiparasitic activity [55]. The synthesis yield is low at room temperature but by increasing the temperature up to 80 or 90ºC the reaction yield increases to 87 and 74% with bmim [PF_6] and bmim [BF_4], respectively. Tests performed with bmim [PF_6] demonstrate that the IL can be recovered and efficiently reused.

This section could not be completed if the works of Zang et al., on sonochemical organic synthesis in IL media were not mentioned. Several compounds including 2-aryl-4,5-diphenyl imidazoles [56], pyrazole derivatives [57] and oximes [58] with pharmaceutical interest. For the synthesis of these compounds, the authors combined bmim [OH] [58], emim [OAc] [56] or hmim [SO_4] [57] with ultrasonic irradiation. In all the tested conditions, the reaction time was reduced and the synthesis occurred efficiently with high yields, under mild conditions. Similarly, Heravi managed to synthesize quinolines derivatives, with pharmacological activity, through the combination of hbim [BF_4] and ultrasound, resorting to methanol as co-solvent [59].

To finish this section, two applications with immense potential to be implemented in pharmaceutical synthesis must be highlighted due to their importance on the safety and therapeutics perspective. Firstly, the oxidation of ketones, largely applied to the synthesis of antibiotics and steroids, was already performed in IL media resorting to imidazolium ILs as an alternative to the commonly used dichloromethane, chloroform and acetonitrile [112]. The most promising ILs were bmim [BF_4] and 1-methylimidazolium acetate (mim [OAc]) providing yields of lactone and ester synthesis between 65 and 95%.

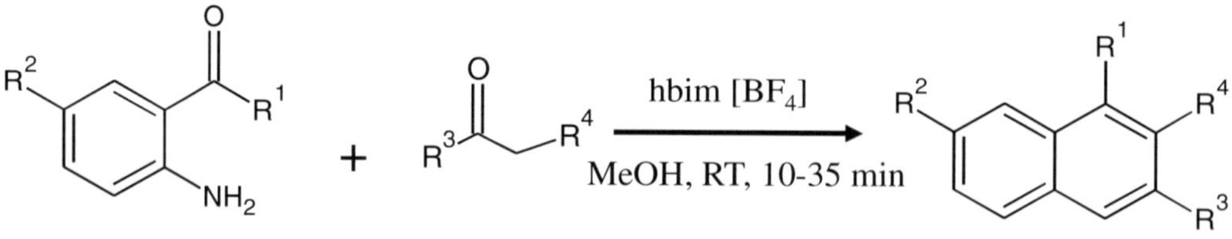

Scheme 5. Condensation of ketones to produce quinolone derivatives in the presence of hbim [BF_4] [59].

Again, ILs could be recycled 3 times without loss of yield. Finally, we highlight the major importance of the work of Hwang and co-workers regarding the formation of amyloid from α-synuclein in IL media. There are a number of amyloid related diseases that largely depend on the development of novel therapeutic strategies based on amyloid formation. However, amyloid formation is usually a very slow process which hinders more effective developments on this field. In this work, a set of imidazolium ILs and one pyridinium IL were tested and in all cases the ILs acted as stimulators of amyloid formation with decrease of the lag time of the process. Amongst the tested compounds, emim [NTf_2] is a stronger promoter when compared with other ILs.

3.2. ILs as Drug Vehicles

Solubility is one of the most important properties during drug development. It limits the concentration that a dosage form can incorporate and the rate at which the molecule dissolves from the solid form. It is then a key property for gastro-intestinal absorption of orally administered drugs. In fact, poor aqueous solubility of drugs is one of the major causes for low systemic exposure and, consequently, lack of *in vivo* activity [2, 113]. From the total amount of new chemical entities developed in the pharmaceutical industry, 40% fail to reach higher development levels due to solubility problems [114, 115]. The importance of solubility during drug development is illustrated by the *Biopharmaceutics Classification System (BCS)* which is basically a guide for the prediction of the intestinal drug absorption, provided by the U.S. FDA [116]. In this classification system, prediction relies in only two major parameters: solubility and intestinal permeability. In this context, it is not surprising that the search for novel strategies to increase the solubility of new drug candidates is one of the biggest concerns of both industry and academic researchers. Various techniques are used for the enhancement of the solubility of poorly soluble drugs which include physical and chemical modifications of drug and other methods like particle size reduction, crystal engineering, salt formation, solid dispersion, use of surfactant, complexation, among others. The selection of the strategy to be adopted is highly dependent on drug´s property, site of absorption, and characteristic of the dosage form [117].

ILs are well-known for their solvation properties that make them able to dissolve a wide range of organic and inorganic compounds [118]. This property has been associated with the fact that ILs form a highly solvating and little coordinating media. Additionally, the short range repulsive interactions between anions and cations enhance the solvent power of ILs [119]. In the pharmaceutical field, it has been demonstrated that ILs are capable of dissolving several drugs that are insoluble or poorly soluble in water and in the solvents commonly used in these applications [8, 13, 51, 111, 120-126]. Additionally, their inherent tunability associated with the possibility of designing a specific IL for a specific application are major attractions of these compounds in this field. Generally, there is an increase of the solvency of ILs with the increase of the alkyl side chain of the cation [127].

The studies with nucleoside derivatives of Kumar and co-workers illustrate the importance of solubility studies in a stage prior to synthesis [52, 53, 111]. As described in section 1.3.1.5, imidazolium ILs present better solvent properties for both deoxyribo- and ribonucleosides than the commonly used VOCs. These findings opened new perspectives regarding the synthesis of nucleoside derivatives in IL media.

The pioneering researches of Professor´s A.T. Florence group boosted the pharmaceutical utilization of ILs as solvents or drug reservoirs [121-123]. This application is mainly related with the ability of ILs to dissolve poorly water-soluble molecules drugs like albendazole and danazol [122], dexamethasone [121, 123], penicillin V, progesterone and dehydro-epiandrosterone [123]. For instance, the solubility of albendazole increased more than 20,000 times in hmim [PF_6] and more than 10,000 in bmim [PF_6]. The issue of IL´s water miscibility was also explored during this evaluation considering the preferential utilization of aqueous systems during drug´s formulation. The water miscibility of poorly water-miscible ILs, like bmim [PF_6], can be enhanced through the addition of a more water miscible IL with higher effects for more hydrophilic ILs. The variation of danazol solubility is far more significant with very good results in the water miscible IL bmim [BF_4]. Hexafluorophosphate ILs of the imidazolium group were tested as solvents and reservoirs of progesterone, dehydroepiandrosterone, dexamethasone, sucrose and penicillin V [123]. It should be highlighted that the solubility decreases with the increase of the alkyl chain length from bmim to omim. All the ILs are adequate for the entrapment and release of solubilized drugs. Distinctly, the same ILs were utilized to evaluate the release rate of solubilized hydrophilic and hydrophobic drugs by means of the passage of an electric current through the immiscible ILs [121]. The application of electric current to the ILs increases the release of both solubilized drugs with the highest variation being registered for omim [PF_6]. Similarly, Smith et al., studied the solubility of ibuprofen and paracetamol in bmim [PF_6] and hmim [PF_6] at different temperatures. Even though both ILs are good solvents of both drugs, the solubilities are higher in hmim [PF_6] than in bmim [PF_6]. ILs with trifluoromethanesulfonate and NTf_2 anions were selected to dissolve isoniazide [124], N-acetyl-L-cysteine, coumarin and 4-hydroxycoumarin [125]. Trifluoromethanesulfonate ILs are better solvents for isoniazide [124], N-acetyl-L-cysteine and 4-hydroxycoumarin than NTf_2 ones [125]. The inverse was observed for coumarin [125]. The solubility of isoniazide was also studied in ammonium ILs, together with other antibiotic drug pyrazinecarboxamide [128]. Amongst the tested compounds, didecyldimethylammonium nitrate is the best solvent for both drugs with isoniazid exhibiting higher solubility than pyrazinecarboxamide in all the tested ILs.

ILs can also be design to provide tunable and adequate hydrophilic-lipophilic balance to enhance water solubility of poorly soluble APIs [126]. This was evidenced recently on a study with amphotericin B and itraconazole, two complex, amphiphilic drugs with reduced water solubility. The selection of the anions and cations took in consideration not only the desired hydrophilic-lipophilic balance but also safety issues keeping in mind that they must be approved for consumption. Thus, acetate ILs based on short-chain fatty amines were synthesized (by choosing complementary functionality in one or both ions) and their effectiveness as solvents was evaluated. The adopted strategy proved to be a good option not only for the solubilization of drugs in the ILs but also to solubilize and maintain the drug in aqueous media.

Also on the perspective of the use of ILs as solvents of pharmaceuticals, there are already informations on the influence on drug´s properties that determine their pharmaceutical profile and *in vivo* behavior [129]. These informations were gathered during the study of imidazolium ILs as solvents of nimesulide, a widely used NSAID with reduced water solubility. The binding of nimesulide to human serum albumin (HSA) as well as its interaction with micelles of hexadecylphosphocholine to calculate partition coefficients, were studied in the presence and in the absence of emim [Ms] and emim [TfMs] 1%. As in aqueous media, in the presence of the selected ILs, nimesulide binds strongly to HSA by means of spontaneous interactions. The studied drug-IL systems exhibit properties that favor the interaction with biological membranes, evidenced by the high Kp values. Thus, there are good perspectives for the effective absorption and distribution *in vivo*.

ILs can also be incorporated in IL-in-oil microemulsions to be used as carriers of poorly soluble drugs [7-10]. Microemulsions are self-assembled structures widely recognized as vehicles for encapsulation, stabilization, and delivery of pharmaceutical drugs [130]. They are thermodynamically stable colloidal dispersions formed by water and oil and stabilized by an interfacial film or surfactant. The initial report on this topic was elaborated by Professor´s Goto group and involved a four component system composed of dimethylimidazolium dimethylphosphate (dmim $[(CH_3O)_2PO_2]$), isopropyl myristate, polyoxyethylene sorbitan (Tween-80) and sorbitan laurate (span-20) for the delivery of acyclovir [8, 9]. The IL was selected as the most adequate compound on solubility studies that were performed with seven imidazolium ILs. The increase of IL content on the micelle conducts to an increase of micelle volume and the incorporation of acyclovir results on a slight decrease of the mean diameter. The efficiency of the proposed delivery system was studied *in vitro* recurring to a porcine skin model using Franz-type diffusion cells. Skin permeability and the transdermal permeation of acyclovir increases when IL based microemulsions are used. The insertion of IL 4%wt in the microemulsion leads to a reduction of less than 20% in cell viability (human epidermal model LabCyte™ EPI-MODEL 12). With these pioneering studies, the authors confirmed the adequacy of these microemulsions to perform transdermal drug delivery. In continuation, the authors studied a set of imidazolium ILs as dispersed phase [7]. It was noticed that hydrophilic ILs with coordinating anions can be solubilized on the core of Tween-80/ span-20/ isopropyl myristate micelles and form stable microemulsion droplets more effectively. It is important to highlight that the solubility of the ILs in the microemulsion follow the same tendency as in water. The potentialities of the delivery system were further explored with methotrexate and dantrolene, two poorly water soluble drugs. The solubility of both drugs is enhanced in microemulsions with dmim $[(CH_3O)_2PO_2]$ with further increase by higher IL concentration. Similarly, Dobler and co-workers prepared oil-in-water and water-in-oil microemulsions containing hmim [Cl] and bmim $[PF_6]$ and evaluated the influence of the ILs on the properties of the micellar systems [131]. The ILs can be successfully incorporated in the micelles and due to their antimicrobial activity can have the effect of a preservative. As related above, skin permeation is enhanced in the presence of ILs.

Other microemulsion systems were explored as vehicles of insoluble drugs such as curcumin [132], dopamine and acetylcholine [133]. The hydrophilic IL bmim $[BF_4]$ increases about 7 times the association of curcumin with CTAB and induces a decrease of association in the case of cationic surfactants like sodium dodecylsulfate (SDS), acting like a modulator of drug´s solubilization in micellar systems. On a distinct perspective, the micellization and surface behavior of 1-tetradecyl-3-methylimidazolium bromide (tmim [Br]) was studied in the presence of dopamine and acetylcholine aiming the evaluation of the applicability of the micellar system in drug delivery [133]. The selected IL exhibits better surface activity and better drug carrier performance than the structurally similar cationic surfactant, tetradecyltrimethylammonium bromide. Imidazolium ILs are also good solvents for the release of APIs from styrene-divinylbenzene polymeric vehicles as demonstrated by Ramos-Rodriguez and co-workers [127]. The studies conducting to this conclusion were based on atomistic and mesoscopic simulations that evaluated solubility parameters and the diffusion of albendazole from the polymeric vehicle to an aqueous medium modified with bmim [BF_4], bmim [PF_6] or hmim [Br]. The diffusion of albendazole was related with the solvent power of the studied ILs, with bmim [PF_6] inducing higher release from the vehicle than bmim [BF_4] or hmim [Br] due to the higher polarity of [PF_6]. The evolution of the research on this field has been only foreclosed by one of the most delicate issues that concerns with ILs: toxicity. Further developments on the use of ILs as carriers or solvents of drugs are largely dependent on the preparation of safer ILs with reliable toxicity data that could be readily accepted as GRAS and used as excipients in pharmaceutical formulations.

3.3. ILs as Active Pharmaceutical Ingredients

Despite the technological advances of the last decades, nowadays there is still a significant rate of failure during drug development in the pharmaceutical industry. A study conducted between 1990 and 2002 revealed that only 40% of the drugs that pass phase I and II clinical trials will reach the market [134]. Failure was associated with lack of efficacy (50%), safety concerns (30%) or both (20%). The issue of efficacy is quite intriguing considering that this is the major concern of phase II clinical trials. Drug´s efficacy is tightly associated with bioavailability that in turns is dependent on fundamental properties such as solubility, dissolution rate and lipophilicity. Besides economical and judgmental reformulations, the strongest possibilities to overcome this failure rates are to profile drug-like properties as early as possible and work on lower-risk compounds.

Typically, the pharmaceutical industry is based on solid and crystalline forms of APIs with claimed advantages in terms of purity, solubility and thermal stability [135, 136]. It is also commonly accepted that solid forms of APIs facilitate isolation, manufacturing and product handling with reduction of the costs of the overall production process. This tendency is further encouraged by a vast amount of guidelines, most of them provided by FDA, concerning the manufacturing and quality control of solid pharmaceuticals with a notorious lack of guiding principles for liquid dosage forms. However, solid forms of APIs often suffer from low solubility and polymorphic conversion which can influence negatively the bioavailability of the drug and ultimately its therapeutic effect [136].

Polymorphism is the ability of a compound to exist as two or more crystalline phases that exhibit distinct arrangements and/or conformations of the molecule in the crystal lattice [135, 137, 138]. Thus, polymorphs display different physical, thermodynamic, spectroscopic, interfacial and mechanical properties. Moreover, the different lattice energies of polymorphs can originate different solubilities and dissolution rates, affecting the bioavailability of pharmaceutical drugs. Polymorphism is one of the major concerns of the pharmaceutical industry during drug development due its impact during formulation and on pharmacological effect, as the conversion can occur throughout drug manufacturing or storage [138]. During drug development, if polymorphs are detected the most active and stable is selected. As examples, acetaminophen (paracetamol) can be presented in two forms with distinct compression properties [137] and the antiviral drug ritonavir has an inactive polymorph that exhibits the ability to inactive the active polymorph [139]. Ranitidine hydrochloride has been subject of deep discussion and legal litigation due to an attempt to approve a more active polymorph than the patented and commercialized by GSK. However and despite all the concerns on this topic, problems with commercialized drugs can still occur.

Particle size is another problem related to solid forms of pharmaceuticals, so that there are rigid guidelines concerning this aspect. It is known that particle size influences several parameters that determine drugs´ bioavailability and stability [140].

In this context one cannot also forget the problems of solubility of the majority of the novel drug candidates that hinder their future commercialization [113]. It is then important to control critically polymorphism, solubility and particle size during the development of solid APIs to avoid failures in posterior phases.

3.3.1. Control and Modulation of Physico-Chemical and Biological Properties of Pharmaceutically Active ILs

The most common strategy of the pharmaceutical industry to overcome inadequate physico-chemical properties is salt formation [13, 135, 136]. Recent estimations indicate that around 50% of all drugs utilized in the pharmaceutical industry are salts with improved properties regarding the corresponding neutral molecules. As for neutral drugs, FDA has got lists of approved conterions which include organic and inorganic ones. Even though this strategy hinders some of the abovementioned problems, the possible combinations are relatively small and the industry recurs systematically to the same ions.

In this context, an IL approach, in the design of novel APIs, appears to be appropriate as it enables uncountable possible cation-anion combinations while providing distinctive properties unreachable in solid salts namely enhanced solubility and absence of polymorphic forms [12, 14, 141]. In this pharmaceutically active ILs (IL-APIs), the cation and the anion can both present biological activity or the counterion can have the ability to reduce the adverse effects of the active ingredient or change its solubility, for example. The possibility of combining more than two ions in the same compound cannot be forgotten as it can provide an additional biological activity or property. In these cases, the total charge of the cations is balanced by the total charge of the anions [142, 143]. Even though not predictable, synergistic or additive effects must also be considered as they may provide novel treatment possibilities and distinct drug delivery options. Furthermore, the assertive combination of appropriate anions and cations empowers the chemical manipulation of the compounds with specific objectives related with the manufacturing process, the stability of the formulations, their bioavailability and eventual adverse effects. In this perspective, it is important to highlight that small modification of an API can result on dramatic changes of its properties and eventually on its classification in the BCS. Thus, the IL approach can be profitable not only from the technological perspective but also on the pharmacological point of view since the possibility of modulate the abovementioned properties can facilitate the interaction of the compounds with biological membranes with consequent increase in their bioavailability and higher probabilities to actuate *in vivo.*

3.3.2. Synthesis and Profiling of Active Pharmaceutical Ingredients with IL Properties

The synthesis of APIs with IL properties is based on the evident similarity between IL-forming cations and APIs and on the assumption that the majority of bioactive ions present properties to form low melting salts, namely adequate size, diffuse charge and asymmetry. The initial developments on this field demanded a deep knowledge not only of the biological action of the ions but also of their ability to form ILs. Even though a critical ion selection is not a guarantee for IL formation, the most determining aspect during IL-APIs´ synthesis is definitely ion selection. Independently from the objective, the ions must be GRAS compounds with properties to be incorporated in pharmaceutical formulations either as APIs or excipients.

At this point, it is notorious that the research in this field is still mainly focused on the synthesis of novel compounds through the association of ions of distinct pharmacological groups or with diverse excipient properties. This is related with the relatively easy implementation of synthetic procedures based on metathesis reactions after the selection of ions with pharmaceutical interest. Moreover, the intensive investment on novel APIs with IL properties is still motivated by the enthusiasm related with the possibility of associating ions with the objective of modulate the pharmacological activity and physico-chemical properties while obtaining dual activity compounds.

The research group of Professor Rogers developed pioneer work regarding the synthesis and characterization of APIs with IL properties with a significant amount of papers and patents on this topic [12, 13, 141-147]. The first experiences on this field were reported on an interesting article exploring the so called *third evolution of ILs* [14]. Table 2 summarizes the most significant IL-APIs synthesized in the last years as well as their main properties and pharmacological activity.

Ranitidine docusate, lidocaine docusate and didecyldimethylammonium ibuprofenate can be synthesized through simple metathesis reactions to overcome issues of polymorphism, solubility and bioactivity [14]. The compounds exhibit dual activity since they incorporate well-known and widely used pharmaceutical ions and counterions with emollient (docusate) or antibacterial properties (didecyldimethylammonium). The association of docusate to lidocaine results on a hydrophobic API with increased thermal stability and enhanced bioactivity from topical application, possibly due to a slower release profile and higher skin permeability associated to the presence of the emollient. The three IL-APIs present low surface tension and low contact angles due to weak cation-anion interactions [161].

The IL-API approach enables the synthesis of compounds based on specifically selected ions. For instance benzalkonium and didecyldimethylammonium can form a dual active IL by combination with ibuprofen with applicability for the treatment of inflammation caused by bacterial infections [12]. Benzalkonium can also be used to form ILs with anions with similar activity, such as sulfacetamide. The result is an IL-API with enhanced antibiotic activity that can potentially reduce bacterial resistance. Similarly acetylsalicylic acid and salicylic acid can be combined with antimicrobial (cetylpyridinium, benzethonium, benzalkonium and hexetidinium), analgesic (tramadolium), local anesthetic (lidocainium and procainium) and antiarrhythmic (procainiumamide) cations to obtain dual activity IL-APIs with advantages in terms of solubility and functionality [145]. These combinations result in low melting or liquid salts at room temperature. However, acetylsalicylate IL-APIs suffer from low stability and their therapeutic application can only be guarantee if moisture is rigorously controlled. From the selected combinations, salicylate IL-APIs incorporating antibacterial actions, such as benzalkonium salicylate and cetylpyridinium salicylate exhibit promising physical and chemical properties that foresee their utilization in antimicrobial applications. This was further confirmed through the study of protein binding affinity, partition coefficients and surfactant properties of benzalkonium salicylate, cetylpyridinium salicylate and emim salicylate [148].

Table 2. Pharmacological activity and main features of IL-APIs

IL-API	Anion	Cation	Pharmacological activity	Main features	Ref.
Ranitidine docusate	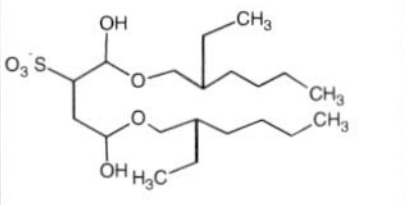	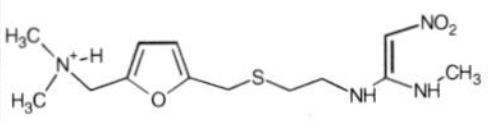	Histamine H_2 receptor antagonist Emollient	- Absence of polymorphs - Low surface tension - Low contact angles	[14]
Lidocainium docusate	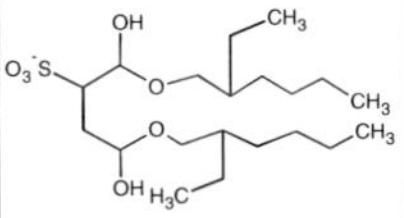	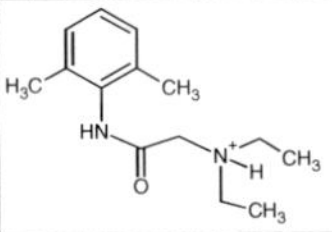	Local anesthetic Emollient	- Higher skin permeation	
Didecyldimethyl-ammonium ibuprofenate	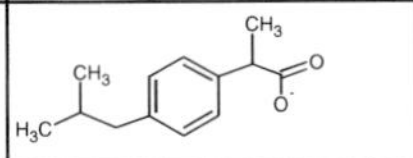	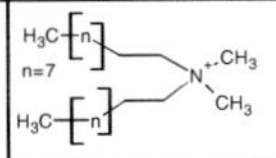	Antibacterial Anti-inflammatory	- Anion with primary biological activity	
Benzalkonium ibuprofenate	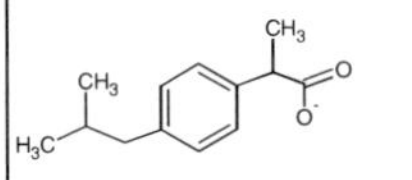	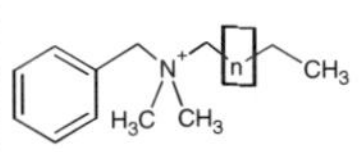	Antibacterial Anti-inflammatory	- Appropriate to combat inflammation associated to bacterial infections	[12]
Didecyldimethyl-ammonium ibuprofenate	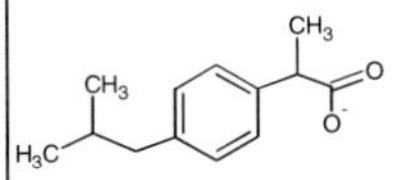	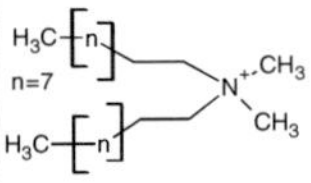	Antibacterial Anti-inflammatory	- Appropriate to combat inflammation associated to bacterial infections	
Benzalkonium sulfacetamide	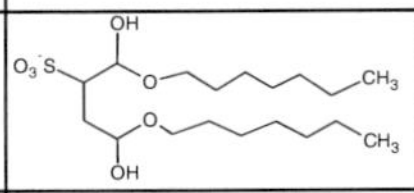	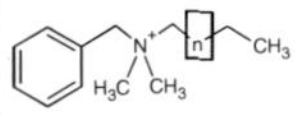	Antibacterial Anti-acne	- Reduction of bacterial resistance	
Cetylpyridinium salicylate	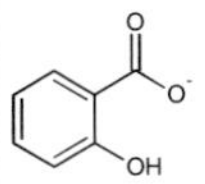	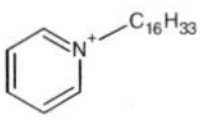	Antibacterial Anti-inflammatory	- Improved solubility - Strong bind to HSA - High partition coefficient - Slightly toxic (aquatic toxicity)	[145, 148, 149]
Cetylpyridinium acetylsalicylate	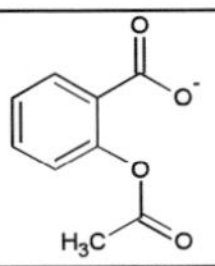	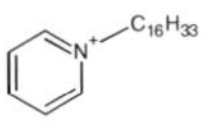	Antibacterial Anti-inflammatory	- Water sensitive - Low stability	[145]
Benzethonium salicylate	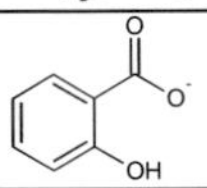	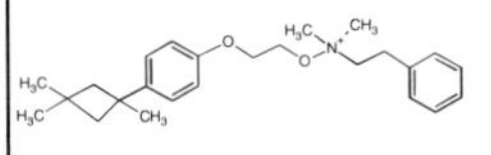	Antibacterial Anti-inflammatory	- Improved solubility	[145, 149]
Benzethonium acetylsalicylate	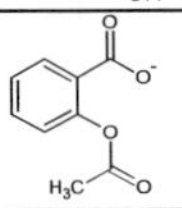	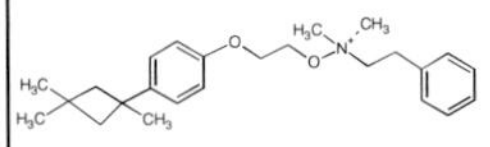	Antibacterial Anti-inflammatory	- Water sensitive - Low stability	[145]
Benzalkonium salicylate	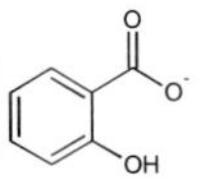	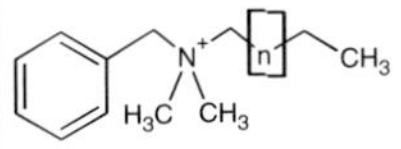	Antibacterial Anti-inflammatory	- Improved solubility - Strong bind to HSA - High partition coefficient - Slightly toxic (aquatic toxicity)	[145, 148, 149]

Table 2. (Continued)

IL-API	Anion	Cation	Pharmacological activity	Main features	Ref.
Benzalkonium acetylsalicylate	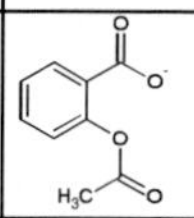	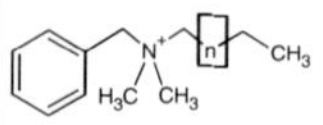	Antibacterial Anti-inflammatory	- Water sensitive - Low stability	[145]
Hexetidinium salicylate	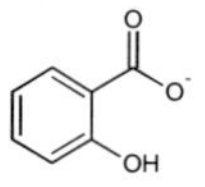	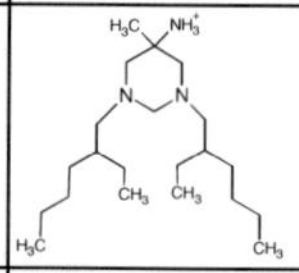	Antibacterial Anti-inflammatory	- Improved solubility	
Lidocainium salicylate	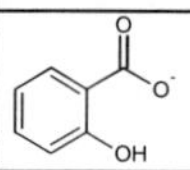	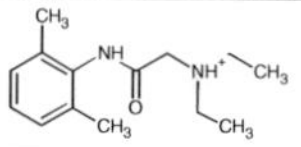	Local anesthetic Anti-inflammatory	- Improved solubility	
Procainium salicylate	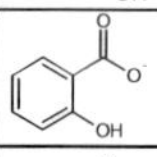	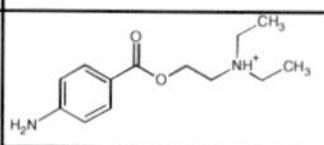	Local anesthetic Anti-inflammatory	- Improved solubility	
Procainamidium salicylate	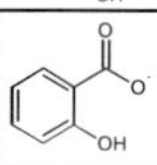	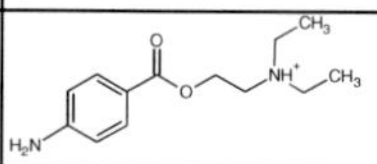	Antiarrhythmic Anti-inflammatory	- Improved solubility	
Tramadolium salicylate	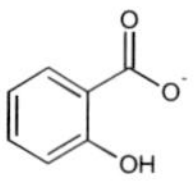	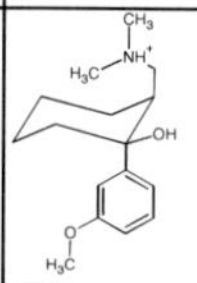	Analgesic Anti-inflammatory	- Improved solubility	
Tramadolium acetylsalicylate	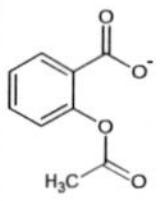	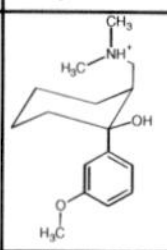	Analgesic Anti-inflammatory	- Water sensitive - Low stability	
Emim salicylate	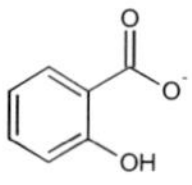	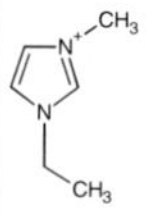	Low antibacterial activity Anti-inflammatory	- Improved solubility - Strong bind to HSA - Higher partition coefficient - Practically harmless (aquatic toxicity)	[145, 148, 149]
Emim ampicillin	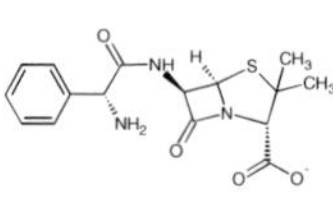	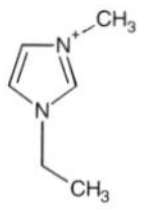	Antibacterial	- Improved solubility -Higher thermal stability Higher partition coefficient	[150-152]
C_2OHmim ampicillin	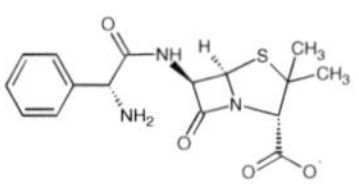	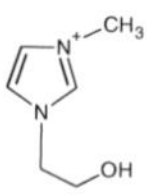	Antibacterial	-Improved solubility -Higher thermal stability Higher partition coefficient	
Choline ampicillin	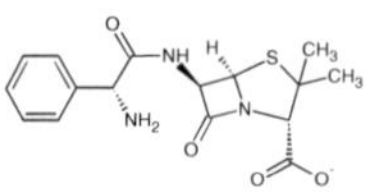	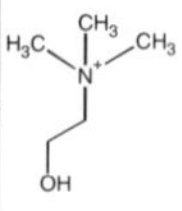	Antibacterial	- Low toxicity - Improved solubility -Higher thermal stability - Higher partition coefficient	

IL-API	Anion	Cation	Pharmacological activity	Main features	Ref.
TEA ampicillin	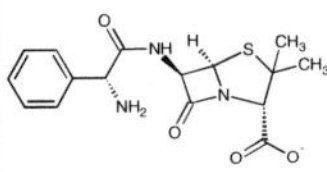	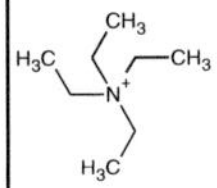	Antibacterial	- Improved solubility -Higher thermal stability - Higher partition coefficient	
Cetylpyridinium ampicillin	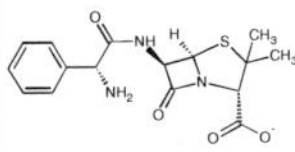	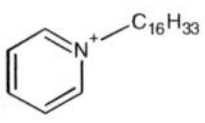	Antibacterial	-Higher thermal stability - Higher partition coefficient	[150-152]
T3htdph ampicillin	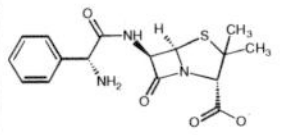	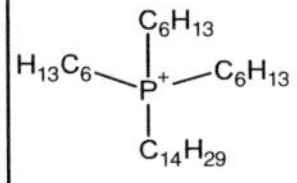	Antibacterial	-Higher thermal stability - Higher partition coefficient	
Tetracycline docusate	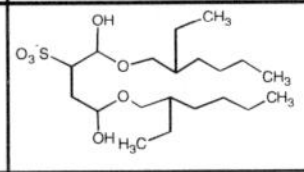	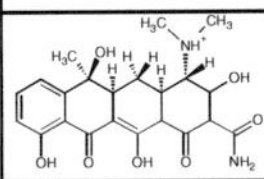	Antibacterial Emollient	- Lower solubility - Higher partition coefficient	[153]
Benzethonium saccharinate	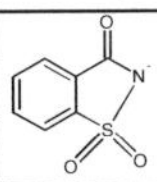	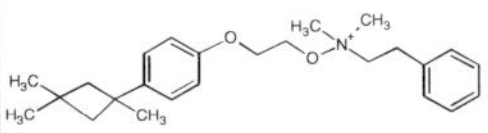	Antibacterial		[154]
Propantheline acesulfamate	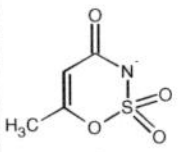	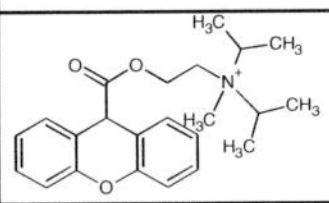	Muscarinic acethylcoline receptor antagonist	- Crystalline solids with low melting points - Absence of polymorphs	[154, 155]
Propantheline *p*-toluenesulfonate	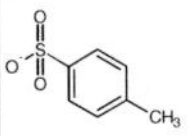	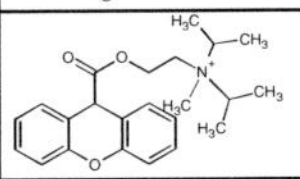	Muscarinic acethylcoline receptor antagonist		
Lidocainium ibuprofenate	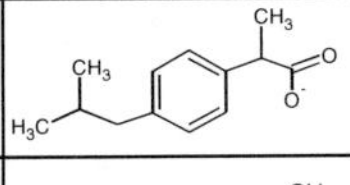	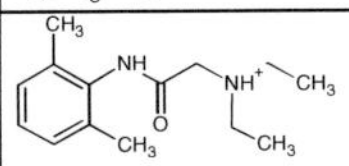	Local anesthetic Anti-inflammatory	- API-SILPs - Fast release - Controlled solubility	[146]
Tetrabutylphos-phonium ibuprofenate	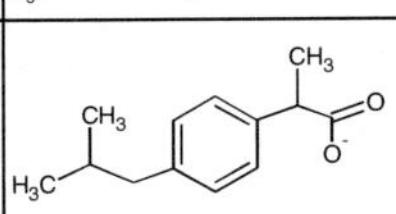	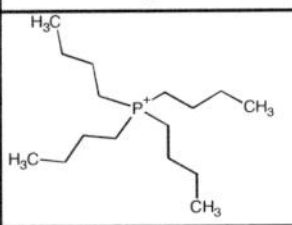	Antibacterial Anti-inflammatory		
Tuaminoheptane benzoate	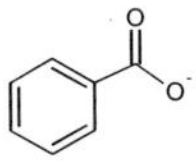	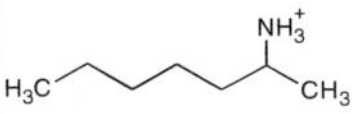	Nasal decongestant Preservative	- Solid or liquid and water soluble or insoluble depending on the ration	
Amantadine benzoate	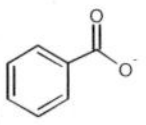	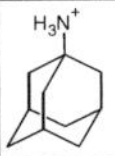	Antiviral/antipar-kinson Preservative	-Higher solubility (but still poor water soluble)	
Tuaminoheptane salicylate	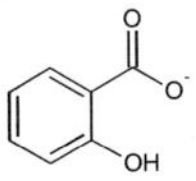	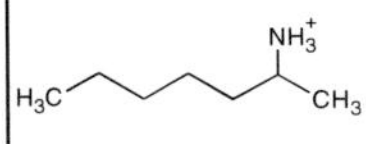	Nasal decongestant Anti-inflammatory Keratolytic	- Water soluble or insoluble depending on the ratio	[156]
Amantadine salicylate	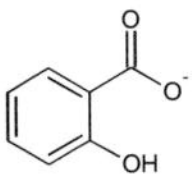	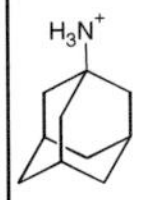	Antiviral/antipar-kinson Anti-inflammatory Keratolytic	- Water insoluble	
2-Pyrrolidino-ethanol salicylate	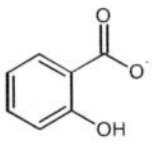	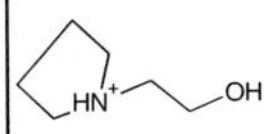	Anti-inflammatory Keratolytic	- Water soluble or insoluble depending on the ratio	
Tuaminoheptane gentisate	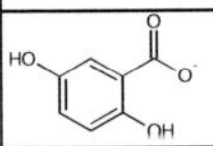	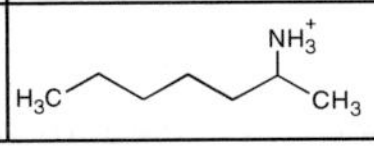	Nasal decongestant Analgesic	- Solid - Water soluble	

Table 2. (Continued)

IL-API	Anion	Cation	Pharmacological activity	Main features	Ref.
Amantadine gentisate	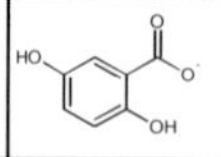	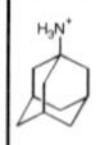	Antiviral/antiparkinson Analgesic	- Water insoluble	
2-Pyrrolidino-ethanol gentisate	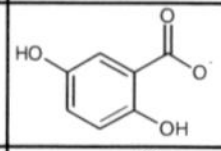	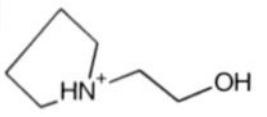	Analgesic	- Higher solubility than gentisic acid	
Tuaminoheptane salicylate	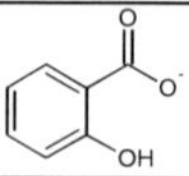	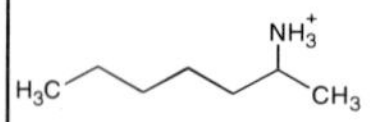	Nasal decongestant Anti-inflammatory Keratolytic	- Rapid membrane permeation	[157]
Bromohexinium ibuprofenate	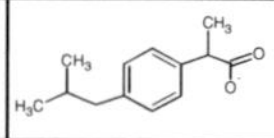	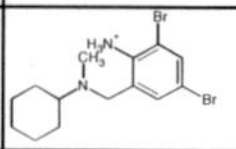	Mucolytic Anti-inflammatory		
Lidocainum etodolac	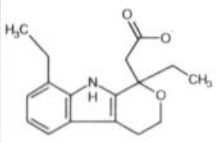	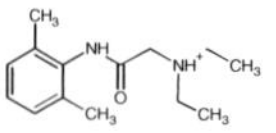	Local anesthetic Anti-inflflmmatory	- Higher skin permeability - Safe and well tolerated (results of clinical trials)	[158-160]
Benzethonium docusate	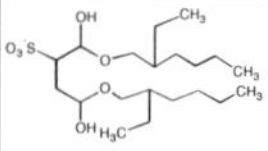	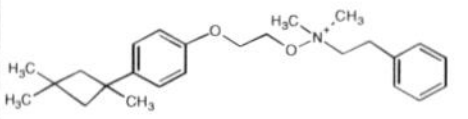	Antibacterial Emollient	- Water insoluble - Moderately toxic (aquatic toxicity)	[149]
Benzethonium bistriflimide		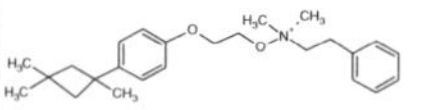	Antibacterial	- Water insoluble - Moderately toxic (aquatic toxicity)	
Trihexyltetradecyl-phosphonium bistriflimide	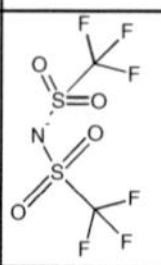	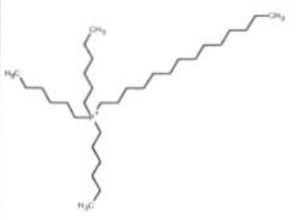	Antibacterial	- Water insoluble	
Trihexyltetradecyl-phosphonium docusate	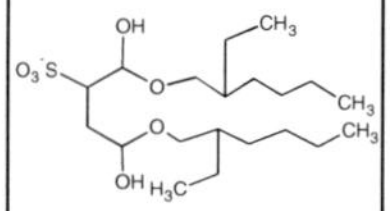	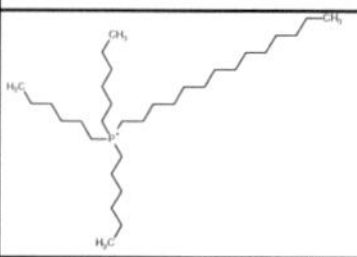	Antibacterial Emollient	- Water insoluble	
Trihexyltetradecyl-phosphonium salicylate	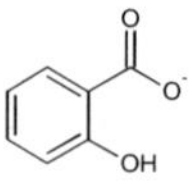	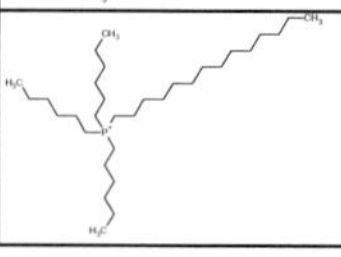	Antibacterial Anti-inflammatory	- Water insoluble	

The three compounds bind strongly and spontaneously to human serum albumin and exhibit the ability to form micelles. Moreover, the determined partition coefficients are up to 6 times higher than for sodium salicylate. These features indicate that these IL-APIs can exert direct effect on cell membranes and present adequate properties to be incorporated in antimicrobial pharmaceutical formulations, mainly topical ones.

Also in the field of antimicrobials, the thermal stability, solubility and melting point of ampicillin can be dramatically changed by association with organic cations such as emim, 1-hydroxy-ethyl-3-methylimidazolium (C_2OHmim), choline, tetraethylammonium (TEA), cetylpyridinium and trihexyltetradecylphosphonium [150-152]. IL-APIs can be obtained through the neutralization of moderately basic ammonia solutions of ampicillin with the mentioned cations. The obtained compounds present more adequate partition properties which result in enhanced membrane affinity and permeation [151]. Choline ampicillin is the most interesting compound resulting from this association as it presents adequate solubility and low toxicity. In terms of bacterial resistance reversion, the compound with highest potential is cetylpyridinium ampicillin [152]. Another interesting approach, explored since the first experiences with IL-APIs through the use of docusate [14], is to couple a bioactive ion with a well-known excipient as counterion. A good example is the synthesis of tetracycline docusate to modulate important biopharmaceutical properties of tetracycline [153]. Even though the solubility of the drug decreases to half on the IL format, both octanol-water and liposome-water partition coefficients show a favorable trend concerning *in vivo* activity. This approach is also very well illustrated by the synthesis of stable IL-APIs based on benzethonium and propantheline combined with artificial sweeteners such as acesulfamate, saccharinate and *p*-toluenesulfonate, via metathesis reaction [144, 154, 155]. The obtained compounds overcome issues of polymorphism, typical for instance of propantheline bromide and modify the thermal properties and solubility of the compounds. The utilization of sweeteners can be also interesting to camouflage the unpleasant taste of specific APIs being good options to integrate oral formulations.

Distinctly, the advantages of the liquid properties of an IL can be combined with the benefits of solid state by means of the immobilization of IL-APIs on mesoporous silica [146]. The study of the sorption, stability and release profile of tetrabutylphosphonium ibuprofenate and lidocainium ibuprofenate confirm that this strategy can be a tool for drug delivery with the ability to control the release of the adsorbed compound by manipulating the IL properties. There are evidences of total and fast leaching of the IL-API in physiological conditions (phosphate buffer saline, pH 7.4, 37ºC) and enhanced thermal stability of the IL-API compared with its liquid form.

The prodrug approach is also a possibility when dealing with pharmaceutically active ILs [147]. Prodrugs are pharmacologically inactive derivatives of a related drug molecule that demand a spontaneous or enzymatic transformation *in vivo* to release the active drug. They can be idealized to overcome pharmaceutical, pharmacokinetic, or pharmacodynamic problems such as chemical instability, reduced solubility, unacceptable taste or odor, insufficient oral absorption, inadequate blood-brain barrier permeability and toxicity, among others [162]. The combination of an IL strategy with the prodrug approach can be an important tool for the development of APIs with improved properties and higher possibilities to actuate efficiently *in vivo*. This was explored through the functionalization of acetaminophen (paracetamol) through esterification with chloroacetyl chloride to obtain a neutral acetoxy derivative [147]. This was subjected to further alkylation with typical IL precursors to form chloride salts or to anion exchange with silver docusate to obtain prodrug docusate ILs. This new prodrugs present low water solubility and are readily hydrolyzed in physiological conditions suggesting good bioactivity.

This section could not be closed without a mention to the importance of protic ILs (PILs) in the pharmaceutical field, namely as APIs. PILs are a subclass of ILs formed by the reaction of a Brønsted acid with a Brønsted base [100]. The main difference between PILs and other ILs is the proton transfer from the acid to the base that originates proton-donor and -acceptor sites, which can be used to build up a hydrogen-bonded network. The simplicity of the synthetic route for the production of PILs makes these compounds very promising for pharmaceutical applications [145, 156, 157, 163]. It is then possible to obtain PIL-APIs from acids and bases commonly used in the pharmaceutical industry. As examples, salicylic, benzoic and gentisic acids can form PIL-APIs through the combination with pharmaceutical bases such as tuaminoheptane, amantadine and 2-pyrrolidinoethanol, ethylamine and diethylamine, among others [156, 164]. The variable properties of the obtained compounds can be explored to modulate their membrane permeation. Studies with model membranes suggest that PIL-APIs with low ionicity can readily cross the lipid bilayer, probably as hydrogen bond complexes [157, 164]. Considering that PIL-APIs exist mainly as ion-pairs their behavior is similar to that of neutral species and thus can cross membranes more easily than ionic species. It is also possible that, proton transfer, at the point of dissolution into the membrane, plays a key role in the transport.

Even though it is controversial, in this group of compounds one can also include oligomeric ILs that result from the combination of non-stoichiometric ratios of the selected acids and bases [165]. This results in the formation of hydrogen-bonded moieties containing both ions and some neutral non-ionized compound. Thus, the composition of these liquids can range from fully ionized to partially ionized or non-ionized species generating intensive debate regarding the classification of these compounds. These are proton transfer IL systems in which the proton is shared by two or more cationic or anionic moieties leading to a controlled reduction of melting point with expansion of the liquid range. Oligomerization can be also applied to the formation of pharmaceutically active compounds as demonstrated by Bica and Rogers through the synthesis of oligomeric tetrabutylphosphonium salicylates and lidocainium salicylate [13, 163].

Recently, MEDRx Co. Ltd. (a Japanese pharmaceuticals company) and IL Pharma Inc. (a subsidiary of MEDRx Co., Ltd.) developed an etodolac-lidocaine patch (MRX-7EAT) for the treatment of pain and inflammation using ILTS® (Ionic Liquid Transdermal System) [158-160]. The patch presents higher skin permeability than the conventional etodolac patch with a consequent increase of the systemic exposure to etodolac. The results of non-clinical and clinical studies confirmed the safety and tolerability of the product [159, 160].

Considering the future pharmaceutical utilization of IL-APIs the toxicity and biodegradability of these compounds must be carefully evaluated and discussed as usually performed with conventional pharmaceutical drugs. Even though these compounds are synthesized from GRAS materials, the evaluation of toxicity must include the study of their effects on humans as well as their environmental impact since the principal pathway for their release is the aquatic environment due to their low volatility. Up to now there is very few information concerning the issues of toxicity and biodegradability and the available data refers to *in vitro* assays based on human carboxylesterase and *Vibrio fischeri* [149]. The effect of the abovementioned salicylates [148] on aquatic bacteria *Vibrio fischeri* ranges from practically harmless to slightly toxic. Due to its antimicrobial activity compounds incorporating the cetylpyridinium cation tend to be more toxic for both humans and the aquatic environment than those incorporating benzethonium and benzalkonium. These salicylates are only slightly more toxic than their original starting materials. This marginally enlarged risk is totally compensated by the benefits that accrue from the dual activity of the compounds. Interestingly, the anion bistriflimide can modulate the toxicity of the IL-APIs. For instance, the toxicity of benzethonium and phosphonium cations is decrease when combined with bistriflimide to form an IL-API. Not surprisingly, the toxicity of IL-APIs resulting from the association of benzethonium or trihexyltetradecylphosphonium with the docusate anion is related almost exclusively to the toxicity of the cations, due to the innocuous nature of this anion.

4. Final Remarks and Future Trends

The pharmaceutical utilization of ILs either as solvents, carriers or APIs has increased in the last years with successful examples both at the industrial and academic level.

The peculiar properties of ILs are of pivotal importance for their application as substitutes of conventional VOCs during the synthesis of pharmaceutical drugs and in other phases of pharmaceutical processing. With adequate compound selection the use of ILs as pharmaceutical solvents can result on a significant reduction of the human and environmental impact of pharmaceutical synthesis while providing higher yield synthetic routes as described for biocatalytic synthesis. The appearance of greener ILs with more attractive properties in terms of toxicity and biodegradability further foments their incorporation in pharmaceutical formulations as part of drug delivery systems with improved properties in terms of drug release profile and pharmacologic activity. This seems to be one of the most promising application of ILs in a near future being a full acceptance of this approach mainly dependent on the abovementioned toxicity concerns and on evidences of adequate pharmaceutical profiles of the involved drugs.

Despite the high resistance of the pharmaceutical industry to this concept, the synthesis of APIs with IL properties is undoubtfully the most exciting and captivating application in this field. The possibility of obtaining dual activity compounds or drugs with improved properties either concerning their efficiency or their undesirable effects opens countless possibilities for drugs with reduced bioavailability and for the treatment of diseases with particular and complex demands in terms of pharmacological intervention. The recent advances with IL-API prodrugs further enhances these possibilities. Studies with pharmaceutically active silica supported ILs offer also promising perspectives in the field of nanotechnology, mainly through the association of IL-APIs with nanoparticles for drug delivery.

The implementation of the IL strategy for the synthesis of novel drugs in the pharmaceutical industry demands only a deep change in mentalities of the involved institutions and the creation of specific tools to perform drug´s purification and manipulation on the liquid state. Legal and approval issues are largely facilitated by the well-known safe nature of the anions and cations used to synthesize novel IL-APIs. In conclusion, the conditions for the insertion of more products based on the IL-API concept in the pharmaceutical market are created and it is expect that once the abovementioned advantages are clearly evidenced by influential authorities the concept will disseminate rapidly.

References

[1] Morgan, S.; Grootendorst, P.; Lexchin, J.; Cunningham, C.; Greyson, D. The cost of drug development: A systematic review. *Health Policy* 2011, 100, 4-17.

[2] Di, L.; Kerns, E. H. Profiling drug-like properties in discovery research. *Curr. Opin. Chem. Biol.* 2003, 7, 402-408.

[3] Alfonsi, K.; Colberg, J.; Dunn, P. J.; Fevig, T.; Jennings, S.; Johnson, T. A.; Kleine, H. P.; Knight, C.; Nagy, M. A.; Perry, D. A.; Stefaniak, M. Green chemistry tools to influence a medicinal chemistry and research chemistry based organization. *Green Chem.* 2008, 10, 31-36.

[4] Sheldon, R. A. Green solvents for sustainable organic synthesis: state of the art. *Green. Chem.* 2005, 7, 267-278.

[5] Plechkova, N. V.; Seddon, K. R. Applications of ionic liquids in the chemical industry. *Chem. Soc. Rev.* 2008, 37, 123-150.

[6] Wasserscheid, P.; Welton, T. *Ionic liquids in synthesis*; 2nd Weinheim: Wiley-VCH Verlag GmbH & Co. KGaA; 2008.

[7] Moniruzzaman, M.; Kamiya, N.; Goto, M. Ionic liquid based microemulsion with pharmaceutically accepted components: Formulation and potential applications. *J. Colloid. Interf. Sci.* 2010, 352, 136-142.

[8] Moniruzzaman, M.; Tahara, Y.; Tamura, M.; Kamiya, N.; Goto, M. Ionic liquid-assisted transdermal delivery of sparingly soluble drugs. *Chem. Commun.* 2010, 46, 1452-1454.

[9] Moniruzzaman, M.; Tamura, M.; Tahara, Y.; Kamiya, N.; Goto, M. Ionic liquid-in-oil microemulsion as a potential carrier of sparingly soluble drug: Characterization and cytotoxicity evaluation. *Int. J. Pharm.* 2010, 400, 243-250.

[10] Moniruzzaman, M.; Goto, M. Ionic Liquids: Future Solvents and Reagents for Pharmaceuticals. *J. Chem. Eng. Jpn.* 2011, 44, 370-381.

[11] Rodriguez, H.; Bica, K.; Rogers, R. D. Ionic Liquid Technology: A Potential New Platform for the Pharmaceutical Industry. *Tropical Journal of Pharmaceutical Research* 2008, 7, 1011-1012.

[12] Hough, W. L.; Rogers, R. D. Ionic liquids then and now: From solvents to materials to active pharmaceutical ingredients. *B. Chem. Soc. Jpn.* 2007, 80, 2262-2269.

[13] Shamshina, J. L.; Barber, P. S.; Rogers, R. D. Ionic liquids in drug delivery. *Expert Opin. Drug Del.* 2013, 10, 1367-1381.

[14] Hough, W. L.; Smiglak, M.; Rodriguez, H.; Swatloski, R. P.; Spear, S. K.; Daly, D. T.; Pernak, J.; Grisel, J. E.; Carliss, R. D.; Soutullo, M. D.; Davis, J. H.; Rogers, R. D. The third evolution of ionic liquids: active pharmaceutical ingredients. *New J. Chem.* 2007, 31, 1429-1436.

[15] Walden, P. Über die Molekulargrösse und elektrische Leitfähigkeit einiger gesehmolzenen Salze *Bulletin de l'Académie Impériale des Sciences de St Petersbourg* 1914, 8, 405-422.

[16] Chum, H. L.; Koch, V. R.; Miller, L. L.; Osteryoung, R. A. Electrochemical scrutiny of organometallic iron complexes and hexamethylbenzene in a room-temperature molten-salt. *J. Am. Chem. Soc.* 1975, 97, 3264-3265.

[17] Gale, R. J.; Gilbert, B.; Osteryoung, R. A. Raman-spectra of molten aluminum-chloride - 1-butylpyridinium chloride systems at ambient-temperatures. *Inorg. Chem.* 1978, 17, 2728-2729.

[18] Pernak, J.; Sobaszkiewicz, K.; Mirska, I. Anti-microbial activities of ionic liquids. *Green Chem.* 2003, 5, 52-56.

[19] Pernak, J.; Smiglak, M.; Griffin, S. T.; Hough, W. L.; Wilson, T. B.; Pernak, A.; Zabielska-Matejuk, J.; Fojutowski, A.; Kita, K.; Rogers, R. D. Long alkyl chain quaternary ammonium-based ionic liquids and potential applications. *Green Chem.* 2006, 8, 798-806.

[20] Walkiewicz, F.; Materna, K.; Kropacz, A.; Michalczyk, A.; Gwiazdowski, R.; Praczyk, T.; Pernak, J. Multifunctional long-alkyl-chain quaternary ammonium azolate based ionic liquids. *New J. Chem.* 2010, 34, 2281-2289.

[21] Kumar, V.; Malhotra, S. V. Study on the potential anti-cancer activity of phosphonium and ammonium-based ionic liquids. *Bioorg. Med. Chem. Lett.* 2009, 19, 4643-4646.
[22] Boltic, Z.; Ruzic, N.; Jovanovic, M.; Savic, M.; Jovanovic, J.; Petrovic, S. Cleaner production aspects of tablet coating process in pharmaceutical industry: problem of VOCs emission. *J. Clean. Prod.* 2013, 44, 123-132.
[23] Kerton, F. M. *Alternative Solvents for Green Chemistry*; Cambridge, UK: RSC Publishing; 2009.
[24] Constable, D. J. C.; Jimenez-Gonzalez, C.; Henderson, R. K. Perspective on solvent use in the pharmaceutical industry. *Org. Process Res. Dev.* 2007, 11, 133-137.
[25] FDA, US Department of Health and Human Services. *Guidance for Industry*. Q3C — Tables and List. 2012.
[26] European Medicines Agency. ICH Topic Q3C (R4). Impurities: *Guideline for Residual Solvents*. 2009.
[27] Welton, T. Room-temperature ionic liquids. Solvents for synthesis and catalysis. *Chem. Rev.* 1999, 99, 2071-2083.
[28] Hallett, J. P.; Welton, T. Room-temperature ionic liquids: solvents for synthesis and catalysis. 2 *Chem. Rev.* 2011, 111, 3508-3576.
[29] Sowmiah, S.; Cheng, C. I.; Chu, Y.-H. Ionic liquids for green organic synthesis. *Curr. Org. Synth.* 2012, 9, 74-95.
[30] Wolan, A.; Zaidlewicz, M. Synthesis of arylboronates by the palladium catalysed cross-coupling reaction in ionic liquids. *Org. Biomol. Chem.* 2003, 1, 3274-3276.
[31] Zhang, Q.; Zhang, S.; Deng, Y. Recent advances in ionic liquid catalysis. *Green Chem.* 2011, 13, 2619-2637.
[32] Chakraborty, B.; Sharma, C. D. A new route to the synthesis of isoxazoline derivatives from dihydropyran via cycloaddition reaction in ionic liquid. *Tetrahedron. Lett.* 2013, 54, 5532-5536.
[33] Kathiravan, S.; Raghunathan, R. Synthesis of pyrrolo 2,3-a pyrrolizidino derivatives through intramolecular 1,3-dipolar cycloaddition in ionic liquid medium. *Synthetic Commun.* 2013, 43, 147-155.
[34] Parmar, N. J.; Patel, R. A.; Parmar, B. D.; Talpada, N. P. An efficient domino reaction in ionic liquid: Synthesis and biological evaluation of some pyrano- and thiopyrano-fused heterocycles. *Bioorg. Med. Chem. Lett.* 2013, 23, 1656-1661.
[35] Liu, S.; Thomson, N.; Pettman, A.; Hyder, Z.; Mo, J.; Xiao, J. Ionic liquids as solvent for regioselective arylation of alpha-substituted allylic alcohols by aryl bromides. *J. Mol. Catal. A-Chem.*, 2008, 279, 210-217.
[36] Monteiro, A. L.; Zinn, F. K.; DeSouza, R. F.; Dupont, J. Asymmetric hydrogenation of 2-arylacrylic acids catalyzed by immobilized Ru-BINAP complex in 1-n-butyl-3-methylimidazolium tetrafluoroborate molten salt. *Tetrahedron-Asymmetr.* 1997, 8, 177-179.
[37] Ngo, H. L.; Hu, A. G.; Lin, W. B. Catalytic asymmetric hydrogenation of aromatic ketones in room temperature ionic liquids. *Tetrahedron Lett.* 2005, 46, 595-597.
[38] Earle, M. J.; McCormac, P. B.; Seddon, K. R. The first high yield green route to a pharmaceutical in a room temperature ionic liquid. *Green Chem.* 2000, 2, 261-262.
[39] Yeung, K. S.; Farkas, M. E.; Qiu, Z. L.; Yang, Z. Friedel-Crafts acylation of indoles in acidic imidazolium chloroaluminate ionic liquid at room temperature. *Tetrahedron Lett.* 2002, 43, 5793-5795.

[40] Li, C.; Liu, W.; Zhao, Z. Efficient synthesis of benzophenone derivatives in Lewis acid ionic liquids. *Catal. Commun.* 2007, 8, 1834-1837.

[41] Earle, M. J.; Hakala, U.; Hardacre, C.; Karkkainen, J.; McAuley, B. J.; Rooney, D. W.; Seddon, K. R.; Thompson, J. M.; Wahala, K. Chloroindate(III) ionic liquids: recyclable media for Friedel-Crafts acylation reactions. *Chem. Commun.* 2005, 903-905.

[42] Chen, Y.; Zu, Y.; Fu, Y.; Zhang, X.; Yu, P.; Sun, G.; Efferth, T. Efficient lewis acid ionic liquid-catalyzed synthesis of the key intermediate of coenzyme q(10) under microwave irradiation. *Molecules* 2010, 15, 9486-9495.

[43] Hongwei, Y.; Jinchuan, W.; Chi Bun, C. Kinetic resolution of ibuprofen catalyzed by Candida rugosa lipase in ionic liquids. *Chirality* 2005, 17, 16-21.

[44] Contesini, F. J.; Carvalho, P. d. O. Esterification of (RS)-ibuprofen by native and commercial lipases in a two-phase system containing ionic liquids. *Tetrahedron-Asymmetr.* 2006, 17, 2069-2073.

[45] Kurata, A.; Kitamura, Y.; Irie, S.; Takemoto, S.; Akai, Y.; Hirota, Y.; Fujita, T.; Iwai, K.; Furusawa, M.; Kishimoto, N. Enzymatic synthesis of caffeic acid phenethyl ester analogues in ionic liquid. *J. Biotechnol.* 2010, 148, 133-138.

[46] Pfruender, H.; Amidjojo, M.; Kragl, U.; Weuster-Botz, D. Efficient whole-cell biotransformation in a biphasic ionic liquid/water system. *Angew Chem. Int. Edit.* 2004, 43, 4529-4531.

[47] Pfruender, H.; Jones, R.; Weuster-Botz, D. Water immiscible ionic liquids as solvents for whole cell biocatalysis. *J. Biotechnol.* 2006, 124, 182-190.

[48] Weuster-Botz, D. Process intensification of whole-cell biocatalysis with ionic liquids. *Chem. Rec.* 2007, 7, 334-340.

[49] Braeutigam, S.; Dennewald, D.; Schuermann, M.; Lutje-Spelberg, J.; Pitner, W.-R.; Weuster-Botz, D. Whole-cell biocatalysis: Evaluation of new hydrophobic ionic liquids for efficient asymmetric reduction of prochiral ketones. *Enzyme Microb. Tech.* 2009, 45, 310-316.

[50] Choi, H. J.; Uhm, K.-N.; Kim, H.-K. Production of chiral compound using recombinant Escherichia coli cells co-expressing reductase and glucose dehydrogenase in an ionic liquid/water two phase system. *J. Mol. Catal. B-Enzym.* 2011, 70, 114-118.

[51] Uzagare, M. C.; Sanghvi, Y. S.; Salunkhe, M. M. Application of ionic liquid 1-methoxyethyl-3-methyl imidazolium methanesulfonate in nucleoside chemistry. *Green Chem.* 2003, 5, 370-372.

[52] Prasad, A. K.; Kumar, V.; Malhotra, S.; Ravikumar, V. T.; Sanghvi, Y. S.; Parmar, V. S. 'Green' methodology for efficient and selective benzoylation of nucleosides using benzoyl cyanide in an ionic liquid. *Bioorg. Med. Chem.* 2005, 13, 4467-4472.

[53] Kumar, V.; Malhotra, S. V. Synthesis of nucleoside-based antiviral drugs in ionic liquids. *Bioorg. Med. Chem. Lett.* 2008, 18, 5640-5642.

[54] Kumar, V.; Malhotra, S. V. Ionic liquid mediated synthesis of 5-halouracil nucleosides: key precursors for potential antiviral drugs. *Nucleos. Nucleot. Nucl.* 2009, 28, 821-834.

[55] Zhang, X.; Li, X.; Li, D.; Qu, G.; Wang, J.; Loiseau, P. M.; Fan, X. Ionic liquid mediated and promoted eco-friendly preparation of thiazolidinone and pyrimidine nucleoside-thiazolidinone hybrids and their antiparasitic activities. *Bioorg. Med. Chem. Lett.* 2009, 19, 6280-6283.

[56] Zang, H.; Su, Q.; Mo, Y.; Cheng, B.-W.; Jun, S. Ionic liquid EMIM OAc under ultrasonic irradiation towards the first synthesis of trisubstituted imidazoles. *Ultrason. Sonochem.* 2010, 17, 749-751.

[57] Zang, H.; Su, Q.; Mo, Y.; Cheng, B. Ionic liquid under ultrasonic irradiation towards a facile synthesis of pyrazolone derivatives. *Ultrason. Sonochem.* 2011, 18, 68-72.

[58] Zang, H.; Wang, M.; Cheng, B.-W.; Song, J. Ultrasound-promoted synthesis of oximes catalyzed by a basic ionic liquid bmIm OH *Ultrason. Sonochem.* 2009, 16, 301-303.

[59] Heravi, M. R. P. An efficient synthesis of quinolines derivatives promoted by a room temperature ionic liquid at ambient conditions under ultrasound irradiation via the tandem addition/annulation reaction of o-aminoaryl ketones with alpha-methylene ketones. *Ultrason. Sonochem.* 2009, 16, 361-366.

[60] Funel, J.-A.; Abele, S. Industrial applications of the diels-alder reaction. *Angew Chem. Int. Edit.* 2013, 52, 3822-3863.

[61] Dugger, R. W.; Ragan, J. A.; Ripin, D. H. B. Survey of GMP bulk reactions run in a research facility between 1985 and 2002. *Org. Process Res. Dev.* 2005, 9, 253-258.

[62] Lee, C. W. Diels-Alder reactions in chloroaluminate ionic liquids: Acceleration and selectivity enhancement. *Tetrahedron Lett.* 1999, 40, 2461-2464.

[63] Yeom, C.-E.; Kim, H. W.; Shin, Y. J.; Kim, B. M. Chiral bis(oxazoline)-copper complex catalyzed Diels-Alder reaction in ionic liquids: remarkable reactivity and selectivity enhancement, and efficient recycling of the catalyst. *Tetrahedron Lett.* 2007, 48, 9035-9039.

[64] Doherty, S.; Goodrich, P.; Hardacre, C.; Luo, H. K.; Rooney, D. W.; Seddon, K. R.; Styring, P. Marked enantioselectivity enhancements for Diels-Alder reactions in ionic liquids catalysed by platinum diphosphine complexes. *Green Chem.* 2004, 6, 63-67.

[65] Doherty, S.; Goodrich, P.; Hardacre, C.; Knight, J. G.; Nguyen, M. T.; Parvulescu, V. I.; Paun, C. Recyclable copper catalysts based on imidazolium-tagged bis(oxazolines): A marked enhancement in rate and enantioselectivity for Diels-Alder reactions in ionic liquid. *Adv. Synth. Catal.* 2007, 349, 951-963.

[66] Torborg, C.; Beller, M. Recent applications of palladium-catalyzed coupling reactions in the pharmaceutical, agrochemical, and fine chemical industries. *Adv. Synth. Catal.* 2009, 351, 3027-3043.

[67] Crisp, G. T. Variations on a theme - recent developments on the mechanism of the Heck reaction and their implications for synthesis. *Chem. Soc. Rev.* 1998, 27, 427-436.

[68] Bellina, F.; Chiappe, C. The Heck reaction in ionic liquids: progress and challenges. *Molecules* 2010, 15, 2211-2245.

[69] Soloway, A. H.; Tjarks, W.; Barnum, B. A.; Rong, F. G.; Barth, R. F.; Codogni, I. M.; Wilson, J. G. The chemistry of neutron capture therapy. *Chem. Rev.* 1998, 98, 1515-1562.

[70] Nolan, S. P.; Clavier, H. Chemoselective olefin metathesis transformations mediated by ruthenium complexes. *Chem. Soc. Rev.* 2010, 39, 3305-3316.

[71] Tomasek, J.; Schatz, J. Olefin metathesis in aqueous media. *Green Chem.* 2013, 15, 2317-2338.

[72] Moss, T. A. A ring-closing metathesis approach to heterocycle-fused azepines *Tetrahedron Lett.* 2013, 54, 993-997.

[73] Kim, M.; Gajulapati, K.; Kim, C.; Jung, H. Y.; Goo, J.; Lee, K.; Kaur, N.; Kang, H. J.; Chung, S. J.; Choi, Y. A facile synthetic route to diazepinone derivatives via ring

closing metathesis and its application for human cytidine deaminase inhibitors. *Chem. Commun.* 2012, 48, 11443-11445.

[74] Ishigai, K.; Fuwa, H.; Hashizume, K.; Fukazawa, R.; Cho, Y.; Yotsu-Yamashita, M.; Sasaki, M. Total synthesis and biological evaluation of (+)-gambieric acid a and its analogues. *Chem.-Eur. J.* 2013, 19, 5276-5288.

[75] Pradere, U.; Clavier, H.; Roy, V.; Nolan, S. P.; Agrofoglio, L. A. The shortest strategy for generating phosphonate prodrugs by olefin cross-metathesis - application to acyclonucleoside phosphonates. *Eur. J. Org. Chem.* 2011, 7324-7330.

[76] Muthusamy, S.; Azhagan, D. Efficient synthesis of 19-31 membered macrocyclic tetralactones via ring closing metathesis in ionic liquids. *Tetrahedron* 2010, 66, 8196-8202.

[77] Thomas, P. A.; Marvey, B. B. C18:1 methyl ester metathesis in bmim x type ionic liquids. *Int. J. Mol. Sci.* 2009, 10, 5020-5030.

[78] Buijsman, R. C.; van Vuuren, E.; Sterrenburg, J. G. Ruthenium-catalyzed olefin metathesis in ionic liquids. *Org. Lett.* 2001, 3, 3785-3787.

[79] Audic, N.; Clavier, H.; Mauduit, M.; Guillemin, J. C. An ionic liquid-supported ruthenium carbene complex: A robust and recyclable catalyst for ring-closing olefin metathesis in ionic liquids. *J. Am. Chem. Soc.* 2003, 125, 9248-9249.

[80] Yao, Q. W.; Sheets, M. An ionic liquid-tagged second generation Hoveyda-Grubbs ruthenium carbene complex as highly reactive and recyclable catalyst for ring-closing metathesis of di-, tri- and tetrasubstituted dienes. *J. Organomet. Chem.* 2005, 690, 3577-3584.

[81] Clavier, H.; Audic, N.; Guillemin, J. C.; Mauduit, M. Olefin metathesis in room temperature ionic liquids using imidazolium-tagged ruthenium complexes. *J. Organomet. Chem.* 2005, 690, 3585-3599.

[82] Mayo, K. G.; Nearhoof, E. H.; Kiddle, J. J. Microwave-accelerated ruthenium-catalyzed olefin metathesis. *Org. Lett.* 2002, 4, 1567-1570.

[83] Rueping, M.; Nachtsheim, B. J. A review of new developments in the Friedel-Crafts alkylation - From green chemistry to asymmetric catalysis. *Beilstein Journal of Organic Chemistry* 2010, 6.

[84] Kazemi, S.; Peters, C. J.; Kroon, M. C. Phase behavior of the ternary system acetylferrocene, the ionic liquid 1-butyl-3-methylimidazolium bis(trifluoromethylsulfonyl)imide, and carbon dioxide to be applied in friedel-crafts acylation reactions. *J. Chem. Eng. Data* 2013, 58, 951-955.

[85] Boon, J. A.; Levisky, J. A.; Pflug, J. L.; Wilkes, J. S. Friedel crafts reactions in ambient-temperature molten-salts. *J. Org. Chem.* 1986, 51, 480-483.

[86] Pan, A. X.; He, T.; Xu, L. L.; Fang, Y.; Li, H. M.; Xie, J. M.; Chen, M. Efficient synthesis of 1-acetylpyrene using bmim Cl-$FeCl_3$ ionic liquid as dual catalyst and solvent. *Int. J. Chem. React. Eng.* 2013, 11, 1-7.

[87] Zhang, L.-R.; Yi, F.-P.; Zou, J.-Z.; Zhang, X.; Wang, Z. Regioselective Friedel-Crafts acylation of indoles catalysed by zinc oxide in an ionic liquid. *J. Chem. Res.* 2012, 600-602.

[88] Phuong Hoang, T.; Duus, F.; Thach Ngoc, L. Friedel-Crafts acylation using bismuth triflate in BMI PF6. *Tetrahedron Lett.* 2012, 53, 222-224.

[89] Rasor, J. P.; Voss, E. Enzyme-catalyzed processes in pharmaceutical industry. *Appl. Catal. A-Gen.* 2001, 221, 145-158.

[90] Clouthier, C. M.; Pelletier, J. N. Expanding the organic toolbox: a guide to integrating biocatalysis in synthesis. *Chem. Soc. Rev.* 2012, 41, 1585-1605.
[91] Klibanov, A. M. Enzymatic catalysis in anhydrous organic-solvents. *Trends Biochem. Sci.* 1989, 14, 141-144.
[92] Hudson, E. P.; Eppler, R. K.; Clark, D. S. Biocatalysis in semi-aqueous and nearly anhydrous conditions. *Curr. Opin. Biotech.* 2005, 16, 637-643.
[93] Klibanov, A. M. Improving enzymes by using them in organic solvents. *Nature* 2001, 409, 241-246.
[94] Klibanov, A. M. Asymmetric enzymatic oxidoreductions in organic solvents. *Curr. Opin. Biotech.* 2003, 14, 427-431.
[95] van Rantwijk, F.; Sheldon, R. A. Biocatalysis in ionic liquids. *Chem. Rev.* 2007, 107, 2757-2785.
[96] Zhao, H. Methods for stabilizing and activating enzymes in ionic liquids - *a review J. Chem. Technol. Biotechnol.* 2010, 85, 891-907.
[97] Muginova, S. V.; Galimova, A. Z.; Polyakov, A. E.; Shekhovtsova, T. N. Ionic liquids in enzymatic catalysis and biochemical methods of analysis: Capabilities and prospects. *J. Anal. Chem.* 2010, 65, 331-351.
[98] Yang, Z.; Pan, W. B. Ionic liquids: Green solvents for nonaqueous biocatalysis. *Enzyme Microb. Tech.* 2005, 37, 19-28.
[99] Park, S.; Kazlauskas, R. J. Biocatalysis in ionic liquids - advantages beyond green technology. *Curr. Opin. Biotech.* 2003, 14, 432-437.
[100] Greaves, T. L.; Drummond, C. J. Protic ionic liquids: Properties and applications. *Chem. Rev.* 2008, 108, 206-237.
[101] Feher, E.; Major, B.; Belafi-Bako, K.; Gubicza, L. On the background of enhanced stability and reusability of enzymes in ionic liquids. *Biochem. Soc. T* 2007, 35, 1624-1627.
[102] Kaftzik, N.; Wasserscheid, P.; Kragl, U. Use of ionic liquids to increase the yield and enzyme stability in the beta-galactosidase catalysed synthesis of N-acetyllactosamine *Org. Process Res. Dev.* 2002, 6, 553-557.
[103] Kragl, U.; Eckstein, M.; Kaftzik, N. Enzyme catalysis in ionic liquids. *Curr. Opin. Biotech.* 2002, 13, 565-571.
[104] Erbeldinger, M.; Mesiano, A. J.; Russell, A. J. Enzymatic catalysis of formation of Z-aspartame in ionic liquid - An alternative to enzymatic catalysis in organic solvents. *Biotechnol. Progr.* 2000, 16, 1129-1131.
[105] Moon, Y. H.; Lee, S. M.; Ha, S. H.; Koo, Y. M. Enzyme-catalyzed reactions in ionic liquids. *Korean J. Chem. Eng.* 2006, 23, 247-263.
[106] Cull, S. G.; Holbrey, J. D.; Vargas-Mora, V.; Seddon, K. R.; Lye, G. J. Room-temperature ionic liquids as replacements for organic solvents in multiphase bioprocess operations. *Biotechnol. Bioeng.* 2000, 69, 227-233.
[107] Liu, B. K.; Wang, N.; Chen, Z. C.; Wu, Q.; Lin, X. F. Markedly enhancing lipase-catalyzed synthesis of nucleoside drugs' ester by using a mixture system containing organic solvents and ionic liquid. *Bioorg. Med. Chem. Lett.* 2006, 16, 3769-3771.
[108] Milner, S. E.; Maguire, A. R. Recent trends in whole cell and isolated enzymes in enantioselective synthesis. *Arkivoc.* 2012, 321-382.

[109] Kurbanoglu, E. B.; Zilbeyaz, K.; Kurbanoglu, N. I.; Ozdal, M.; Taskin, M.; Algur, O. F. Continuous production of (S)-1-phenylethanol by immobilized cells of Rhodotorula glutinis with a specially designed process. *Tetrahedron-Asymmetr.* 2010, 21, 461-464.

[110] Braeutigam, S.; Bringer-Meyer, S.; Weuster-Botz, D. Asymmetric whole cell biotransformations in biphasic ionic liquid/water-systems by use of recombinant Escherichia coli with intracellular cofactor regeneration. *Tetrahedron-Asymmetr.* 2007, 18, 1883-1887.

[111] Kumar, V.; Parmar, V. S.; Malhotra, S. V. Enhanced solubility and selective benzoylation of nucleosides in novel ionic liquid. *Tetrahedron Lett.* 2007, 48, 809-812.

[112] Chrobok, A. The Baeyer-Villiger oxidation of ketones with Oxone (R) in the presence of ionic liquids as solvents. *Tetrahedron* 2010, 66, 6212-6216.

[113] Avdeef, A. Physicochemical profiling (solubility, permeability and charge state). *Curr. Top Med. Chem.* 2001, 1, 277-351.

[114] Naseem, A.; Olliff, C. J.; Martini, L. G.; Lloyd, A. W. Effects of plasma irradiation on the wettability and dissolution of compacts of griseofulvin. *Int. J. Pharm.* 2004, 269, 443-450.

[115] Patel, N. D.; Patel, K. V.; Panchal, L. A.; Shukla, A. K.; Shelat, P. K. An emerging technique for poorly soluble drugs: self emulsifying drug delivery system. *Int. J. Pharm. Biol. Arch.* 2011, 2, 621-629.

[116] Amidon, K. S.; Langguth, P.; Lennernas, H.; Yu, L.; Amidon, G. L. Bioequivalence of Oral Products and the Biopharmaceutics Classification System: Science, Regulation, and Public Policy. *Clin. Pharmacol. Ther.* 2011, 90, 467-470.

[117] Savjani, K. T.; Gajjar, A. K.; Savjani, J. K. Drug solubility: importance and enhancement techniques. *ISRN pharmaceutics* 2012, 1-10.

[118] Blanchard, L. A.; Brennecke, J. F. Recovery of organic products from ionic liquids using supercritical carbon dioxide. *Ind. Eng. Chem. Res.* 2001, 40, 287-292.

[119] Aerov, A. A.; Khokhlov, A. R.; Potemkin, I. I. Why ionic liquids can possess extra solvent power. *J. Phys. Chem. B.* 2006, 110, 16205-16207.

[120] Smith, K. B.; Bridson, R. H.; Leeke, G. A. Solubilities of pharmaceutical compounds in ionic liquids. *J. Chem. Eng. Data* 2011, 56, 2039-2043.

[121] Jaitely, V.; Mizuuchi, H.; Florence, A. T. Current-stimulated release of solutes solubilized in water-immiscible room temperature ionic liquids (RTILs). *J. Drug Target* 2010, 18, 787-793.

[122] Mizuuchi, H.; Jaitely, V.; Murdan, S.; Florence, A. T. Room temperature ionic liquids and their mixtures: Potential pharmaceutical solvents. *Eur. J. Pharm. Sci.* 2008, 33, 326-331.

[123] Jaitely, V.; Karatas, A.; Florence, A. T. Water-immiscible room temperature ionic liquids (RTILs) as drug reservoirs for controlled release. *Int. J. Pharm.* 2008, 354, 168-173.

[124] Forte, A.; Melo, C. I.; Bogel-Lukasik, R.; Bogel-Lukasik, E. A favourable solubility of isoniazid, an antitubercular antibiotic drug, in alternative solvents. *Fluid Phase Equilibr.* 2012, 318, 89-95.

[125] dos Santos, A. D.; Morais, A. R. C.; Melo, C.; Bogel-Lukasik, R.; Bogel-Lukasik, E. Solubility of pharmaceutical compounds in ionic liquids. *Fluid Phase Equilibr.* 2013, 356, 18-29.

[126] McCrary, P. D.; Beasley, P. A.; Gurau, G.; Narita, A.; Barber, P. S.; Cojocaru, O. A.; Rogers, R. D. Drug specific, tuning of an ionic liquid's hydrophilic-lipophilic balance to improve water solubility of poorly soluble active pharmaceutical ingredients. *New J. Chem.* 2013, 37, 2196-2202.

[127] Ramos-Rodriguez, D.-A.; Rodriguez-Hidalgo, M.-d.-R.; Soto-Figueroa, C.; Vicente, L. Molecular and mesoscopic study of ionic liquids and their use as solvents of active agents released by polymeric vehicles. *Mol. Phys.* 2010, 108, 657-665.

[128] Melo, C. I.; Bogel-Lukasik, R.; da Ponte, M. N.; Bogel-Lukasik, E. Ammonium ionic liquids as green solvents for drugs. *Fluid Phase Equilibr.* 2013, 338, 209-216.

[129] Azevedo, A. M. O.; Ribeiro, D. M. G.; Pinto, P. C. A. G.; Lucio, M.; Reis, S.; Saraiva, M. L. M. F. S. Imidazolium ionic liquids as solvents of pharmaceuticals: Influence on HSA binding and partition coefficient of nimesulide. *Int. J. Pharm.* 2013, 443, 273-278.

[130] Gupta, S.; Moulik, S. P. Biocompatible microemulsions and their prospective uses in drug delivery. *J. Pharm. Sci.* 2008, 97, 22-45.

[131] Dobler, D.; Schmidts, T.; Klingenhofer, I.; Runkel, F. Ionic liquids as ingredients in topical drug delivery systems. *Int. J. Pharm.* 2013, 441, 620-627.

[132] Patra, D.; Barakat, C. Unique role of ionic liquid bmim BF4 during curcumin-surfactant association and micellization of cationic, anionic and non-ionic surfactant solutions. *Spectroc. Acta Pt. A-Molec. Biomolec. Spectr.* 2011, 79, 1823-1828.

[133] Mahajan, S.; Sharma, R.; Mahajan, R. K. An investigation of drug binding ability of a surface active ionic liquid: micellization, electrochemical, and spectroscopic studies. *Langmuir* 2012, 28, 17238-17246.

[134] Gordian, M.; Singh, N.; Zemmel, R.; Elias, T. Why products fail in Phase III. *In vivo* 2006, April, 1-8.

[135] Cains, P. W. Solid-state chemistry in drug development. *Chem. Today* 2007, 25, 10-12.

[136] Variankaval, N.; Cote, A. S.; Doherty, M. F. From form to function: Crystallization of active pharmaceutical ingredients. *Aiche. Journal* 2008, 54, 1682-1688.

[137] Brittain, H. G. Drugs and the pharmaceutical sciences. *Polymorphism in pharmaceutical solids;* New York: Marcel Dekker, Inc; 1999.

[138] Sarma, B.; Chen, J.; Hsi, H.-Y.; Myerson, A. S. Solid forms of pharmaceuticals: Polymorphs, salts and cocrystals. *Korean J. Chem. Eng.* 2011, 28, 315-322.

[139] Chemburkar, S. R.; Bauer, J.; Deming, K.; Spiwek, H.; Patel, K.; Morris, J.; Henry, R.; Spanton, S.; Dziki, W.; Porter, W.; Quick, J.; Bauer, P.; Donaubauer, J.; Narayanan, B. A.; Soldani, M.; Riley, D.; McFarland, K. Dealing with the impact of ritonavir polymorphs on the late stages of bulk drug process development. *Org. Process Res. Dev.* 2000, 4, 413-417.

[140] Chow, K.; Tong, H. H. Y.; Lum, S.; Chow, A. H. L. Engineering of pharmaceutical materials: An industrial perspective. *J. Pharm. Sci.* 2008, 97, 2855-2877.

[141] Rogers, R. D.; Daly, D. T.; Swatloski, R. P.; Hough, W. L.; Davis, J. H.; Smiglak, M.; Pernak, J.; Spear, S. K. *Multi- functional ionic liquid compositions for overcoming polymorphism and imparting improved properties for active pharmaceutical, biological, nutritional, and energetic ingredients.* WO2007044693. A2. The University of Alabama. 2007.

[142] Rogers, R. D.; Rijksen, C.; Daly, D. T.; Caldwell, K.; Caldwell, G.; Hough-Troutman, W. L.; Bica, K. *Compounds comprising two or more biologically functional ions and*

method of treating Parkinson disease. WO2010078258. A1. The board of trust of the University of Alabama. 2010.

[143] Rogers, R. D.; Daly, D. T.; Gurau, G.; MacFarlane, D. R.; Turanjanin, J.; Dean, P. M.; L., S. J.; Bica, K.; Seddon, K. R. *Dual functioning ionic liquids and salts thereof*. WO2010078300. A1. The board of trust of the University of Alabama. Monash University. Queen´s University of Belfast. 2010.

[144] Stoimenovski, J.; MacFarlane, D. R.; Bica, K.; Rogers, R. D. Crystalline vs. Ionic Liquid Salt Forms of Active Pharmaceutical Ingredients: A Position Paper. *Pharm. Res.* 2009, 27, 521-526.

[145] Bica, K.; Rijksen, C.; Nieuwenhuyzen, M.; Rogers, R. D. In search of pure liquid salt forms of aspirin: ionic liquid approaches with acetylsalicylic acid and salicylic acid. *Phys. Chem. Chem. Phys.* 2010, 12, 2011-2017.

[146] Bica, K.; Rodriguez, H.; Gurau, G.; Cojocaru, O. A.; Riisager, A.; Fehrmann, R.; Rogers, R. D. Pharmaceutically active ionic liquids with solids handling, enhanced thermal stability, and fast release. *Chem. Commun.* 2012, 48, 5422-5424.

[147] Cojocaru, O. A.; Bica, K.; Gurau, G.; Narita, A.; McCrary, P. D.; Shamshina, J. L.; Barber, P. S.; Rogers, R. D. Prodrug ionic liquids: functionalizing neutral active pharmaceutical ingredients to take advantage of the ionic liquid form. *Med. Chem. Comm.* 2013, 4, 559-563.

[148] Pinto, P. C. A. G.; Ribeiro, D. M. G. P.; Azevedo, A. M. O.; De la Justina, V.; Cunha, E.; Bica, K.; Vasiloiu, M.; Reis, S.; Saraiva, M. L. M. F. S. Active pharmaceutical ingredients based on salicylate ionic liquids: insights into the evaluation of pharmaceutical profiles. *New J. Chem.* 2013, 37, 4095-4102.

[149] Costa, S. P. F.; Justina, V. D.; Bica, K.; Vasiloiu, M.; Pinto, P. C. A. G.; Saraiva, M. L. M. F. S. Automated evaluation of pharmaceutically active ionic liquids' (eco)toxicity through the inhibition of human carboxylesterase and *Vibrio fischeri*. *J. Hazard. Mater.* 2014, 265, 133-141.

[150] Ferraz, R.; Branco, L. C.; Marrucho, I. M.; Araujo, J. M. M.; Rebelo, L. P. N.; da Ponte, M. N.; Prudencio, C.; Noronha, J. P.; Petrovski, Z. Development of novel ionic liquids based on ampicillin. *Med. Chem. Comm.* 2012, 3, 494-497.

[151] Florindo, C.; Araujo, J. M. M.; Alves, F.; Matos, C.; Ferraz, R.; Prudencio, C.; Noronha, J. P.; Petrovski, Z.; Branco, L.; Rebelo, L. P. N.; Marrucho, I. M. Evaluation of solubility and partition properties of ampicillin-based ionic liquids. *Int. J. Pharm.* 2013, 456, 553-559.

[152] Ferraz, R.; Teixeira, V.; Rodrigues, D.; Fernandes, R.; Prudencio, C.; Noronha, J. P.; Petrovski, Z.; Branco, L. C. Antibacterial activity of Ionic Liquids based on ampicillin against resistant bacteria. *Rsc. Advances* 2014, 4, 4301-4307.

[153] Alves, F.; Oliveira, F. S.; Schroder, B.; Matos, C.; Marrucho, I. M. Synthesis, characterization, and liposome partition of a novel tetracycline derivative using the ionic liquids framework. *J. Pharm. Sci.* 2013, 102, 1504-12.

[154] Dean, P. M.; Turanjanin, J.; Yoshizawa-Fujita, M.; MacFarlane, D. R.; Scott, J. L. Exploring an Anti-Crystal Engineering Approach to the Preparation of Pharmaceutically Active Ionic Liquids. *Cryst. Growth Des.* 2009, 9, 1137-1145.

[155] Hough-Troutman, W. L.; Smiglak, M.; Griffin, S.; Reichert, W. M.; Mirska, I.; Jodynis-Liebert, J.; Adamska, T.; Nawrot, J.; Stasiewicz, M.; Rogers, R. D.; Pernak, J. Ionic

liquids with dual biological function: sweet and anti-microbial, hydrophobic quaternary ammonium-based salts. *New J. Chem.* 2009, 33, 26-33.

[156] Stoimenovski, J.; Dean, P. M.; Izgorodina, E. I.; MacFarlane, D. R. Protic pharmaceutical ionic liquids and solids: Aspects of protonics. *Faraday Discuss.* 2012, 154, 335-352.

[157] Stoimenovski, J.; MacFarlane, D. R. Enhanced membrane transport of pharmaceutically active protic ionic liquids. *Chem. Commun.* 2011, 47, 11429-11431.

[158] Hamamoto, H.; Ishibashi, M.; Matsumura, S.; Yamasaki, K.; Hidetoshi, H.; Masaki, I.; Sueko, M.; Keiko, Y. *External preparation* e.g., *a tape, having antiinflammatory and analgesic activity, comprising etodolac and local anesthetic.* MEDRX CO LTD (MEDR-Non-standard) NIPPON SHINYAKU CO LTD (NNSH) MEDREX KK (MEDR-Non-standard) HAMAMOTO H (HAMA-Individual) ISHIBASHI M (ISHI-Individual) MATSUMURA S (MATS-Individual) YAMASAKI K (YAMA-Individual) MEDOREKKUSU KK (MEDO-Non-standard). 2005.

[159] Miwa, Y.; Hamamoto, H.; Hikake, S.; Kuwabara, Y. A phase I, randomized, open-label, cross-over study of the pharmacokinetics, dermal tolerability, and safety of MRX-7EAT Etodolac-Lidocaine Topical Patch in healthy volunteers. *J. Pain* 2013, 14, S72-S72.

[160] Kuwabara, Y.; Hamamoto, H.; Hikake, S.; Miwa, Y. A randomized, multi-Center, double-blind, placebo-controlled phase II/III trial to evaluate the efficacy, tolerability and safety of MRX-7EAT Etodolac-Lidocaine Topical Patch in the treatment of pain. *J. Pain* 2013, 14, S73-S73.

[161] Restolho, J.; Mata, J. L.; Saramago, B. Peculiar surface behavior of some ionic liquids based on active pharmaceutical ingredients. *J. Chem. Phys.* 2011, 134, 074702-1 – 074702-7.

[162] Ettmayer, P.; Amidon, G. L.; Clement, B.; Testa, B. Lessons learned from marketed and investigational prodrugs. *J. Med. Chem.* 2004, 47, 2393-2404.

[163] Bica, K.; Rogers, R. D. Confused ionic liquid ions-a "liquification" and dosage strategy for pharmaceutically active salts. *Chem. Commun.* 2010, 46, 1215-1217.

[164] Tantishaiyakul, V.; Phadoongsombut, N.; Wongpuwarak, W.; Thungtiwachgul, J.; Faroongsarng, D.; Wiwattanawongsa, K.; Rojanasakul, Y. ATR-FTIR characterization of transport properties of benzoic acid ion-pairs in silicone membranes. *Int. J. Pharm.* 2004, 283, 111-116.

[165] Johansson, K. M.; Izgorodina, E. I.; Forsyth, M.; MacFarlane, D. R.; Seddon, K. R. Protic ionic liquids based on the dimeric and oligomeric anions: (AcO)(x)H(x-1) (-) *Phys. Chem. Chem. Phys.* 2008, 10, 2972-2978.

In: Ionic Liquids: Synthesis, Characterization and Applications ISBN: 978-1-63117-716-3
Editor: Anson Brooks

Chapter 2

IONIC LIQUID APPLICATION FOR LIGNIN EXTRACTION AND DEPOLYMERIZATION

Raquel Prado, Xabier Erdocia and Jalel Labidi
Environmental and Chemical Engineering Department,
University of the Basque Country, Plaza Europa,
Donostia-San Sebastián, Spain

ABSTRACT

Today, in the chemical industry, solvents are used in large quantities. Therefore, solvents establish a significant part of the environmental performance of a process and also impact on safety and health issues. In order to get greener processes especially in the case of new biorefineries it is essential the use of green solvents. Ionic liquids as alternative solvents and catalyst are investigated in the area of biomass deconstruction to obtain valuable chemicals, in order to minimize the impact of the use of volatile organic solvents in the environment.

Until now, the ionic liquids were generally applied in the cellulose solubility and its conversion into a biofuel among others applications. However, lignin is also a key target to take into a count as a renewable source of chemicals and several studies are gaining in importance in the last years. The lignin being the most abundant residue in the paper industries, due to its structure can be developed as aromatics source.

With this purpose, the traditional delignification processes like alkali and organosolv were compared with ionosolv in order to obtain superior quality lignin, suitable to be depolymerised. It was found that ionosolv lignin has greater purity compared to organosolv and alkali lignins, avoiding the attached hemicelluloses and sulphate salts in case of alkali lignins. The recyclability of the ionic liquid was also studied to assess the economic cost of the process. In addition, they were applied as solvent and catalyst in order to depolymerize lignin to valuable products, as vanillin and syringaldehide.

INTRODUCTION

Biomass is mainly constituted by cellulose, lignin, hemicelluloses, and some extractives. Concretely, lignin constitute the second main component of lignocelluloses (cellulose, lignin and hemicelluloses), which in general consist approximately 30% of dry weight of softwoods and around 20% in hardwoods and grasses, though, this concentration also depends on each species (1).

Lignin is water insoluble polymer associated with the development of vascular system in plants, providing water-proofing, structural reinforcement and resistance to biological and physical attack. Lignin is formed by dehydrogenative polymerization of three phenylpropanoid precursors. It is biosynthesized from L-phenylalanine via cinnamic acids: p-coumaril alcohol (trans-3-(4-hydroxyphenyl)-2-propen-1-ol), coniferyl alcohol (trans-3-(4-methoxy-4-hydroxy-phenyl)-2-propen-1-ol) and sinapyl alcohol (trans-3-(3,5-dimethoxy-4-hydroxyphenyl)-2-propen-1-ol) (Figure 1), in a process promoted by H_2O_2 system. Lignin formation reaction mechanism consist on the removal of the phenolic hydrogen atom from the precursors leading to phenoxyl radicals, which by non-enzymatic radical-radical couplings produce a three dimensional amorphous polymer which form a randomized structure in a three dimensional network inside the cell wall (2). Once incorporated into the lignin polymer, the precursor subunits are modified and identified by their aromatic ring structure, becoming guaiacyl, syringyl and p-hydroxyphenyl subunits (the most repetitive units into lignin complex structure). Despite having randomized structure, there are some aspect that are repetitive, for example; more than two-thirds of the linkages in lignin are ether linkages, and there are some common linkages that are commonly repeated in the lignin structure β-O-4-aryl ether bonds, β-5-phenylcoumaran, 5–5' biphenyl, 4-O-5-diarylether, β-1-(1,2-diarylpropane), α-O-4-aryl ether, and β–β'-resinol linkages which are depicted in Figure 2.

Depending on the source of lignin whether hardwood softwood or grasses, the composition of lignin is different. Softwood lignin consists almost exclusively of guayiacyl units while hardwood one contains also large units of syringyl units and grasses lignin contains also minor amounts of p-hydroxyphenyl groups. Table 1 shows the proportion among different precursors and linkages depending on the source of the lignin (3). The proportion among linkages also depends on the source of lignin, being around β-Aryl ether (β-O-4) linkage 46% on softwood and 60% on hardwood, and biphenyl (5-5') linkage around 10% on softwood and 5% on hardwood (1).

Table 1. Lignin structure proportions

%	Softwood	Hardwood	Grasses
Coumaryl			5
Coniferyl	90-95	50	75
Sinapyl	5-10	50	25
β-O-4	46	60	
α-O-4	6-8	6-8	
4-O-5	3.5-4	6.5	
β-5	9-12	6	
β-1	7	7	
β-β'	2	3	
5-5'	9.5-11	4.5	

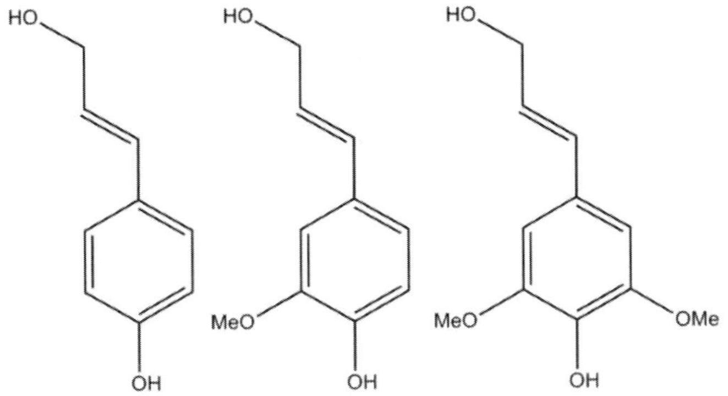

Figure 1. Phenylpropanoids precursors a) p-coumaryl alcohol, b) coniferyl alcohol and c) sinapyl alcohol.

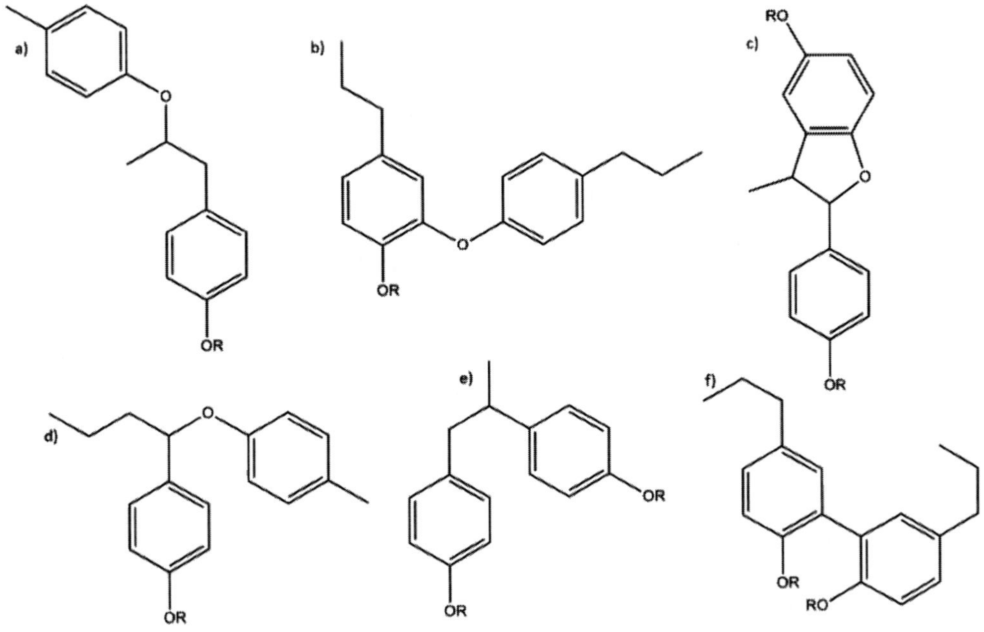

Figure 2. Lignin linkages a) α-Aryl ether (α-O-4), b) Diphenyl ether (4-O-5), c) Phenylcoumaran (β-5), d) β-Aryl ether (β-O-4), e) Diaryl propane (β-1) and f) Byphenyl (5-5').

However, all lignin structure data are approximate, it is very difficult to elucidate the real chemical structure of lignin, because it does not only depends on its randomize synthesis and specie but also on the process used for its extraction which makes practically impossible to know its starting structure. As example, guayacil units are more likely C-C cross-link at the C-5 position of the ring; these cross-links can be formed during lignification as well as delignification processes. The C-5 position is substituted in the syringyl units therefore it cannot participate in substitution reactions. The C-C cross-link cannot be hydrolyzed by acid or base, making the delignification process of softwood more difficult than for hardwood.

1.1. Lignin Applications

In chemical pulping processes, around 100 million tons per year lignin are generated as a byproduct most of which is burned to produce the energy required by the process. Only a very small fraction of the lignin (ca. 1.2%) is utilized as material in industrial processes. In addition, it is necessary to purify the lignin obtained by traditional methods, in order to obtain revalorized products. Therefore, there is an increasing demand for new delignification processes that could provide new means to use this renewable resource in a more efficient manner, not only as fuel but also as building blocks for producing commodities and fine chemicals (4).

The applicability of lignin has been studied by many authors. The diversity of functional groups present in lignin allows its use as dispersant in cement and gypsum blends (5), as emulsifier or chelating agent for removing heavy metals from industrial effluents (6). Besides, lignin has been used as polymer additives (7), surfactant, as substitute for phenolic powder resins, as polyol in polyurethane foams, as epoxy functionalized lignin in epoxy resins and soda lignin has been also used as dispersant (8). Lignin is used as substitute for phenol in phenol formaldehyde resins, polyurethanes, polyvinylchlorides (PVC) and, to a lesser extent, in polypropylenes, polyethylenes, polyhydroxyalkanoates and polylactides (9).

Moreover, lignin has antioxidant properties, acting as free radical scavengers. This activity allows the use of lignin for cosmetic formulations such as topical applications (10). Košiková et al. studied the potential medicinal application of lignin obtained from pulp and paper production (kraft and prehydrolysis processes), and found that it provided a protective effect on DNA and inhibited mutagenicity processes, resulting in a potential antimutagenic and anticarcinogenics agent (11). Recent works have confirmed that lignin can be used in cosmetics and pharmaceutical preparations without harmful effects in human cells. These studies have also demonstrated that the origin of the lignin strongly influences their antiradical activity (12).

2. Delignification Process

The main aim of delignification processes was the obtaining of good quality cellulose, discarding other components of biomass as hemicelluloses and lignin. With the time, those components have gained in importance, because their applicability in different and valuable fields. So delignification processes have evolved in order to achieve a complete deconstruction of biomass maintaining the properties of all its components. This chapter is focused on lignin properties and applicability related to ionic liquids. In this section, the evolution of traditional delignification processes will be briefly explained, to finish with the emerging use of ionic liquids on biomass processing.

2.1. Sulphite Process

The first delignification process used in the pulp and paper industry was the sulphite process. This process produces wood pulp which is almost pure cellulose fibers by using various salts of sulphurous acid to extract the lignin from wood chips in large pressure vessels called digesters. In this pulping process sulphites (SO_3^{2-}), bisulphites (HSO_3^-), sulphur dioxide (SO_2) and sulphuric acid (H_2SO_4) are added as sulphur species, their final concentration on the reaction vessel depend on the pH of the system, that is controlled by the acid amount. The cation of the sulphite and bisulphite salts can be sodium (Na^+), calcium (Ca^{2+}), potassium (K^+), magnesium (Mg^{2+}) or ammonium (NH_4^+). With this process sulphonation of lignin occurs and subsequently the hydrophobicity of lignin increased. The importance of sulphite pulping process has dramatically decreased, and only about 10% of the pulp is currently produced by this method. It was substituted by the Kraft process in the paper industry (13).

2.2. Kraft Process

Kraft process is the dominant global pulping process with about 90% share of the total production capacity. In this process, cellulose fibers are isolated through dissolution of lignin and hemicelluloses of the wood chips. In a conventional Kraft cooking, an aqueous solution of sodium hydroxide and sodium sulphide, also known as white liquor, is reacted with the wood chips in a large pressure vessel called a digester. The white liquor and the chips are heated to a cooking temperature of about 170 °C and are allowed to cook at that temperature for about two hours. During this treatment, the hydroxide and hydrosulphide anions react with the lignin, causing the fragmentation of the polymer into smaller water/alkali-soluble fragments.

Delignification in the Kraft cook proceeds in three distinct phases: the initial phase, the bulk phase, and the final or residual phase. The initial phase of delignification takes place up to a temperature of about 150 °C, and it is controlled by diffusion. The bulk phase includes the heating period from about 150-170 °C and the cooking treatment at 170 °C. The rate of delignification in the bulk phase is controlled by chemical reactions. Most of the lignin is removed in this phase. The residual or final phase, in which the rate of delignification significantly decreases, begins when about 90% of the lignin has been removed and marks the end of the cook. The selectivity in this phase is poor, and further pulping could result in significant degradation of carbohydrates (14).

2.3. Sulphur Free Alkaline Process

Biomass pulping by alkaline process accounts for nearly 5% of the total pulp production. The effect of the alkaline delignification depends on lignin content of the raw materials, in this process lignin and hemicelluloses are dissolved. They utilize lower temperature and pressure than other delignification processes. Sodium, potassium, calcium and ammonium hydroxides are suitable as alkaline delignification agents. Sodium hydroxide is the most investigated, despite calcium hydroxide has been shown to be an effective delignification agent and is the cheapest of the hydroxides. In addition, it is possible to recover calcium from the reaction like calcium carbonate and regenerate the hydroxide. The main drawbacks of alkaline delignification are that it is highly depend on the temperature and the presence of oxygen, high residence times are required at ambient conditions, and there are formed inorganic salts that are difficult to separate from the products (15).

2.4. Organosolv Process

Organosolv process is the less destructive delignification procedure, lignin structure remains practically as its primary structure. In the organosolv process, an organic or aqueous organic solvent mixture is used to break the internal lignin and hemicelluloses bonds. The solvents most commonly used in the process are methanol, ethanol, acetone, ethylene glycol, triethylene glycol and tetrahydrofurfuryl alcohol. Organic acids such as oxalic, acetic, acetylsalicylic and salicylic can also be used in organosolv process. It involves simultaneous prehydrolysis and delignification of lignocellulosic biomass supported by organic solvents and diluted aqueous acid solutions. The concentration of the solvent can vary between 40-80%, the temperature between 140-200 °C and normally the pressure of this pulping process is high. The solvents used in the process can be recycled by evaporation and condensation, but the cost is high (16).

3. Ionic Liquid and Lignin

As it has been mentioned before, this chapter is mainly focused on lignin reactivity with IL and its applicability. However, at the beginning as it had occurred with the traditional methods the main aim of the studies was the delignification of the biomass in order to obtain cellulose with good properties for paper and fibers applications. Nevertheless, it also evolves to a complete deconstruction and separation of biomass components maintaining the properties of all of them. The first studies realized to obtain good cellulose are extremely useful also for the lignin applicability since to achieve the isolation of cellulose the lignin behavior and solubility into ILs was tested, and first to try the delignification some authors tried the solubility of the lignin into several ionic liquids. Therefore, the same ILs can be used to check the reactivity of the lignin.

Among the different ionic liquids that had been checked for the same purpose, most of them are based on imidazolium ring, so their study deserves a complete section in this chapter.

3.1. Imidazole Based Ionic Liquids

The studied ionic liquids are mostly based on imidazolium cation ionic liquid; because they are particularly suitable for lignin solubilization as the cation aromatic ring can favorably interact with the lignin phenyl rings via aromatic interactions. With the variation of the alkyl chain (methyl-, ethyl-, allyl-, butyl-, hexyl- and benzyl) and combined with different anions, in order to establish the most suitable combination for having good lignin solubility.

3.1.1. Lignin solubility and Ionosolv process

The ionosolv process consists on delignification of biomass by dissolving the lignin by IL as solvent, whereas the cellulose remains solid as a pulp without change on its crystallinity. The ionosolv process has been studied by many authors using imidazolium cation based ILs as solvents; the influence of the anion, substituents and conditions were studied. It was observed by many of them that sulphate and sulphonate based anions were the most appropriate to dissolve lignin, and the halides more specifically fluorides the worst candidates. Because of the high variety of methods discovered to achieve good delignification yields, in the following paragraph a brief description of different studies will be detailed.

Pu et al. studied the solubility of softwood kraft lignin on different ionic liquids and temperatures (1, 3-dimethylimidazolium methylsulphate ($m_2imMeSO_4$), 1-hexyl-3-methylimidazolium trifluoromethanesulphonate ($hmimCF_3SO_3$), 1-butyl-2,3-dimethylimidazolium tetrafluoroborate (bm_2imBF_4), 1-butyl-4-methylpyridinium hexafluoro-phosphate ($bmpyPF_6$), 1-butyl-3-methylimidazolium chloride (bmimCl), 1-butyl-3-methylimidazolium bromide (bmimBr)). The solubility of lignin depended strongly on the anion rather than the cation, $MeSO_4$>CF_3SO_4>Cl-Br>BF_4>>>PF_6. In addition, it was also affected by the temperature (17). Lee et al. also worked on the solubility of kraft lignin on different ILs. $M_2imMeSO_4$, 1-butyl-3-methylimidazolium hexafluoro phosphate ($bmimPF_6$), bm_2imBF_4, $bmimCF_3SO_3$, 1-ethyl-3-methylimidazolium acetate ($emimCH_3COO$), 1-alil-3-methylimidazolium chloride (amimCl), bmimCl and 1-benzil-3-methylimidazolium Chloride (bzmimCl) were checked at 90 °C during 24 h. Their observations were in agreement with those of Pu et al. about the sulphate base anions were better solvents than ILs with halogenure anions, lignin solubility varied for the different IL, around 500 kg/L for $mimMeSO_4$ and $bmimCF_3SO_3$, 300 Kg/L for $emimCH_3COO$ and amimCl, 100 kg/L for bmimCl and bzmimCl, 40 kg/L $bmimBF_4$ and insoluble in $bmimPF_6$ (18). Lateef et al. checked the solubility of mixtures between synthetic lignin and cellulose, into 3-(2-Cyanoethyl)-1-methyl-1H-imidazol-3-ium bromide (CNmimBr), 1-propyl-3-methylimidazolium bromide (prmimBr) and bmimCl. The solubility was higher in CNmimBr, around 95.3 mg/g, and 83-97% of lignin was recovered (19). Tan et al. achieved the delignification of sugarcane by ethylmethylimidazolium alkylbenzenesulphonate (emimABS) mixed with xylanesulfonate; the best delignification performance was observed after 2h at 190 °C and atmospheric pressure, achieving 93% of lignin extraction (20). Alkali (NaOH) and organosolv lignins solubility into $bmimMeSO_4$ was checked, it was observed that organosolv lignin was more soluble than alkali lignin, 90% for organosolv and 27% for alkali, however, when the reaction was enhanced by microwave radiation the solubility increased until 97% and 73% respectively. The lignin recovery yields were 75% for organosolv lignin and 64% for alkali lignin under microwave radiation. The main impurities that were present on lignins obtained by those pulping methods were the hemicelluloses in both cases and inorganic salts that were formed during lignin precipitation for alkali pulping. When lignins were reprecipitated from the $bmimMeSO_4$ a considerable decrease on hemicelluloses content was observed, in addition the inorganic salts are insoluble on IL, so the alkali lignin purity enhanced significantly. Due to the good performance on lignin dissolution showed by $bmimMeSO_4$ it was tried to extract directly the lignin from the biomass. The delignification power of the $bmimMeSO_4$ process was compared with organosolv and alkali pulping processes, to determine the performance of the processes obtaining good quality lignin. The purity of ionosolv lignin was higher compared to organosolv and alkali lignin, being negligible the hemicelluloses amount attached to the lignin, whereas was around 6% for organoslv and 22% for alkali lignins. Besides, the inorganic salt impurity was also avoided when $bmimMeSO_4$ was used (21). Brant et al. successfully achieved the delignification of miscanthus, pine and willow, by water-IL mixtures; they compared 1-butyl-3-methylimidazolim methylsulphonate ($bmimMeSO_3$), 1-

butyl-3-methylimidazolium hydrogensulphate ($bmimHSO_4$) and a mixture $bmimMeSO_4/HSO_4$ mixed with water. With the addition of water the cellulose solubility was also avoided, the treatment was carried out at 120 ºC and after 22 h more than a half of the biomass was dissolved for miscanthus. The optimum water content was determined as 20 % w/w. For the miscanthus, they achieved 92.8% of deligninfication with $bmimHSO_4$ and 67.9% for $bmimMeSO_3$. The pine and willow were also treated with $bmimHSO_4$ and 20% water; with this process 85% and 65.5% delignification rates respectively were obtained (22).

3.1.2. Wood solubility and selective separation

After the lignin solubility, direct extraction from different kinds of biomass (softwood, hardwood and grass) was assessed. In order to dissolve wood it was observed that the particle size was determinant for the solubility, due to the diffusion problems that present some ionic liquids because of their high viscosity, being the most soluble ball-milled power and following the order sawdust>TMP fibers>>wood chips. The presence of water in the reaction media is also extremely important issue. The dissolution of the cellulose precise the absence of water, because the mechanism of dissolution implies the formation of hydrogen bonds, when water is present in the media those hydrogen bonds are formed between water and IL decreasing solubility of wood. So that, the reaction has to be done on inert atmosphere, and both, the ILs and the wood, have to be dried before. In general in order to have a successful solubility of wood into ionic liquids, presence of water, particle size, and reaction time are the main factors to take into account. However, there is not a unified procedure for the biomass deconstruction, similar performances can be obtained by different ways, it depends on the kind of biomass that is treated, and the used ionic liquid. Reaction time can be severally reduced by microwave radiation. BmimCl and $etmimCH_3COO$ are the most studied ILs for biomass deconstruction; the first one because it presents good performance dissolving cellulose (23) (24) (25), and the second one because it is supposed to be greener. The first ionic liquid utilized for the wood dissolution was bmimCl. Fort et al. studied the dissolution of both hardwood and softwood into bmimCl; they observed that the softwood was more easily dissolved than hardwood, but the solubility not only depended on the type but on the specie, the oak was the less soluble, though after 24 h of treatment at 100 ºC the yields are similar for all species. Kilpeläinen et al. also used BmimCl among others (amimCl, bzlmimCl, 1-benzyl-3-methylimidazolium dicyanamide ($bzmimCN_2$) and 1-m-methoxybenzyl 1-methylimidazolium chloride (MeObzmimCl)) as solvent for wood, concretely for Norway spruce and pine. Complete wood dissolution was achieved until saturation up to 8% w/w of solute with amimCl and amimCl as solvents, stirring the mixture of sawdust and IL at 80-120 ºC; however the dissolution was not completely clear, and the viscosity was quite high. It was found that the anion Cl^- disrupt the hydrogen bonds present in the wood. However, the interaction with the aromatic part of the wood, the lignin, was poor in those ILs, which explains why the dissolution was not completely clear. In order to increase the solvation of the lignin bzmimCl was introduced as solvent, the handicap of this IL was its high viscosity which was decreased adding amimCl. The mixture of both ILs leads to a clear light brown solution of wood, after the solution the components were selectively recovered (4). The main drawback of bmimCl is that at high temperatures can be discomposed and releases halogenides that are toxic; in general all the ionic liquids with halogen as anion can decompose to halogenides. As it has been mentioned before in this section, $emimCH_3COO$ ionic liquid is considered greener than the other ILs, because of its possibility of being biodegraded. In the search for greener processes for the biomass decomposition emimCH_3COO has been extensively investigated. Lee et al. tried the solubility of maple wood flour into emimCH_3COO; they optimized the temperature and reaction time. They found that the lignin solubility resulted higher at 130 °C where 63% of the lignin was extracted. In order to optimize the reaction time the temperature was fixed at 90 °C being the optimum time 70 h, with 86% of lignin extraction (18). Sun et al. also studied emimCH_3COO in order to dissolve wood, both, softwood (southern yellow pine) and hardwood (red oak), achieving around 93% of wood solubility in the case of softwood and around of 98% of solubility in the case of hardwood (26). Magalhães da Silva et al. tried the solubility of grasses (wheat straw) on emimCH_3COO; they compose a complete biomass deconstruction scheme, achieving the selective separation of cellulose, hemicelluloses and lignin with good yields, including the recovery of emimCH_3COO. The wheat straw was dissolved into emimCH_3COO at 5% w/w at 140 °C stirring during 6 h. The selective separation was achieved in 3 steps by adding 3 different antisolvents, the cellulose was isolated adding 3% aqueous solution of NaOH, the hemicelluloses by the addition of EtOH and the lignin with the adjusting of pH at 2 adding acidified water (HCl). It was achieved the recovery of 91% of cellulose, 96% of hemicelluloses and 97% of lignin (27).

3.1.3. Non imidazole cation based ionic liquids

Ionic liquids derived from polycyclic amidine bases have successfully dissolved sulphur free lignin and direct extract of lignin from aspen wood chips (28). For the dissolution of lignin 1,8-diazobicycle [5.4.0]undec-7-enium (HDBU) cation combined with several anions, as chloride Cl, acetate (CH_3COO), formate (HCOO), hydrogensulfate (HSO_4), trifluoroacetate TFA, methanesulfonate ($MeSO_3$), lactate (lac), tosylate (tos), Saccharinate (Sacch) and bis[(trifluoromethane)sulfonyl]imide, were checked achieving around 93% of yield. The reaction temperature was 100 °C for most of them with the exception of HDBUCl (150 °C) and HDBUlac (50 °C). In all cases, long reaction time were needed, longer than 24 h. 8-butyl 1,8-diazobicycle [5.4.0]undec-7-enium Chloride, 8-octyl-1,8-diazobicycle [5.4.0]undec-7-enium chloride, HDBU HCOO, HDBU Cl and HDBU CH_3COO were able to dissolve partially wood chips, achieving nearly complete dissolution with the HDBU HCOO, HDBU Cl, at 100 °C after 48 h.

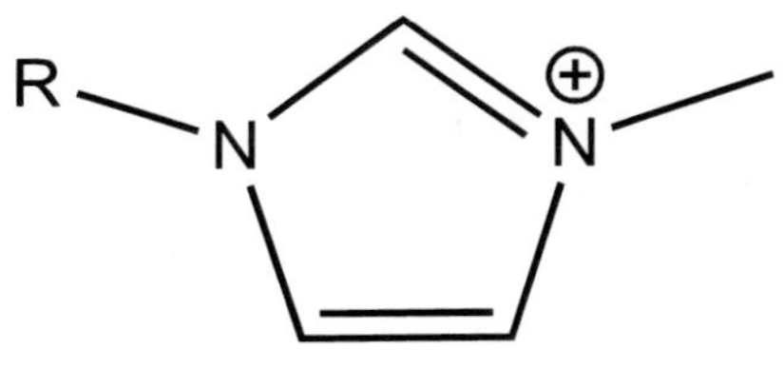

Figure 3. methyl imidazolium cation.

4. Lignin Depolymerization

Lignin conversion into simple aromatic products is a difficult task due to lignin recalcitrant structure. There are some depolymerization processes that are already under research namely pyrolysis of the isolated lignins, catalytic hydrogenolysis, alkaline hydrolysis, supercritical water and solvent depolymerization. The C–C linkages of lignin are significantly more resistant towards cleavage and the majority of them survive regardless of the depolymerization process. Thus, depolymerization of the isolated lignins results in a complex mixture of aromatic compounds in which the individual mass fraction of each compound barely exceeds few percent. In addition, the depolymerization of lignin depends on the source of lignin. During the pulping processes, a chemical modification of lignin occurs, with the formation of different structures depending on the pulping properties: for example is extremely hard to depolymerize kraft lignin due to the percentage of refractory C–C linkages between the propyl phenol monomers which is significantly higher than in the proto lignin initially presented in the wood (14) (29). Organosolv lignin structure is the most similar to the wood lignin structure, and it is the most chosen target to be catalytically depolymerized because of its lower molecular weight and sulphur free structure (30). The different depolymerization techniques are briefly described in the following sections.

4.1. Lignin Pyrolisis

Thermogravimetric studies have revealed that lignin thermal degradation occurs over a wider range of temperatures (160–900 °C) compared to polysaccharides (220–400 °C), due to its complex and diverse structure. The amount of solid residue (char) from isolated lignins is significantly higher than the residue from carbohydrates or no fractionated biomass. There are reported data of pyrolysis depolymerization at different temperatures from 500 to 560 °C, the oil yield is improved with the temperature from 16 to 52% (31) (32) (33). In any case, the selectivity of pyrolysis depolymerization is very low. The oils obtained from the pyrolysis of isolated lignins are complex mixtures of different structures (aromatic and no aromatic compounds) with a wide range of MW (from hydrocarbons and aromatic monomers to oligomers), each of which typically constitutes less than 1% of the total.

4.2. Catalytic Hydropyrolysis

Dry lignin can be catalytically transformed into a liquid mixture at temperature significantly lower than that of pyrolysis with the aid of hydrogen. In this case, the obtained oil is also a complex mixture of different products, resulted of a wide range of reactions that occurs in the medium. It has been observed that the oil yield is hardly affected by the pressure; comparing pressures from 50 to 140 bar, the oil yield increases from 28 to 80 with the same catalyst NiMo, Cr_2O_3 (34).

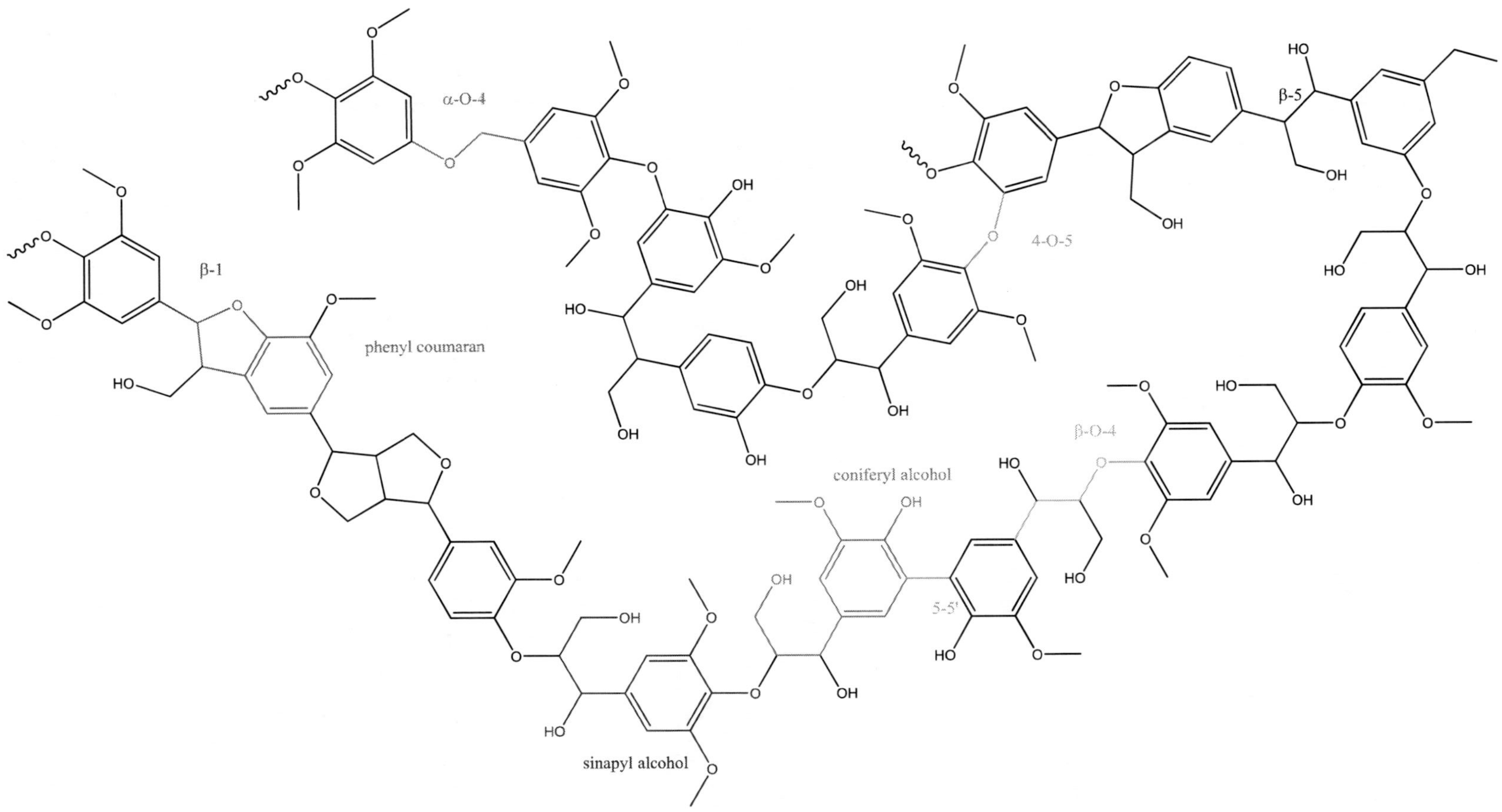

Figure 4. Lignin structure and typical linkages.

4.3. Sub- and Supercritical Water Treatment

Processing of lignin in sub and supercritical water can produce the breakage of the ether linkages and produce larger fragments trough cross linking between the reactive fragments, predominantly by Friedel-Craft mechanism. Compared with the dry methods, due to a low concentration of lignin in the water, the yield of undesirable side condensation reactions is considerably reduced. Besides, the addition of alkaline salts increased the oil yield until a maximum (35) (36) and the addition of p-cresol or phenols improves the selectivity of the process (34).

4.4. Supercritical Solvents

Generally ethanol and methanol are used as supercritical solvents; the mixtures of alcohol and water have been also studied. These reactions are carried out at temperature of 200 and 350 ºC, and at high pressures.

Lignin solvolysis processes can be categorized into two general groups: base-catalyzed depolymerization and hydrogenolysis. The hydrogenolysis processes in supercritical solvents can be divided in other three subgroups: fragmentation with pressurized H_2 gas, fragmentation with in situ hydrogen production from a hydrogen-donor solvent such as formic acid which is thermally decomposed during the depolymerization reaction and fragmentation with hydrogen produced from the partial reforming of the solvent in presence of a metal catalyst. The use of catalyst leads to a decrease of the char yield and the monomer concentration in the extracted oil reached 5% (37).

4.5. Lignin Reactivity on Ionic Liquids

Before studying reactivity of lignin on ionic liquids, is important to determine how its structure is affected by different cation-anion combinations. The lignin structure is affected by the pretreatment, which caused chemical changes in its structure and fragmentation, due to the chemical agents, temperature and pressure. This fact makes especially important how the lignin is affected by the IL that it is using for its selective modification since ILs can acts as catalyst. For example, Tan et al. observed that after delignification of sugarcane waste using emimABS as IL, the obtained lignin had lower molecular weight than the autocatalysis lignin. Both lignin were acetylated in order to carry out a proper comparison. Acetylated lignin from emimABS had lower molecular weight (3690 g/mol) than acetylated autohydrolysis lignin (19300 g/mol), polydispersity was also lower for the emimABS treated lignin 1.66 and 11.4 respectively (20). George et al. studied the effect of different ILs (emim n-hexylsulfate ($emimHxSO_4$), emim diethylphosphate ($emimEt_2P$), emim n-butylsulfate ($emimBuSO_4$), emim dimethylphophate ($emimMe_2P$), emim lactate emimlac, emimCl, $emimCH_3COO$, bmimCl, Cyphos 101, Cyphos 165 and 1-methyl-2-pyrrolidone Chloride mpyrCl) into lignins from different sources (organosolv, alkali (NaOH) and slightly sulfonated alkali lignin). It was observed that the anion had more influence in the lignin structure than the cation, the activity of the anions modifying the molecular weight was in the following order:

sulfate>lactate>acetate>chloride>phosphate. It has to be mention that all lignins had not the same behavior, organosolv lignin is more suitable to have changes in its structure, following by the slightly sulphonated lignin, and the less affected was the alkali lignin (38).

Due to its complex structure, there are no many works that use lignin as starting material to study its reactivity. However, there are many authors that have studied lignin model compounds in order to elucidate the reaction mechanism and the proper conditions to extent the reaction to the whole lignin. The depolymerization of lignin model compounds and lignin on IL has been studied under reductive and oxidative conditions. The reductive conditions were achieved by adding Lewis and Brönsted acids as catalyst; however, these reactions were briefly studied because of the low reactivity showed by the lignin despite the good yields obtained on lignin model compounds. The oxidation has been more extensively studied, in most of the cases coupling a metal catalyst as Fe, Mn, Co, V and oxidant agent as O_2 or H_2O_2 has been used with many different ILs in different conditions. The role of ILs was as both, reaction solvent and/or catalyst (39).

4.5.1. Reduction

Binder et al. worked on reductive depolymerization of organosolv lignin at moderate temperatures and with catalysts. First of all, Eugenol reactivity was studied at moderate temperatures with Brönsted acid catalyst into ionic liquids (emim triflate (emimOTF) and emimCl), to check its conversion to guaiacol, 4-propylguaiacol and isoeugenol. The reactions were done at 180 °C during 2 h, with several catalyst; 100% of eugenol conversion was achieved, but the obtained products were not always the desired targets. With Co/silica, Cu/alumina, $RuCl_3$, $RhCl_3$, $H_3PMo_{12}O_{40}$, $H_4SiW_{12}O_{40}$, nafion and several Pt catalyst 100% of conversion was achieved. But the best 4-propylguaiacol yield (77.5%) was obtained with 0.5% Pd/alumina (C3677) emimOTF tandem, despite eugenol conversion was 99.6% and the best guaiacol yield was achieved with Nafion emimOTF (11.6%). Then other model compounds as 4-ethylguaiacol and 2-phenilethyl phenyl ether were treated under the optimum conditions obtained for eugenol. 4-ethylguaiacol did not show any reactivity and 2-phenilethyl phenyl ether was reduced to phenol, obtaining the best conversion yield with nafion 62.9% and 6.7% of phenol as product. When organosolv lignin is treated with Nafion under the same reaction conditions as eugenol, guaiacol was not obtained (30).

4.5.2. Oxidation

Zakzeski et al. tried the depolymerization of lignin through the oxidation using $CoCl_2 \cdot 6H_2O$ salt at mild conditions (80 °C and 0.5 MPa of O_2) dissolved in emim diethylphosphate (emimDEP). Both alcell organosolv lignin and alkali lignin (NaOH) were soluble into emimDEP, and were oxidised under the same conditions for 4 h; the products were extracted by ethyl acetate. GC chromatogram indicated that there were not monomers from the depolymerisation. However, the IR-ATR spectra showed changes in chemical structure of lignin; the lignin was oxidised, but the catalyst was not strong enough to disrupt the linkage and depolymerize it. In order to understand what was happening in the reaction media, two model compounds (5–5'model compound, 3,3'-dimethoxy-5,5'-dimethyl-1,1'-biphenyl-2,2'-diol and β-*O*-4 model compound 1-hydroxy-1-(4-hydroxy-3-methoxyphenyl)-2-(2,6-dimethoxyphenoxy)ethane) and veratryl alcohol were treated under the same conditions. It was observed that 5-5', β-*O*-4 and phenolic alcohols remains intact after the reaction, while benzyl and other alcohols were oxidized. From the cinnamyl alcohol, cinnamaldehyde, cinnamic acid, benzoic acid and epoxides were obtained, and from veratryl alcohol, only benzyl alcohol was oxidized giving as a product vanillin (40). Veratryl alcohol was previously treated under Co salen complexes in water under alkali conditions with good results by other authors (41). So that, Sonar et al. associated salen complexes with task specific IL (TSIL) (Figure 5); this TSIL was used as ligand on Co salen complexes allowing the selective oxidation of veratryl alcohol to veratraldehyde obtaining yields between 24-56%. Furthermore, no further oxidation was obtained. The reaction was done under alkali conditions at pH 12.5 and air as oxygen source at 80 ºC (42).

Stärk et al. worked on the oxidation of lignin into IL. Manganese, iron and cooper salts were used as catalyst into 4 different ionic liquids: emimCFSO$_3$, m$_2$imMeSO$_4$, emimEtSO$_4$ and emimMeSO$_3$. The reaction conditions were 10% w/w of lignin in the IL and 2% of catalyst, at 100ºC and 80 x 10^5 Pa in a batch reactor. It was observed that $Mn(NO_3)_2$ salt was the most effective. After the reaction, the possible monomers were extracted by 3 different solvents, toluene, dichlorometane and ethylacetate. The best performance was observed into emimCFSO$_3$, achieving 54% of lignin conversion. When catalyst concentration was increased to 20%, 63% of conversion was achieved. Finally DCM was chosen as extracting solvent for the monomers because GC chromatogram showed the highest range of monomer compounds, syringol, vanillin, 2,6-dimethoxy-1,4-benzoquinone DMBQ and syringaldehyde were determined as reaction products. Extracted oil yield was also increased with the increasing of catalyst concentration from 27% to 32%. The DMBQ was determined as the main product and 11.5% was finally selectively isolated (43).

Lahtinen et al. studied the reactivity of lignin model compounds (2,6-dimethoxyphenol andconyferyl alcohol) on amimCl by the enzyme M. Albomyces laccase. The reaction was carried out during 20 h at pH 6, achieved by Na-citrate buffer, varying the concentration of IL in water from 0 to 40 mol%, and NaN_3 was used as enzyme inhibitor. The oxidation of 2,6-dimethoxyphenol and conyferyl alcohol was achieved resulting in the formation of β-5 4-(3-(hydroxymethyl)-7-methoxy-5-methyl-2,3-dihydrobenzofuran-2-yl)2-methoxyphenol, β-β 4,4'-(tetrahydro-1H,3H-furo[3,4-c]furan-1,4-diyl)bis(2-methoxypehol), β-O-4 1-(4-(hydroxyl-3-methoxyphenyl)-2-(2-methoxy-4-methylphenoxy)propane-1,3-diol, and α-C=O/ β-O-4 dimmers 3-hydroxy-1-(4-hydroxy-3-methosyphenyl)-2-(2-methoxy-4-methylphenoxy) propan-1-one, α-O-4/β-O-4 trimmers 4-(3-hydroxy-1,2-bis(2-methoxy-4-methylphenoxy) propyl)-2-methoxyphenol and coniferyl alcohol dehydropolymer. It was observed that the presence of amimCl affected to the kind of linkages which are formed in the polymer enhancing the quantity of β-β structures and decreasing β-O-4 and β-5 links (44).

Jiang et al. studied the selective oxidation of primary alcohols into aldehydes or acids. For the study, metoxybenzylalcohol was used previously to check the optimum conditions in order to obtain aldehydes and then lignin model compounds as veratryl alcohol and vanillyl alcohol were oxidized under the same conditions to establish if the behavior of the catalyst can be extended to other phenolic structures. The influence of the cocatalyst and IL into $VO(acac)_2$ catalyst activity was studied in order to selectively oxidize primary alcohols; under aerobic conditions several cooper salts as cocatalyst and both imidazolium and pyridinium type ionic liquids as solvent, were added to 2 mol% of $VO(acac)_2$ and 6 mol% of 1,4-diazabicyclo[2.2.2]octane (DABCO), in order to check metoxybenzylalcohol reactivity into aldehyde or acid. All the reactions were done at 90 ºC for 15 h. It was observed that the aldehyde was 99% selectively obtained without adding the cocatalyst and using bmimPF6 as IL achieving 90% of yield. To obtain the acid selectively Cu(II) 2-ethylhexanoate was added as cocatalyst and hmimOTf was use as solvent, but only 87% of selectivity and 77% of yield was obtained. It was observed that not only the co-catalyst affected reaction selectivity but also the IL used. With pyridinium type ILs a complex mixture of products was obtained (45). The oxidation of the lignin model compounds was done under 5 mol % $VO(acac)_2$, 10 mol % DABCO, and $bmimPF_6$ as solvent at 80 °C for 12 h, in the vanillyl alcohol the oxidation selectively occurs in the primary alcohol and not the phenolic one. For both veratryl and vanillyl alcohols, selectively the oxidations stops in the aldehyde with 96% and 94% of selectivity and 94%, 82% yield respectively (46).

It has been demonstrated that transition metal nanoparticles have higher catalytic activity than the bulk material; it is explained by its higher specific surface. So that, the use of nanoparticles as catalyst was studied on lignin model compounds and lignin oxidation by Zhu et al., dispersing metal nanoparticles in the IL. Palladium nanoparticles as catalyst, pyridinium salt of iron bis(dicarbollide) as cocatalyst, O_2 as oxidant and $bmimMeSO_4$ and $bmimPF_6$ as support ILs were used for the oxidation of benzyl alcohol with different substituents (benzylalcohol, (4-methoxyphenyl)methanol, (2-methoxyphenyl)methanol, (3,4-dimethoxyphenyl)methanol and (3-phenoxyphenyl)methanol) and lignin. Substituted benzyl alcohols were selectively oxidized to its corresponding aldehydes, benzaldehyde, 4-methoxybenzaldehyde, 2-methoxybenzaldehyde, 3,4-dimethoxybenzaldehyde and 3-phenoxybenzaldehyde were obtained with 93%, 86%, 77%, 84% and 80% aldehyde isolated yields respectively, under the optimum conditions that correspond to alcohol:nanopaladium:pyridinium salt 100:10:1 ratio and 4 bar of O_2 into a mixture of $BmimMeSO_4$:$BmimPF_6$ 1:2 at 120 °C for 18 h. Lignin was treated under the same conditions, achieving 72% of lignin conversion, the main products were identified as syringaldehyde, vanillin, p-hydroxybenzaldehyde mixed with a slight amount of 2,6-dimethoxy-1,4-benzoquinone (47).

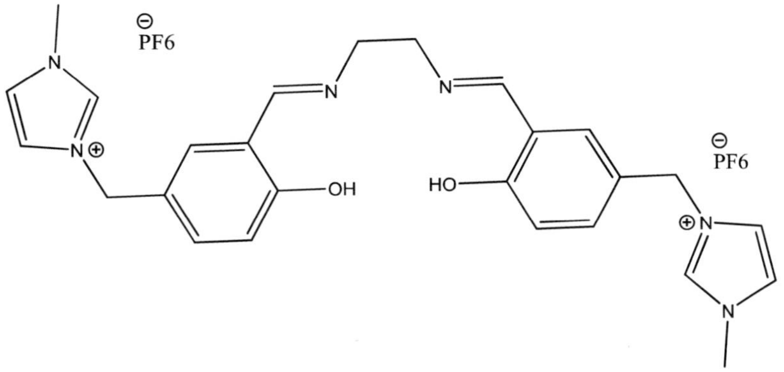

Figure 5. TSIL used as complex for Co.

Liu et al. worked directly with the lignin and its oxidation conditions were optimized; using different ILs (m_2imMe_2P, methylpyridium Me_2P, methyltriethylaminium Me_2P and 4,4-dimethylmorpholin-4-ium Me_2P), 2.5 g of lignin were treated into 47.5 g of solvent with 0.025 g of $CuSO_4$ at 2.5 MPa O_2 pressure at 175 °C during 1.5 h, adding methylisobutylketone to the reaction media. All the solvents were treated in the same conditions and good performances were obtained, being the best option the use of the best m_2imMe_2P as solvent. It was also tried without adding methylisobutylketone, but the reaction yields decreased. Finally, was determined that lignin was successfully oxidized achieving conversion yield nearly to 100%. As lignin derived aromatic aldehydes, vanillin, syringaldehyde and p-hydroxybenzaldehyde were obtained as products with total yield of 29.7% (48).

ILs are highly used on electrochemistry because their high conductivity, taking advantage of this property, electro-catalytic oxidation of lignin into IL was studied by Reichert et al.. The authors prepared an electrode coated by $Ru_{0.25}V_{0.05}Ti_{0.7}Ox$ and Ag as reference electrode; the potential was varied between 0.7 and 1.7 V to check the best conditions. Triethylamonium methylsulphonate (Et_3NMeSO_3) was chosen as solvent for the lignin due to its good solvation performance. Electrolysis was achieved, and the products were extracted by diethyl ether. The best product yield (6%) was obtained with 1.5 V, from the obtained products benzaldehyde, 3-furaldehyde, m-tolualdehyde, vanillin and acetovanillone, were identified by GC-MS (49).

Conclusion

Solubility of lignin into ILs is highly affected by the anion, being the sulphates or sulphonates those which exhibit the best performance. The cation is not so important for lignin extraction; however, most of the ILs used are based on imidazolium ring with different substituents.

Reactivity of several lignin model compounds were studied to establish the optimum conditions, nevertheless, not always these conditions worked in the same way with lignin as with model compounds, even sometimes lignin was not affected by the reaction media, as it occurred in the intent of reduction reaction. The oxidation reactions was probed to be most effective to depolymerize lignin, once again the yields obtained by model compounds were generally much more better than those obtained with lignin. This behavior is principally due to complex structure of lignin that is not completely understood and to the internal interactions that convert lignin into a very stable polymer.

The treatment with ILs resulted in wide range of building blocks and fine chemical with high added value such the DMBQ is developed as a component of anti-tumoral drug.

References

[1] Lanzalunga, O. (2000). Photo- and radiation chemical induced degradation of lignin model compounds. *J. Photochem. Photobiol., B*, 85-108.

[2] García, A. (2009). Characterization of lignins obtained by selective precipitation. *Sep. Purif. Technol., 68*, 193-198.

[3] Azadi, P. (2013). Liquid fuels, hydrogen and chemicals from lignin: A critical review. *Renewable Sustainable Energy Rev., 21*, 506-523.

[4] Kilpeläinen, I. (2007). Disolution of wood in Ionic liquids. *J. Agric. Food Chem., 55*, 9142-9148.

[5] Yang, D. (2007). Properties of sodium lignosulfonate as dispersant of coal water slurry. *Energy Convers. Manage., 48*, 2433–2438.

[6] Sena-Martins, G. (2008). Eco-friendly new products from enzymatically modified. *Ind. Crops Prod.*, *27*, 189-195.

[7] Sivasankarapillai, G. (2012). Lignin valorization by forming toughened lignin-co-polymers: Development of hyperbranched prepolymers for cross-linking. *Biomass and Bioenergy*, *47*, 99–108.

[8] Lora, J.H. (2002). Recent Industrial Applications of Lignin: A Sustainable Alternative to Nonrenewable Materials. *Journal of Polymers and the Environment*, *10*, 39-48.

[9] Sahoo, S. (2011). Characterization of industrial lignins for their utilization in future value added applications. *biomass and bioenergy*, *35*, 4230-4237.

[10] Vinardell, M.P. (2008). Potential applications of antioxidant lignins from different sources. *industrial crops and products*, *27*, 220-223.

[11] Košiková, B. (2002). Reduction of carcinogenesis by bio-based lignin derivatives. *Biomass Bioenergy*, *23*, 153-159.

[12] García, A. (2010). Study of the antioxidant capacity of Miscanthus sinensis lignins. *Process Biochemistry*, *45*, 935-940.

[13] Zhu, J. Y. (2009). Sulfite pretreatment (SPORL) for robust enzymatic saccharification of spruce and red pine. *Bioresource Technology*, *100*, 2411-2418.

[14] Chakar, F. S. (2004). Review of current and future softwood kraft lignin process chemistry. *Ind. Crops Prod.*, *20*, 131-141.

[15] Kumar, P. (2009). Methods for pretreatment of lignocellulosic biomass for efficient hydrolysis and biofuel production. *Ind. Eng. Chem. Res.*, *48* (8), 3713-3729.

[16] Zhao, X. (2009). Organosolv pretreatment of lignocellulosic biomass for enzymatic hydrolysis. *Appl. Microbiol. Biotechnol.*, *82*, 815–827.

[17] Pu, Y. (2007). Ionic Liquid as a Green Solvent for Lignin. *J. Wood Chem. Technol.*, *27*, 23-33.

[18] Lee, S. H. (2009). Ionic Liquid-Mediated Selective Extraction of Lignin From Wood Leading to Enhanced Enzymatic Cellulose Hydrolysis. *Biotechnology and Bioengineering*, *105* (5), 1368-1376.

[19] Lateef, H. (2009). Separation and recovery of cellulose and lignin using ionic liquids: a process for recovery from paper-based waste. *J Chem Technol Biotechnol*, *84*, 1818-1827.

[20] Tan, S. S. Y. (2009). Ionic liquids in biomass processing. *Top Curr. Chem.*, *290*, 311-339.

[21] Prado, R. (2012). Lignin extraction and purification with ionic liquids. *J. Chem. Technol. Biotechnol.*, *88* (7), 1248-1257.

[22] Brandt, A. (2011). Ionic liquid pretreatment of lignocellulosic biomass with ionic liquid–water mixtures. *Green Chem.*, *13*, 2489–2499.

[23] Swatloski, R. P. (2002). Dissolution of Cellose with Ionic Liquids. *J. Am. Chem. Soc.*, *124* (18), 4975-4976.

[24] Kim, S. J. (2010). Cellulose pretreatment with 1-n-butyl-3-methylimidazolium chloride for solid acid-catalyzed hydrolysis. *Bioresour. Technol.*, *101*, 8273-8279.

[25] Cao, Y. (2009). Room temperature ionic liquids (RTILs): A new and versatile platform for cellulose processing and derivatization. *Chem. Eng. J.*, *147*, 13-21.

[26] Sun, N. (2009). Complete dissolution and partial delignification of wood in the ionic liquid 1-ethyl-3-methylimidazolium acetate. *Green Chemistry*, *11*, 646–655.

[27] Magalhães da Silva, S. P. (2013). Novel pre-treatment and fractionation method for lignocellulosic biomass using ionic liquids. *RSC Advances*, *3*, 16040–16050.

[28] D'Andola, G. S. (2008). *Patent No. WO 2008/043837.*

[29] J. Zakzeski, J. (2010). The catalytic valorization of lignin for the production of renewable chemicals. *Chem. Rev.*, *110*, 3552-3599.

[30] Binder, J. B. (2009). Reactions of lignin model compounds in ionic liquids. *Biomass and bioenergie*, *33*, 1122-1130.

[31] Nowakowskia, D. J. (2010). Lignin fast pyrolysis: Results from an international collaboration. *J. Anal. Appl. Pyrolysis*, *88*, 53-72.

[32] Karagöz, S. (2005). Comparative studies of oil compositions produced from sawdust, rice husk, lignin and cellulose by hydrothermal treatment. *Fuel*, *84*, 875-884.

[33] Zhang, B. (2012). Structure and Pyrolysis Characteristics of Lignin Derived from Wood Powder Hydrolysis Residues. *Appl Biochem Biotechnol*, *168*, 37-46.

[34] Okuda, K. (2004). Efficient conversion of lignin into single chemical species by solvothermal reaction in water–p-cresol solvent. *J. Phys.: Condens. Matter*, *16*, S1325-1330

[35] Toledano, A. (2012). Organosolv lignin depolymerization with different base catalysts. *J. Chem. Technol. Biotechnol.*, *87*, 1593–1599.

[36] Vigneault, A (2007). Base-catalyzed depolymerization of lignin: separationofmonomers. *Can. J. Chem. Eng.*, *85*, 906-916.

[37] Barta, K. (2010). Catalytic disassembly of an organosolv lignin via hydrogen transfer from supercritical methanol. *Green Chem.*, *12*, 1640-1647.

[38] George, A. (2011). The effect of ionic liquid cation and anion combinations on the macromolecular structure of lignins. *Green Chem.*, *13*, 3375–3385.

[39] Chatel, G. (2013). Review: Oxidation of lignin using ionic liquids-an innovative strategy to produce renewable chemicals. *ACS Sustainable Chem. Eng.*

[40] Zakzeski, J. (2010). Transition metal catalyzed oxidation of Alcell lignin, soda lignin, and lignin model compounds in ionic liquids. *Green Chem.*, *12*, 1225–1236.K.

[41] Kervinen, H. K. (2003). Oxidation of veratryl alcohol by molecular oxygen in aqueous solution catalyzed by cobalt salen-type complexes: the effect of reaction conditions. *J. Mol. Catal. A: Chem.*, *203*, 9–19.

[42] Sonar, S. (2012). Synthesis and application of Co(salen) complexes containing proximal imidazolium ionic liquid cores. *Can. J. Chem.*, *90*, 60-70.

[43] Stärk, K. (2010). Oxidative Depolymerization of Lignin in Ionic Liquids. *Chem. Sus. Chem.*, *3*, 719 – 723.

[44] Lahtinen, M. (2013). On the reactivity of the Melanocarpus albomyces laccase and formation of coniferyl alcohol dehydropolymer (DHP) in the presence of ionic liquid 1-allyl-3-methylimidazolium chloride. *J. Mol. Catal. B: Enzym.*, *85-86*, 169-177.

[45] Jiang, N. (2007). Selective Aerobic Oxidation of Activated Alcohols into Acids or Aldehydes in Ionic Liquids. *J. Org. Chem.*, *72*, 7030-7033.

[46] Jiang, N. (2007). Vanadium-catalyzed selective aerobic alcohol oxidation in ionic liquid [bmim]PF6. *Tetrahedron Lett.*, *48*, 273-276.

[47] Zhu, Y. (2012). An Efficient and Recyclable Catalytic System Comprising Nanopalladium(0) and a Pyridinium Salt of Iron Bis(dicarbollide) for Oxidation of Substituted Benzyl Alcohol and Lignin. *Chem. Open*, *1*, 67-70.

[48] Liu, S. (2013). Process of lignin oxidation in an ionic liquid coupled with separation. *RSC Adv.*, *3*, 5789-5793.

[49] Reichert, E. (2012). Electro-catalytic oxidative cleavage of lignin in a protic ionic liquid. *Phys. Chem. Chem. Phys.*, *14*, 5214-5221.

In: Ionic Liquids: Synthesis, Characterization and Applications ISBN: 978-1-63117-716-3
Editor: Anson Brooks

Chapter 3

IONIC LIQUIDS AS CATALYSTS FOR THE IMINE BASE/ISOCYANATE-MEDIATED (IBI) RADICAL POLYMERIZATION: MECHANISTIC STUDIES AND POLYMERIZATION KINETICS CONTROL

***Ingmar Polenz*[1,*] *and Stefan Spange*[2,†]**
[1]Max-Planck Institute for Dynamics and Self-organization, Goettingen, Germany
[2]Chemnitz University of Technology, Chemnitz, Germany

ABSTRACT

Room temperature Ionic Liquids (ILs) have received high interest in academia within the last decades due to their use as catalysts or co-reactants in various organic and inorganic syntheses as well as in polymer and material science. Based on the inert nature of the Ionic Liquids, in most of the reactions ILs undergo no measurable conversion and, hence, become attractive even for industry. In especially, systems that require solely catalytic IL amounts are potential candidates for further investigations on the industrial sector.

Subject of this chapter is to show the use of Ionic Liquids as catalysts for the recently reported imine base / isocyanate (IBI) mediated radical polymerization of methyl methacrylate (MMA). Investigations on this quaternary reactive system allowed both an enhancement of the polymerization properties of the IBI mixtures and the gain of deeper insights to the reaction mechanism of this novel radical polymerization type. Typical IBI mixtures initiate the radical polymerization at slightly elevated temperatures or by adding high molar amounts of the imine base and the iscyanate compared to the monomer equivalent; the addition of Ionic Liquid traces to the polymerization mixture gives polymerization of the monomer even at room temperature for all IBI mixtures and an enormous increase in the brutto reaction rate.

* Ingmar Polenz: Max-Planck Institute for Dynamics and Self-organization, Am Fassberg 17, D-37077 Goettingen, Germany.
† Stefan Spange: Chemnitz University of Technology, Strasse der Nationen 62, 09111 Chemnitz, Germany.

Compared to the non-IL containing reaction mixture, the average molecular weight of PMMA isolated from the catalyzed polymerization is significantly increased.

There are no experimental observations that the Ionic Liquids are bonded to the resulting polymer chains and ILs can be obtained from the reaction mixture almost quantitatively, thus their interaction as catalyst at the IBI polymerization is indicated. However, the IL component is measured to be part of the IBI brutto polymerization rate law which underlines its significance on the whole polymerization process.

Mechanistic studies on the IL catalyzed IBI polymerization have solidified the understanding of the initial step of this reaction. To monitor the interaction of the Ionic Liquid with MMA and the individual IBI components, band shifts at the attenuated total reflection Fourier transformation middle infrared spectrums (ATR FT MIR) are investigated and comparison with the appropriate Kamlet-Taft Parameters is done that reveal the activation of isocyanate and MMA sites by the IL as a pre-step to the IBI radical formation.

Keywords: Radical IBI Polymerization, FTIR, Kamlet Taft Parameters, Polymerization Kinetics

1. INTRODUCTION

Room temperature Ionic Liquids (ILs) have gained high interest in academia within the last decades due to their use as catalysts or co-reactants in various organic and inorganic syntheses as well as in polymer and material science. [1–10] Based on the inert nature of Ionic Liquids, in most of the reactions ILs undergo no measurable conversion and, hence, become attractive even for industry. In especially, systems that require solely catalytic IL amounts are potential candidates for further investigations on the industrial sector.

Subject of this chapter is to show the use of Ionic Liquids as catalysts for the recently reported imine base / isocyanate (IBI) mediated radical polymerization of methyl methacrylate (MMA); the general reaction course can be seen in Scheme 1. [11–14]

Ionic Liquid addition to this binary reactive system allows both an enhancement of the polymerization properties of the IBI mixtures and gaining of deeper insights to the reaction mechanism of this novel radical polymerization type. Typical IBI mixtures initiate the radical polymerization at slightly elevated temperature or by adding high molar amounts of the imine base and the isocyanate compared to the monomer equivalent, which is disadvantageous. The addition of Ionic Liquid traces to the reaction mixture results in polymerization of the monomer even at room temperature for all IBI mixtures and an enormous increase in the brutto reaction rate v_{Br}, that is a measure of the overall reaction kinetics, is observed.

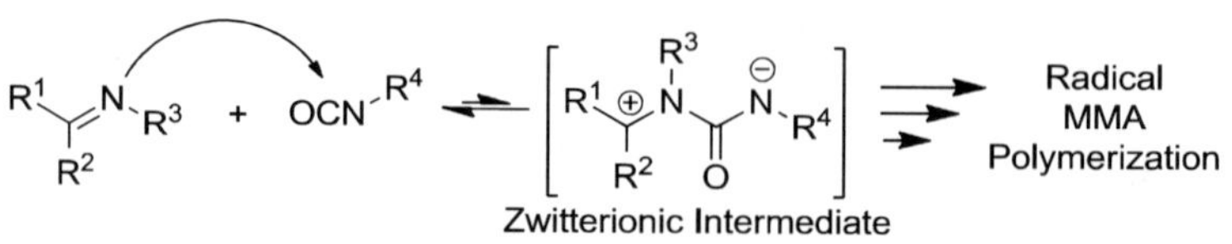

Scheme 1. General reaction scheme of the imine base / isocyanate (IBI) mediated radical polymerization reported in Literature [11–14].

Compared to the non-IL containing reaction mixture, the average molecular weight of PMMA resulting from the catalyzed polymerization is significantly increased, which can be explained by the protected radical mechanism.

There are no experimental observations that the Ionic Liquids are bonded to the resulting polymer chains. ILs can be re-obtained from the reaction mixture quantitatively; therefore, their interaction as a catalyst at the IBI polymerization is indicated. However, the IL component is measured to be part of the IBI brutto polymerization rate law which underlines its significance on the whole polymerization process.

The use of Ionic Liquids at the IBI polymerization enables the generation of designer initiator formulations with well-defined properties for various polymerization techniques and, hence, is of potential interest in industry. From studies on the temperature dependence of the IBI polymerization it is possible to select optimal IL / imine base / isocyanate combinations as promising candidates for material scientific applications.

It is observed for some IBI mixtures the IL to inhibit the formation of side products of the IBI components and consequently to have a positive effect on the initiation properties of the overall polymerization process emphasizing the governing parameters at the selection of an IBI / IL combination for certain applications.

Mechanistic studies on the IL catalyzed IBI polymerization have solidified the understanding of the initial step of this reaction. To monitor the interaction of the Ionic Liquid with MMA and the individual IBI components, band shifts at the attenuated total reflection Fourier transformation middle infrared spectrums (ATR FT MIR) are investigated and comparison with the appropriate Kamlet-Taft Parameters is done that reveal the activation of isocyanate and MMA sites by the IL as a pre-step to the IBI radical formation. It is shown that, compared to common Lewis acids that are used for the enhancement of the electron accepting properties of MMA, interactions of the type IL···MMA as well as the ones with the IBI components comparatively are of weak nature and, hence, predestined at the catalysis of the IBI polymerization.

The results provided here are reprinted from literature 13; completions to the reports are given at certain stages.

2. Interaction of ILs with the IBI Reagents

Recently, it was reported on a novel radical polymerization of vinyl monomers that is initiated by a combination of an imine base and an isocyanate (IBI polymerization). [11, 12]

At this polymerization the formation of the initiating species occurs in the presence of the vinyl monomer; the IBI mixture solely is consumed at the initiation process. The radical character of the initiating species was proved by copolymerization experiments, ESR spectroscopy as well as by radical tracer experiments. It is heretofore suggested that the radical generation results consecutively after formation of a zwitterionic intermediate of the IBI components. IBI combinations can be classified into two categories; those that initiate the polymerization at room temperature (low temperature IBI) and those that are applicable only at elevated temperatures (high temperature IBI). Most of the initiator mixtures have been assigned to the last mentioned type requiring an additional thermal activation. Examples of common imine bases and isocyanates for the IBI polymerization and classified by their IBI reactivity are shown in Scheme 2.

From further kinetic studies of the IBI mixtures the isocyanate reactivity influences the brutto polymerization rate constant $k_{Br,0}$ which is an indicator for the overall polymerization kinetics is concluded. Thus, IBI polymerizations using aryl isocyanates such as phenyl isocyanate (PIC) or 2,4-toluene diisocyanate (TDI) in combination with (2-methylthio)-2-thiazolin (MMT) proceed faster by a factor of 3 compared to polymerizations using alkyl isocyanates such as 1,6-hexamethylene diisocyanate (HDI).

Hence, by taking into account that at radical polymerizations the initiation is the rate determining step it is suggested that at higher concentrations of the IBI zwitterionic species (see Scheme 1) resulting from an increased isocyanate reactivity the overall IBI polymerization kinetics can be accelerated.

Lewis acids $SnCl_4$, $ZnCl_2$ or $AlCl_3$ are suitable electrophilic reagents for the compensation of electron density from electron donation groups. They are known to have a remarkable effect on the enhancement of the electron accepting properties of MMA and other vinyl acceptor monomers and polymerizations carried out in the presence of these Lewisacids proceed significantly faster. [15–19]

However, challenging in this context is the fact that both the imine bases and the isocyanates and for certain reagents even MMA in the combination with Lewis acids are forming a variety of complexes via well-known mechanisms [20–22] and, hence, are not available for polymerization. Addition of lithium trifluoromethansulfonate [Li(OTf)] to a mixture of 2-phenyl-2-oxazoline (POX) / phenyl isocyanate (PIC) and MMA, an IBI system requiring a thermal activation, solely gives poor PMMA yields after 96 h at room temperature (12 %, $M_N = 49{,}600$ g·mol^{-1}, PDI = 1.33).

When adding Ionic Liquid traces to POX / PIC / MMA the mixture gets activated, particularly; immediate polymerization of the vinyl monomer is observed even at room temperature. Table 1 compiles some qualitative experimental results of the IL accelerated IBI polymerizations.

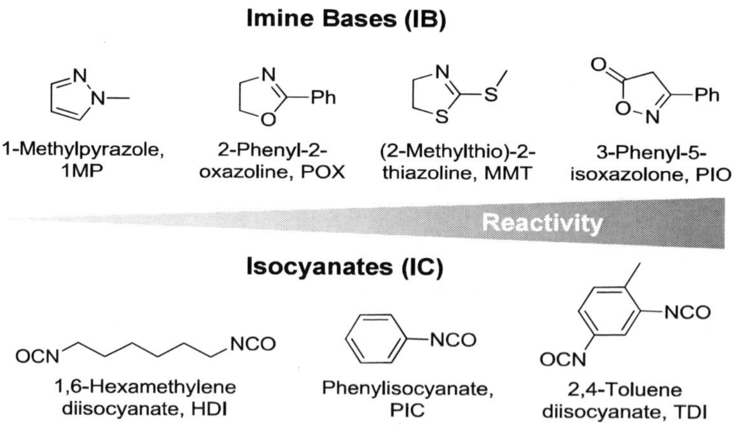

Scheme 2. Common imine bases and isocyanates used for IBI polymerizations sorted by its reactivity. The imine bases (2-methylthio)-2-thiazoline (MMT) and 3-phenyl-5-isoxazolone (PIO) in combination with isocyanates enable the polymerization of MMA at room temperature (low temperature IBI systems); combinations of 2-phenyl-2-oxazoline (POX) or 1-methylpyrazole (1MP) with isocyanates induce MMA polymerizations only with thermal activation (high temperature IBI systems).

Table 1. Reaction conditions and results of the IL accelerated IBI polymerizations of MMA using imine bases commonly requiring a thermal activation. The molecular weights of PMMA are determined using size exclusion chromatography

IB	[IB] / M	IC	[IC] / M	IL	[IL] / M	[MMA] / M	t / h	PMMA yield / %	M_N / g·mol^{-1}	M_W/M_N
POX	0.48	PIC	0.48	–	–	2.85	96	–	–	–
	0.48	PIC	0.48	$[N_{4441}]Ntf_2$	0.05	2.82	35	55	59,800	1.75
	0.48	HDI	0.47	$[N_{4441}]Ntf_2$	0.04	2.76	72	20	336,900	1.83
	0.47	PIC	0.47	$[Py_6]Ntf_2$	0.06	2.81	35	42	65,000	1.96
	0.48	PIC	0.48	[Emim]FAP	0.11	2.85	48	5.5[a]	26,300[a]	2.26[a]
	0.48	PIC	0.48	[Emim]FAP	0.11	2.85	5	13	58,600	1.39
1MP	0.48	PIC	0.48	–	–	2.85	96	–	–	–
	0.48	PIC	0.48	[Emim]FAP	0.21	2.85	5	15	82,000	1.61

All polymerizations are performed in freshly distilled THF at 25 ° C.
[a] Experiment is carried out using styrene as monomer instead of MMA.
Data are reprinted from Lit. [13].

IBI polymerizations at room temperature are feasible using both the POX and the 1MP systems that are stated to be high temperature IBI initiator mixtures. The less reactive HDI induces a slower polymerization but higher molecular weights of the resulting PMMA are observed which evidently shows the smaller initiator concentration at HDI / POX compared to the PIC / POX system.

In contrast, while using styrene instead of MMA, the IBI polymerization proceeds significantly slower indicating the fact that the activation of the monomer component by the IL seems to play a crucial role for the accelerated IBI polymerization. The results of Table 1 also reveal that the activation effect of the Ionic Liquid is not limited on a certain group of ILs; radical MMA polymerization is obtained for different anion and cation types. After intensive washing of the reaction mixture, the IL is extracted quantitatively underlining the fact the IL to be a catalyst at the IBI polymerization.

In general, Ionic Liquids behave as moderately polar solvents with respect to organic solutes. Interactions with other components mainly are of weak nature and can be measured by several spectroscopic methods, namely UV/Vis, NMR, IR and mass spectrometry. [23–25] An established way is to measure the hydrogen bond donating ability α which serves as an indicator for the interaction strength to the lone electron pair of a suitable dye molecule.

Alternatively results can be compared to interactions with reagents having comparable chemical structures. In total, the α values correspond to the acidity of the Ionic Liquid and are measured by UV/Vis spectroscopy using a suitable dye sensitive to the hydrogen bond donating ability. Hydrogen bond donating abilities of Ionic Liquids used in the experiments widely range from 0.36 to 0.83.

Scheme 3 shows the IL chemical structures and corresponding α values. [26]

The α-scale is a suitable tool for characterizing the interaction strength of ionic liquids.

However, it lacks on describing IL interactions with certain sites of a given molecule. To elucidate the effect of the IL on the electronic structure of the IBI component and MMA, Fourier transformation attenuated infrared spectroscopy is used.

Therefore, binary mixtures of the IL and the imine base, the isocyanate and MMA are examined. In Figure 1a–c the IR spectra of [Emim]FAP ($\alpha = 0.83$) mixtures with PIC, MMA and POX are shown.

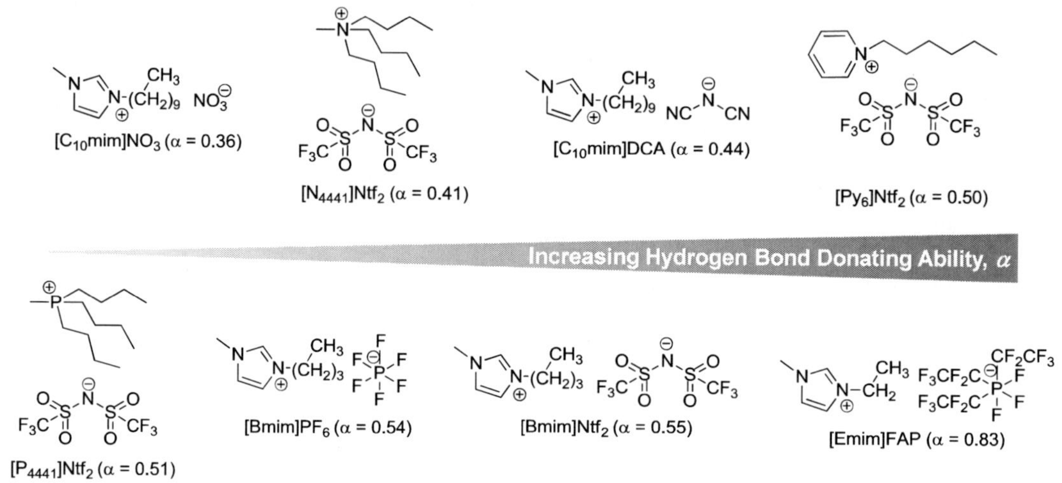

Scheme 3. Ionic Liquids used in this work sorted by their corresponding α values.

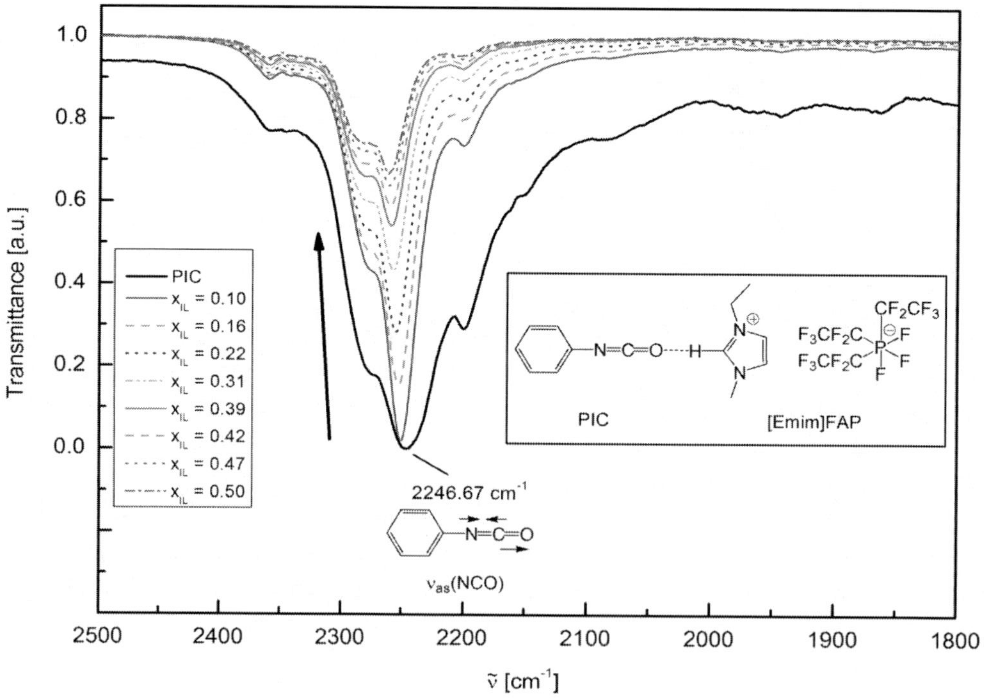

a

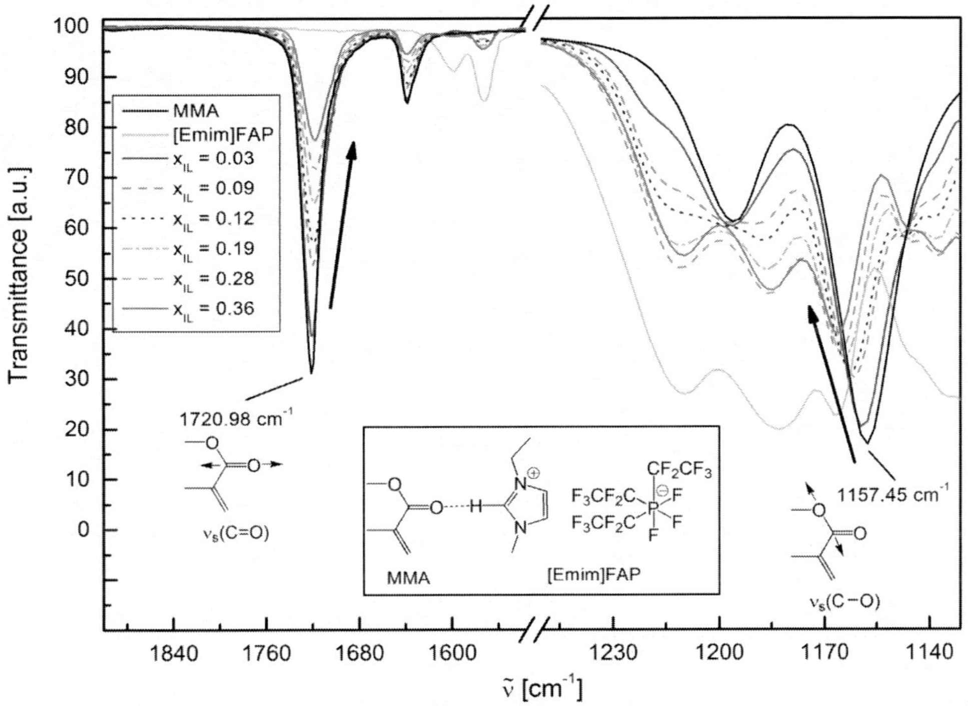

b

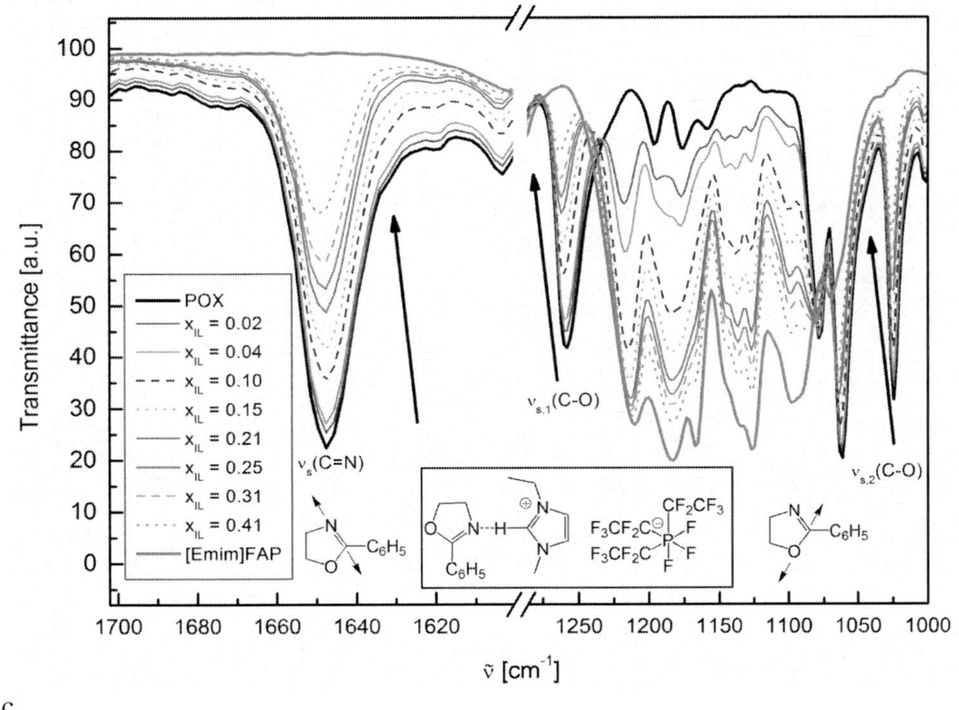

c

Figure 1. (Continued)

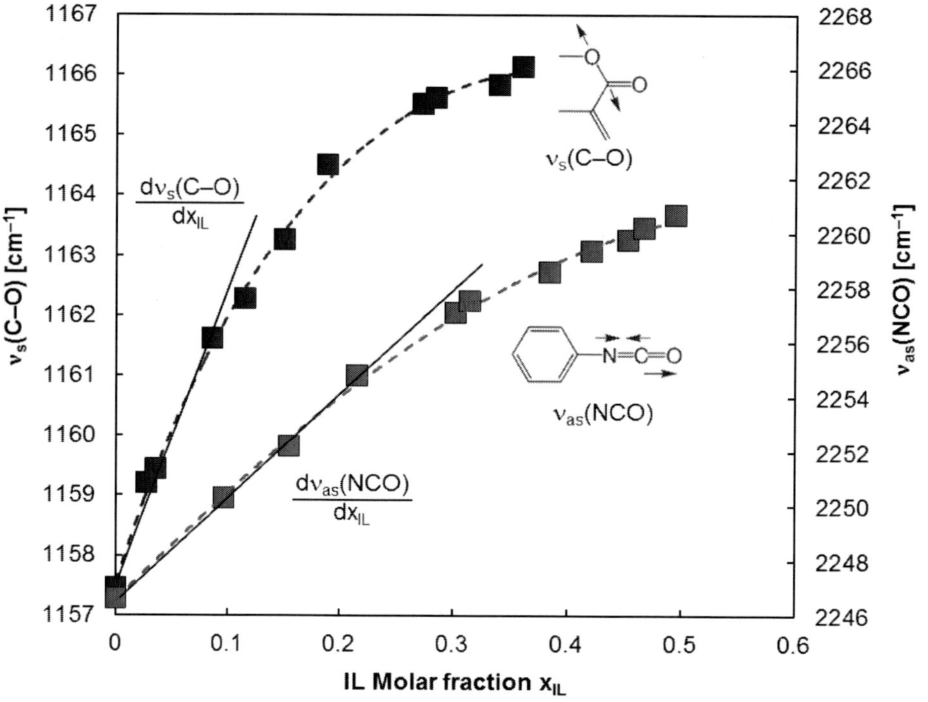

d

Figure 1. IR spectra of binary mixtures PIC / [Emim]FAP (a); MMA / [Emim]FAP (b) and (c) POX / [Emim]FAP using different molar ratios of the Ionic Liquid x_{IL}. Arrows indicate band shifts at increasing IL concentration. (d) Absolute band shifts of the asymmetric stretching ν_{as}(NCO) of PIC and the symmetric stretching ν_s(C–O) of MMA assigned as main interactions at the quaternary mixture of the IL catalyzed IBI polymerization. The corresponding linear fits of the first stage are ν_s(C–O) = $(38.26 \cdot x_{IL} + 1157.85)\mathrm{cm}^{-1}$, $R^2 = 0.96$ and ν_{as}(NCO) = $(34.41 \cdot x_{IL} + 2247.01)\mathrm{cm}^{-1}$, $R^2 = 0.98$.

Significant IR band shifts of characteristic bonds of the IBI components as a function of the IL concentration in all three IBI reagents can be detected. It is known that the imidazolium cation based IL interacts with nucleophilic groups via the C_2–H hydrogen atom, accordingly, certain band shifts at electron bond donating functionalities vibrations of the IBI components IR spectra are measured. This behavior accompanies with the IR band shifts of the symmetrical stretching $\nu_s(C_2\text{–}H)$ of the IL which was also reported in literature. [1–10, 23–26] The interaction of [Emim]FAP with the isocyanate is predominantly considerable by the shift of the IR absorption band at 2246 cm^{-1} originating from the asymmetric stretching of the NCO group, ν_{as}(NCO) (Figure 1a). While increasing the IL amount the ν_{as}(NCO) systematically shifts to higher wavelengths indicating an increase of the electrophilic character of the NCO group. Moreover, at the MMA / [Emim]FAP spectra unambiguous shifts of the ν_s(C=O) to lower wavelengths and ν_s(C–O) to higher wavelengths at increasing IL concentrations are detected. Obviously, the IL attaches upon the C=O group of MMA reducing the electron density at the double bond whereas at the C–O ester group the double bonding character increases. As a result of this interaction electron density at the β-position of the vinyl monomer is removed (partially positively charged) preferentially enabling a nucleophilic addition of the IBI mixture zwitterionic species (see Scheme 1).

Most likely, the information from the IR spectra of the [Emim]FAP interactions with the isocyanate and the monomer can be concluded by the mesomeric species exhibited in Scheme 4.

An interaction of the IL with the imine base POX is measured that is suggested by the IR band shifts of POX symmetric stretching moieties ν_s(C=N), $\nu_{s,1}$(C–O), and $\nu_{s,2}$(C–O) which is shown in Figure 1c. However, the absolute band shifts in the measured range ($\Delta\nu_s$ = 2–3 cm^{-1}) are significantly smaller compared to the ones of the IL–PIC and IL–MMA mixtures (Figure 1d) implying a weaker interaction of [Emim]FAP with 2-phenyl-2- oxazoline. Considering these aspects and mechanistic features so far known of the IBI polymerization [11] the role of the IL at the IBI polymerization catalysis can be stated as shown in Scheme 5.

Measurements of IR spectra of the binary mixtures styrene–[Emim]FAP give no measurable band shifts. Thereof, no mentionable interaction with functional groups of the IL with this monomer occurs. As mentioned, attempts on the styrene polymerization using the combination POX / PIC and traces of [Emim]FAP (see Table 1) give solely tiny amounts of polystyrene, emphasizing the fact that the monomer activation by the IL among the catalyzed IBI polymerization is a crucial step.

The absolute band shifts of ν_s(C–O) of MMA and ν_{as}(NCO) of PIC at the binary mixtures with [Emim]FAP are shown in Figure 1d. In comparison to infrared frequency shifts of the symmetric stretching ν_s(C=O) and ν_s(C=C) of known Zinc complexes of MMA [15] the absolute values of the band shifts at the binary mixtures of this work are small, indicating the weak nature of the IL interactions with the IBI components (see Table 2).

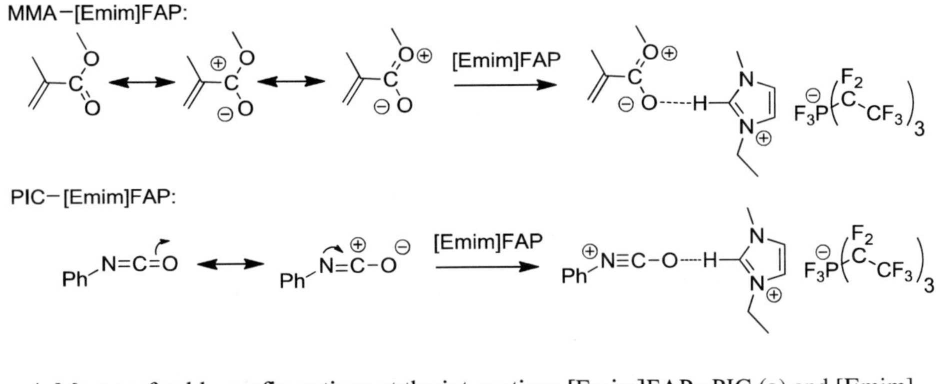

Scheme 4. Most preferable configurations at the interactions [Emim]FAP···PIC (a) and [Emim] FAP···MMA (b) indicated by the IR band shifts revealing the remarkable effect of the IL on the electron structure of the IBI reagents.

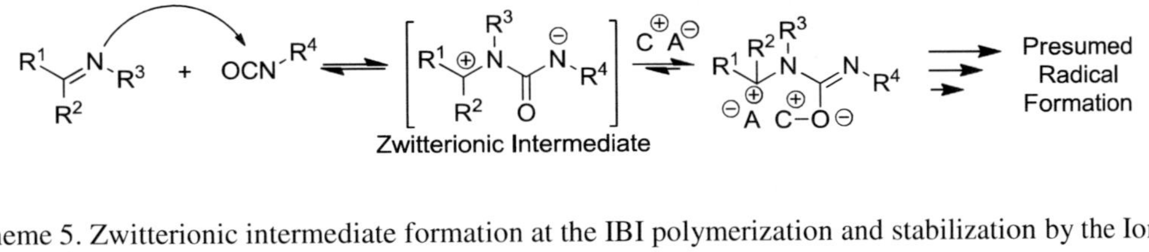

Scheme 5. Zwitterionic intermediate formation at the IBI polymerization and stabilization by the Ionic Liquid C^+A^-.

Table 2. IR band shifts of ν_s(C=O) and ν_s(C=C) of MMA in Zinc complexes and at binary IL / MMA mixtures with the molar fraction x_{MMA}

System	x_{MMA}	$\Delta\nu_s$(C=O) / cm^{-1}	$\Delta\nu_s$(C=C) / cm^{-1}
Zn(MMA)	0.5	–20, –60 [15]	–3
Zn(MMA)2	0.67	–20 [15]	–3
MMA–[Emim]FAP	0.64	–3.9	–2.7
MMA–[Bmim]Ntf_2	0.57	–2.2	–1.3
MMA–[Bmim]PF_6	0.54	–2.8	–1.1
MMA–[C_{10}mim]DCA	0.61	–2.0	–1.5
MMA–[C_{10}mim]NO_3	0.76	–2.6	–0.9

Data are reprinted from Lit. [13].

A specific linear stage of the relations x_{IL} versus the band positions at molar fractions of the IL below 0.2 are detected and plotted in Figure 1d. This effect results from the different solvation moieties of the IL at increasing Ionic Liquid concentration. [27] It is discussed that in binary IL – organic solute mixtures at $x_{IL} < 0.2$ the IL is surrounded by the organic solute; at values $x_{IL} > 0.2$ domain formation of the Ionic Liquid molecules themselves occurs due to the weak solubilizing abilities of the ILs in organic media. To elucidate the changes in the IR spectra at the binary IL MMA mixtures while changing the IL type the slopes at the first stage $d\nu_s/dx_{IL}$ of the ν_s(C–O) MMA band shift relations as a function of the IL concentration are taken as a measure of the relative interaction strength of the IL with MMA. A linear correlation is found when plotting the hydrogen bond donating ability α of the Ionic Liquid versus $d\nu_s/dx_{IL}$ values measured at IR measurements (Figure 2).

The result reveal the slopes $d\nu_s/dx_{IL}$ of the measured imidazolium cation based ILs to be a good alternative measure compared to the literature known Kamlet Taft parameter for monitoring the interaction of the IL with a solute; certain solute molecule moieties are taken into consideration with this approach. For explaining the influence of the Ionic Liquid on the IBI polymerization kinetics both the α values and the slopes $d\nu_s/dx_{IL}$ can be used.

3. Kinetic Studies on the IL Accelerated IBI Polymerization

The effect of the Ionic Liquid on the IBI mixture and the monomer is remarkable. The complexation of the isocyanate and MMA by the IL is of weak nature but obviously leads to their increase in reactivity on the IBI polymerization course (see Table 1).

Even polymerization using IBI combinations commonly requiring a thermal activation such as POX / isocyanate or 1MP / isocyanate in the presence of IL traces can be performed at room temperature at high rates.

For a further characterization of the acceleration effect of the IL, polymerization kinetics of IL catalyzed IBI polymerizations are measured. A feasible quantity in polymer science to describe the polymerization kinetics as a function of a certain component is the brutto polymerization rate $v_{Br,0}$. In a first simple approach the brutto polymerization law has to be verified.

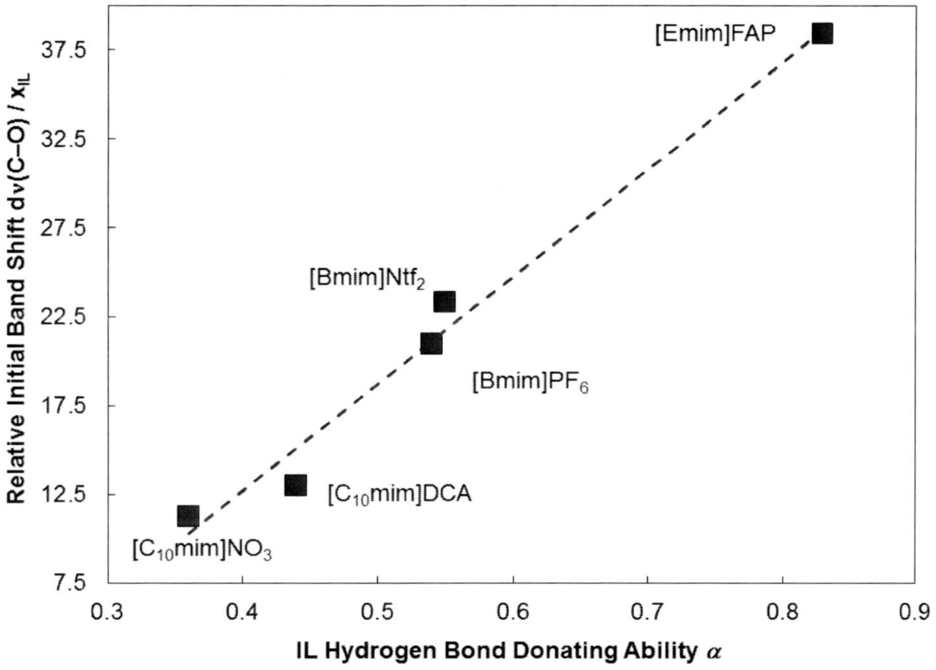

Figure 2. Relationship between the hydrogen bond donating ability α and the slope dv_s/dx_{IL}. The linear fit gives $dv_s/dx_{IL} = 55.24 \cdot \alpha - 7.41$., $R^2 = 0.98$.

From previous studies the shape of the brutto polymerization rate equation $v_{Br,0}$ for the non-catalyzed IBI polymerization is known. Including the IL to the brutto rate law, the correlation gives:

$$v_{Br,0} = k_{Br,0} \cdot [IC]^{0.5} \cdot [IB]^{1} \cdot [MMA]^{1} \cdot [IL]^{x} \quad (1)$$

where $k_{Br,0}$ is the brutto polymerization rate constant and [IC], [IB], [IL] and [MMA] the concentrations of the isocyanate, imine base, Ionic Liquid and methyl methacrylate, respectively. The determination of the reaction order x referring to the Ionic Liquid is done by measuring the brutto polymerization rate $v_{Br,0}$ as a function of the IL concentration like commonly proceeded for radical polymerizations. [28, 29] In Figure 3 the results of selected polymerization kinetic determinations are plotted using POX and PIC as the IBI components

The concentration of the imine base, isocyanate and MMA are kept constant and the quantity of $v_{Br,0}$ is received from the absolute value of the MMA-concentration versus time plot slope at MMA conversions below 5 %. The slope of the double logarithmic plotting of $v_{Br,0}$ versus the initial IL concentrations yields the order x with respect to the Ionic Liquid. A x value close to 0.5 is determined (see Figure 4) that is a similar result to the square root dependency of the initiator species of free radical polymerizations. [28, 29] The fact, that the IL is integrated into the brutto rate law confirms the Ionic Liquid to have a strong impact on the whole polymerization mechanism.

The brutto polymerization rate constant $k_{Br,0}$ value is independent of concentration and is solely a function of the temperature and pressure, thereof, $k_{Br,0}$ allows the comparison of various IL catalyzed IBI polymerizations.

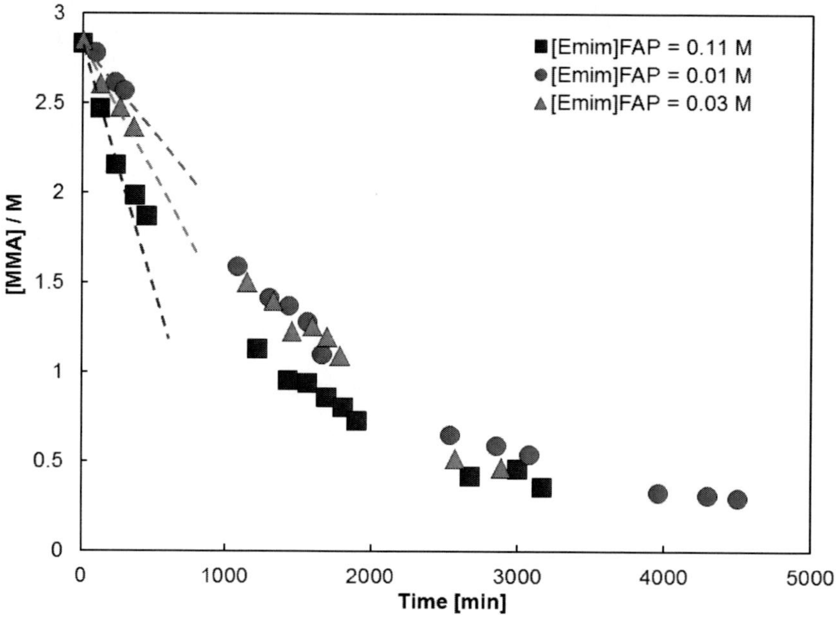

Figure 3. Selected [MMA] versus time plots of the IBI polymerizations of MMA using 2-phenyl-2-oxazoline (POX) and phenyl isocyanate (PIC) using different concentrations of [Emim]FAP. The IL concentrations are 0.01 (●), 0.03 (▲), and 0.11 (■) M; polymerizations are carried out at 25 °C in freshly distilled THF. The corresponding absolute values of the initial slopes are 10.96, 14.07 and 23.25 $M \cdot min^{-1}$, respectively. MMA concentrations are determined using ^1H NMR spectroscopy in $CDCl_3$.

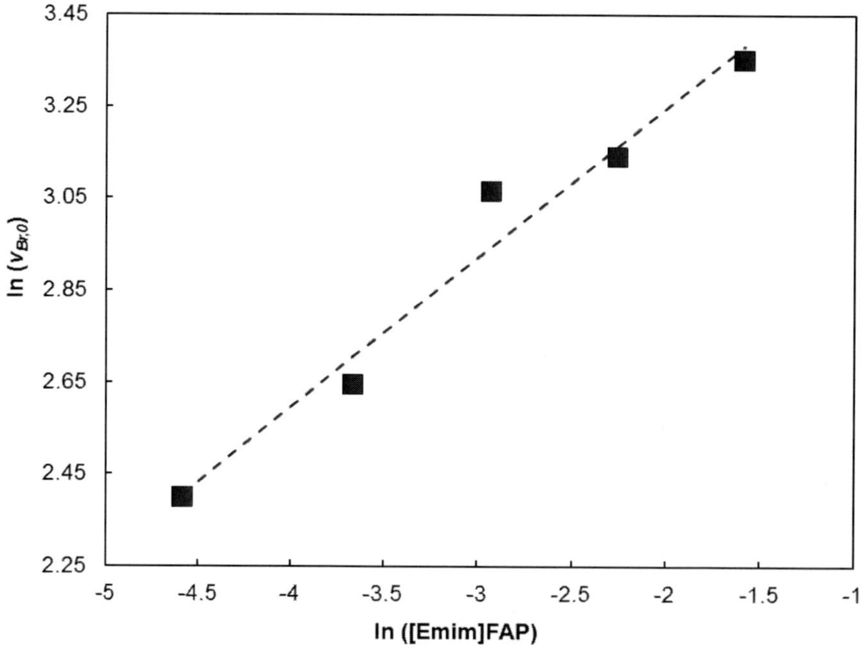

Figure 4. Double logarithmic plot of the brutto polymerization rate $v_{Br,0}$ as a function of the corresponding [Emim]FAP concentrations at the IL catalyzed IBI polymerization of the ternary mixture POX / PIC / MMA. The linear fit of the relation reads ln ($v_{Br,0}$) = 0.33 ln ([Emim]FAP) + 3.9, R^2 = 0.97.

The values of $k_{Br,0}$ of the IL catalyzed IBI polymerization are correlated to the brutto polymerization rate constants of the non-catalyzed IBI polymerizations. In Figure 5a changes in $k_{Br,0}$ of [Emim]FAP catalyzed IBI polymerizations of POX, MMT and 1MP with values reported earlier [11] of the non-catalyzed IBI polymerizations are demonstrated.

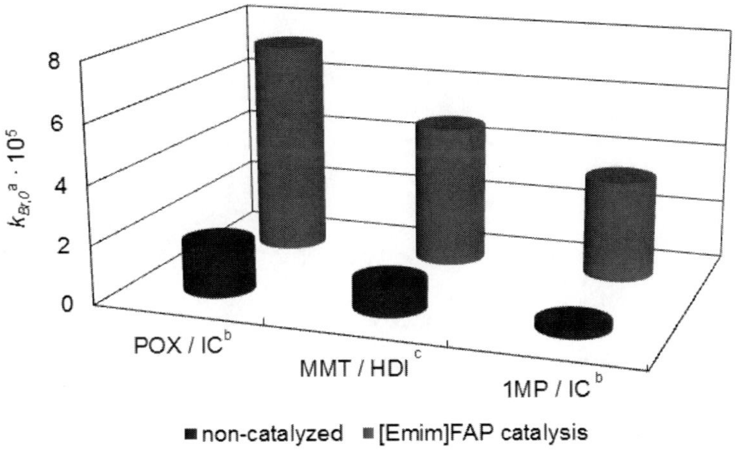

a

[a] Units are: [Emim]FAP catalysis $L^2\ mol^{-2}\ s^{-1}$; non-catalyzed $L^{1.5}\ mol^{-1.5}\ s^{-1}$.

[b] non-catalyzed polymerizations were carried out in substance at 100°C; IL catalyzed at 25°C in tetrahydrofuran.

[c] both polymerizations at 25°C in tetrahydrofuran.

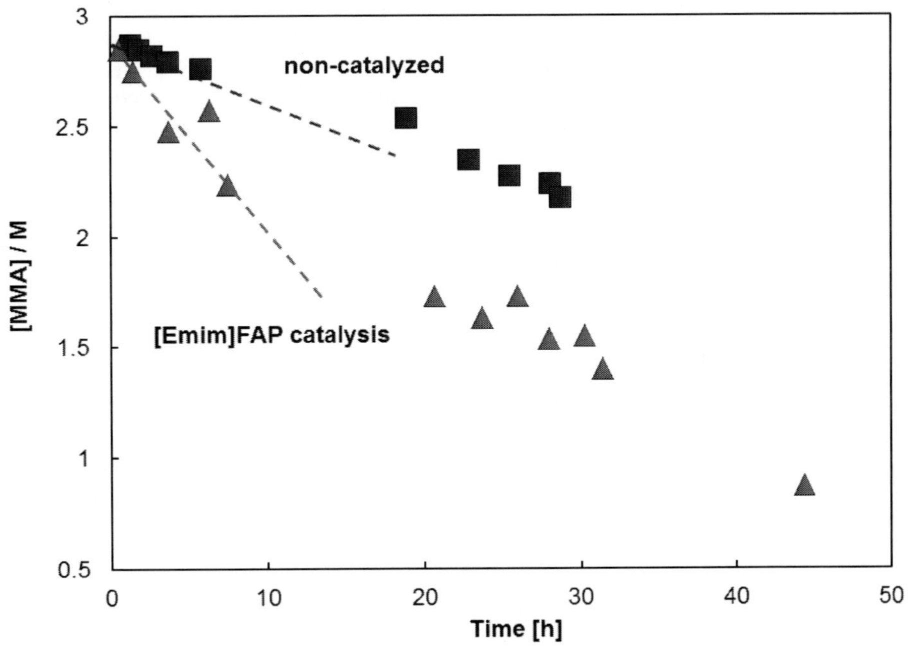

b

Figure 5. (Continued)

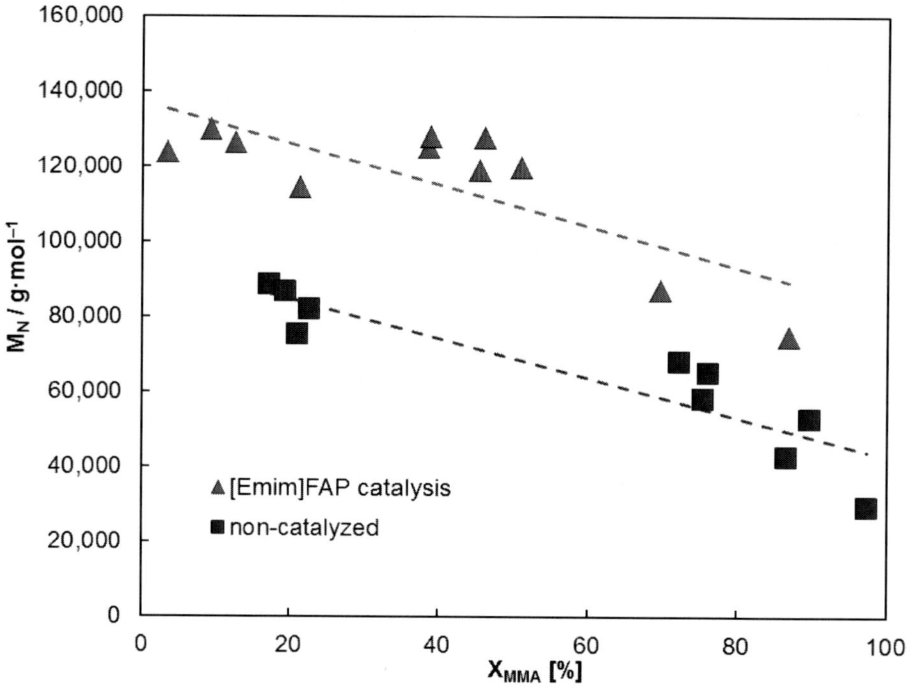

c

Figure 5. (a) Comparison of $k_{Br,0}$ values of IL-catalyzed and non-catalyzed IBI polymerization. (b) MMA concentration versus time plots for the non-catalyzed (■) and [Emim]FAP catalyzed (▲) IBI polymerization using MMT / HDI. The absolute values of the initial slopes are $5.9 \cdot 10^{-2}$ (■) and $2.5 \cdot 10^{-1}$ $M \cdot h^{-1}$ (▲); the corresponding $k_{Br,0}$ values are $0.5 \cdot 10^{-5}$ $L^{1.5} \cdot mol^{-1.5} \cdot s^{-1}$ and $4.8 \cdot 10^{-5}$ $L^{2} \cdot mol^{-2} \cdot s^{-1}$, respectively. (c) Plots of the average number molecular weight (M_N) of the PMMA as a function of the MMA conversion X_{MMA} of the catalyzed (▲) and non-catalyzed (■) IBI polymerization using MMT / HDI / MMA.

Considering literature known values for the IBI polymerization of POX and 1MP, the tremendous influence of the IL on $k_{Br,0}$ can be distinguished. In each example the IL catalysis results in an increase of $k_{Br,0}$ by a factor of approximately 5 to 10.

For a more meaningful visualization of the influence of the IL on the MMA consumption and the molecular weight of the resulting PMMA, the results for the polymerization kinetic determination of the system MMT / HDI / MMA with and without [Emim]FAP traces are shown in Figure 5b and 5c. The whole reaction course is influenced by traces of [Emim]FAP added to the reaction mixture. The addition of 0.11 M of the Ionic Liquid results in an increase of the polymerization rate by a factor of 5. The decrease of the average molecular weight of the polymer with increasing conversion is another indicator for the radical nature of the IL catalyzed IBI polymerization. Even the average molecular weight of the PMMA resulting from IL catalysis is increased by 40 % at comparable molecular weight distributions PDI (both at 1.5–2.2). When taking into account the approach of the kinetic chain length ν [28, 29] as a function of the initiator and monomer concentration for free radical polymerizations, then ν reads:

$$\nu = \frac{k_p}{\sqrt{f \cdot k_t \cdot k_d}} \cdot \frac{[MMA]}{\sqrt{[I]}} \quad (2)$$

where [I] is the initial initiator concentration and k_p, k_t and k_d rate constants of propagation, termination and initiator decay; f is the radical yield (more information see literature [29]). The rate determining step in every radical polymerization is the initiation; the value of k_d typically differs from k_t and k_p by a factor of 10^{-2}–10^{-3}. [30] By assuming a comparable behavior of the IL catalyzed IBI polymerization with kinetics valid in free radical polymerization, the increase of $k_{Br,0}$ indicates an increased concentration of the initiating species (higher k_d) which should therefore lead to shorter chains. However, this is obviously not the case. The results can be explained by the increase of the rate constant of propagation k_p and the decrease of the rate constant of termination k_t that occurs besides the changes in k_d. In literature, the effect of ILs on k_p and k_t at free radical polymerization kinetics are reported and designated as the "protected radical" mechanism. [31-34] Weak H-bonds of the IL with MMA and the growing chain are responsible for both an increase in electrophilicity of the monomer and, hence, an increase in reactivity and k_p and a steric shielding of thc propagating radical ("radical cage") which inhibits the termination reaction by recombination and causes a decrease of k_t. The presence of weak IL·MMA interactions could be confirmed using IR spectroscopy (see chapter A.2) which is in agreement with literature information. However, compared to free radical polymerizations reported by other groups where the IL was used in high amounts as the solvent for the polymerization reaction it is fascinating that at IBI polymerizations solely IL traces can cause these effects.

More considerably, the accelerating effect of the IL on $k_{Br,0}$ is detected by increasing the temperature. Exemplary, for the polymerization of MMA using MMT / HDI with and without catalysis by [Emim]FAP the certain $k_{Br,0}$ values at various temperatures are compared in Figure 6a.

The acceleration of the polymerization with increasing temperature is more significantly notable for the IL catalyzed polymerizations. Rate constants $k_{Br,0}$ of IL catalyzed and the non-catalyzed polymerizations differ by a factor between 5 to 9. Interestingly, compared to the non-catalyzed IBI polymerization (see Literature [11]), the average molecular weight of PMMA increases with increasing temperature (Figure 6b) indicating more drastic changes in k_p and k_t compared to that of k_d.

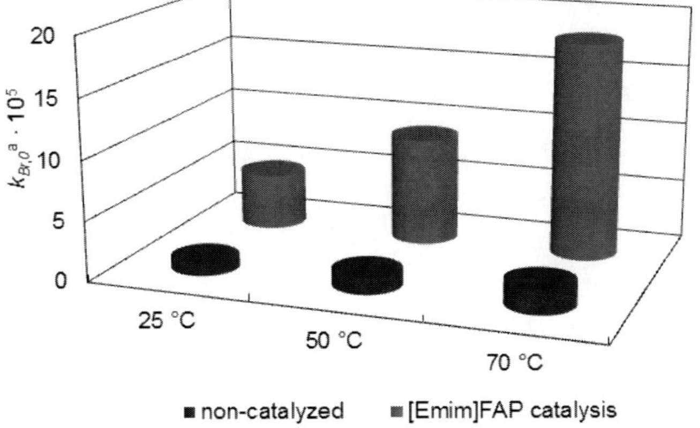

a

[a]Units are: [Emim]FAP catalysis $L^2\ mol^{-2}\ s^{-1}$; non-catalyzed $L^{1.5}\ mol^{-1.5}\ s^{-1}$.

Figure 6. (Continued)

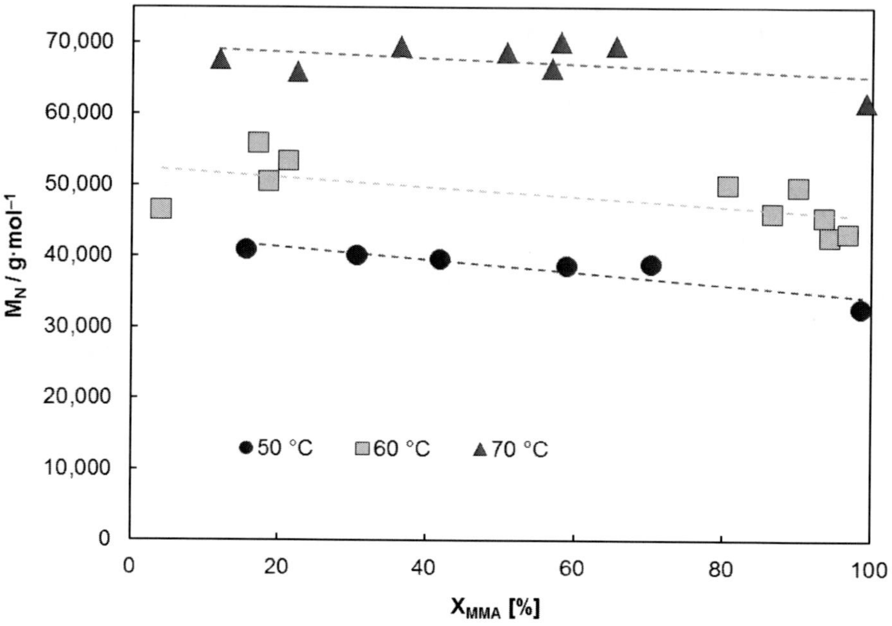

b

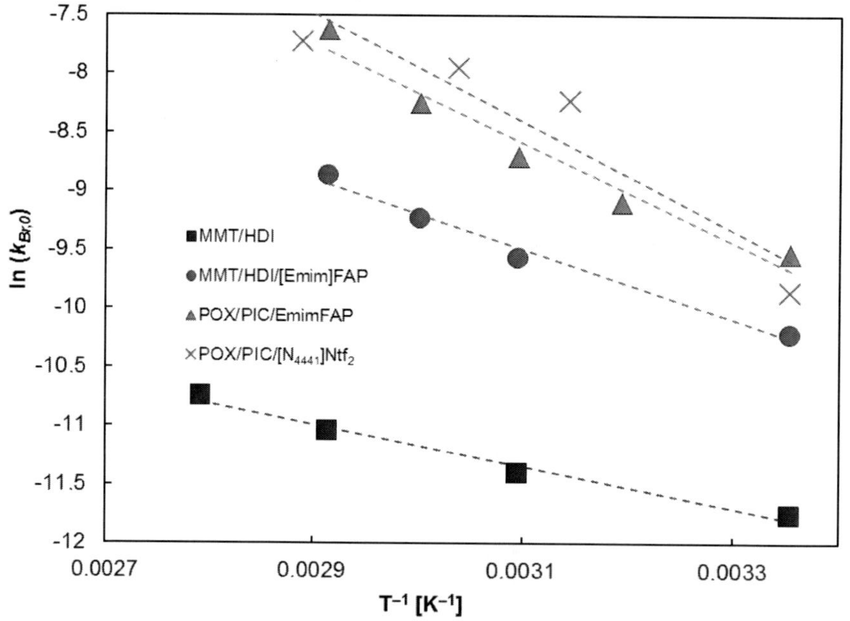

c

Figure 6. (a) Comparison of brutto polymerization rate constants $k_{Br,0}$ of [Emim]FAP catalyzed and non-catalyzed IBI polymerizations using the low temperature combination MMT / HDI at 25, 50 and 70 °C. (b) Average number molecular weights of PMMA at the [Emim]FAP catalyzed MMT / HDI IBI polymerizations as function of the MMA conversion X_{MMA} for 50 (●), 60 (■) and 70 (▲) °C. (c) Arrhenius plots of the non-catalyzed MMT / HDI (■), [Emim]FAP catalyzed MMT / HDI (●) and POX / PIC (▲), as well as $[N_{4441}]Ntf_2$ catalyzed POX / PIC (**x**) IBI polymerizations of MMA. The linear fits are $\ln (k_{Br,0}) = -1773 \cdot T^{-1} - 5.6$, $R^2 = 0.97$; $\ln (k_{Br,0}) = -2964 \cdot T^{-1} - 0.05$, $R^2 = 0.98$; $\ln (k_{Br,0}) = -3838 \cdot T^{-1} - 3.3$, $R^2 = 0.93$ and $\ln (k_{Br,0}) = -4636 \cdot T^{-1} - 6.0$, $R^2 = 0.83$, respectively. The brutto activation energies are 14.74 kJ·mol^{-1}, 24.65 kJ·mol^{-1}, 31.92 kJ·mol^{-1} and 38.55 kJ·mol^{-1}.

The temperature influence on the polymerization kinetics allows the determination of activation parameters. The Arrhenius approach used is shown in Equations 3 and 4.

$$k_{Br,0} = A \cdot \exp(-\frac{E_{A,Br}}{R \cdot T}) \quad (3)$$

$$\ln(k_{Br,0}) = \ln(A) - \frac{E_{A,Br}}{R \cdot T} \quad (4)$$

Herein, A is the pre-exponential factor, R the ideal gas constant, T the absolute temperature and $E_{A,Br}$ the brutto activation energy of the brutto rate constant which is a measure of the overall activation energy of the radical polymerization including the propagation, termination and initiator decay rate constants. For the free radical polymerization, $E_{A,Br}$ can be separated into the three relevant terms propagation (p), termination (t), and initiator decay (d) and so, $E_{A,Br}$ reads [28]

$$E_{A,Br} = E_{A,p} + \frac{1}{2}E_{A,d} - \frac{1}{2}E_{A,t} \quad (5)$$

The shape of the $E_{A,Br}$ equation for the IL catalyzed IBI polymerization is assumed to be similar to Equation 5; solely the coefficients of the separate activation energies are differing. Arrhenius plots of non-catalyzed [11] and IL catalyzed IBI polymerizations are shown in Figure 6c. The activation energies of IL catalyzed IBI polymerizations are ranging from 20 up to 40 $kJ \cdot mol^{-1}$.-

Hence, kinetics of the IL catalyzed IBI polymerization can be classified between polymerizations initiated by photo initiators ($E_{A,Br}$ = 20 $kJ \cdot mol^{-1}$) and free radical polymerizations initiated by thermal decomposable reagents ($E_{A,Br}$ = 80 $kJ \cdot mol^{-1}$). However, when comparing brutto activation energies of non-catalyzed and IL-catalyzed initiator mixtures (MMT / HDI, Figure 6c) the polymerization under IL catalysis has the higher $E_{A,Br}$.

As discussed earlier [11] at increased temperatures (2-methylthio)-2-thiazoline tends to catalyze a side product formation of isocyantes. Thus, a concentration decrease of the initiating species is affected more likely at higher temperatures and the measured activation energy therefore contains an unknown component. Certain IL···isocyanate interactions obviously suppress the side product formations; in the IL catalyzed IBI polymerizations this can be determined by the stronger exponential increase in $k_{Br,0}$ as a function of the absolute temperature. At the kinetics measurements of IL catalyzed IBI polymerization uncertainties occurring from the side product formation are negligible.

The use of ILs allows an improved reactivity comparison of high temperature and low temperature IBI combinations; considering $E_{A,Br}$ values of POX/isocyanate and MMT/ isocyanate systems a distinct higher reactivity of the MMT/isocyanate combination can be stated (Figure 6c).

A significant difference in the brutto activation energies of 6.6 $kJ \cdot mol^{-1}$ can be determined for the IL catalyzed IBI polymerizations POX / PIC / MMA using [Emim]FAP (α = 0.83) and $[N_{4441}]Ntf_2$ ($\alpha = 0.41$), emphasizing the different interaction strengths of various ILs with the IBI components (Figure 6c). In order to verify the influence of the Ionic Liquid type on the IBI polymerization kinetics the specific $k_{Br,0}$ values are measured and set into correlation to the interaction strength.

From IR measurements a linear correlation of α and the certain band shifts is known (see chapter A.2); literature known values of the hydrogen bond donating abilities are used as quantities of IL···IBI component interaction strengths, consequently. In Figure 7 $k_{Br,0}$ values of IL catalyzed POX / PIC / MMA polymerizations at two different temperatures as a function of α are shown.

This result manifests the objective fact that the IBI polymerization acceleration by ILs is not limited to a certain class of Ionic Liquids but rather is transferable to a variety of IL species. Brutto rate constants at 25 °C slightly vary in a range between 5.69±1.4 10^{-5} $L^2 \cdot mol^{-2} \cdot s^{-1}$. Obviously, the IL interaction strength change has no drastic effect on the overall IBI polymerization kinetics at 25 °C.

However, polymerizations carried out at 70 °C exhibit an remarkable dependency of $k_{Br,0}$ from the ILs hydrogen bond donating ability. The stronger the interaction of the IL with the IBI components is (higher α) the slower the IBI polymerization takes place. The inhibition at elevated temperatures potentially is caused by the stronger IL···imine base interaction due to the fact that the side reaction of the isocyanate deactivation is negligible. Thus, the acidity of the Ionic Liquid, expressed by the hydrogen bond donating ability α can be used as a quantity to adjust the kinetics of the IBI polymerization

With these know-how and the previous mentioned tools on hand IL catalyzed IBI mixtures potentially can be used as novel designer initiator mixtures tunable for certain material scientific applications in coatings or in the synthesis of complex polymer architectures.

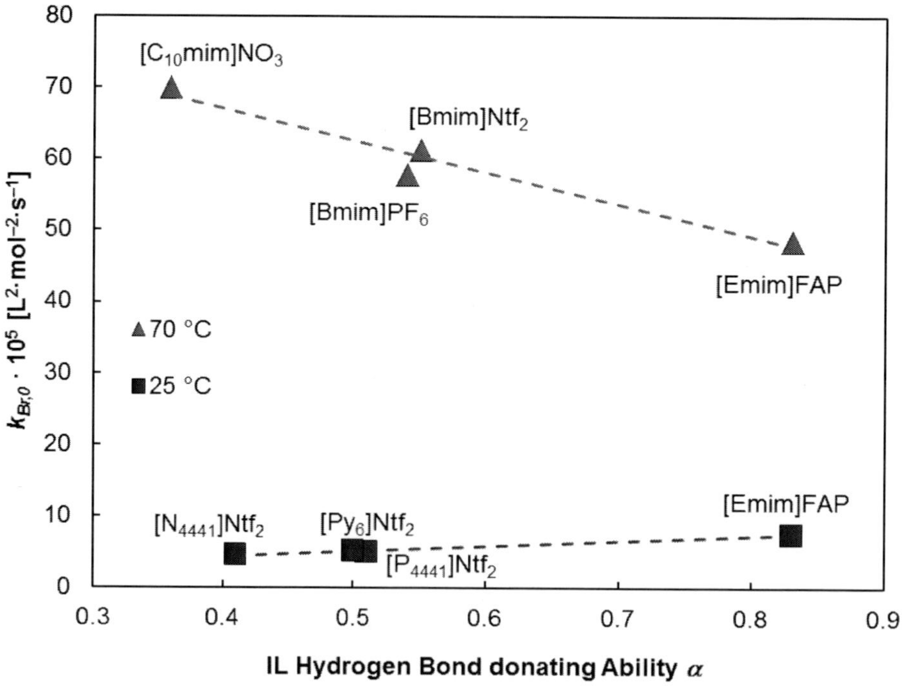

Figure 7. Brutto rate constants $k_{Br,0}$ of IL catalyzed POX / PIC / MMA polymerizations at 25 and 70 °C as a function of the hydrogen bond donating abilities α. The linear fits are $k_{Br,0}$ (70 °C) = $-44 \cdot \alpha + 84$, $R^2 = 0.93$ and $k_{Br,0}$ (25 °C) = $+6 \cdot \alpha + 2.06$, $R^2 = 0.74$.

CONCLUSION

The addition of Ionic Liquid traces to the imine base / isocyanate binary radical source used in polymerization of MMA has notable effects on the whole polymerization process. IL interactions with all IBI components could be detected and quantified by band shifts of certain vibrational modes of the IBI components at the infrared spectroscopy measurements of binary IL / IBI component mixtures. The band shifts correlate linearly with literature known values of the hydrogen bond donating abilities α that are a measure for the acidity of the Ionic Liquid. Interactions of the IL with the isocyanate and the acrylic monomer are concluded to be most distinctive for the initiation process.

Most considerably, IBI polymerizations are accelerated in the presence of IL traces which evidently is shown by the increase of the brutto polymerization rate constant. The IL contributes to the brutto polymerization rate law; the order referring to the Ionic Liquid is 0.5. No indications are found that the IL is converted at the polymerization process or covalently bonds to the polymer formed. Hence, the IL catalyzes the IBI polymerization by an increase of the initiating species concentration by means of stabilizing the zwitterionic intermediate. Polymer chains generated at the IL catalyzed IBI polymerizations are remarkably longer than macromolecules generated by the non-catalyzed IBI polymerizations indicating the "protected radical mechanism" which is in well agreement with results reported at common free radical polymerizations in ILs. [31–34] However, for the IBI polymerization solely IL traces are required to create this notable effect. All IBI initiator systems treated with catalytic IL amounts allow a radical polymerization of acrylic monomers at room temperature. Polymerizations of styrene show a negligible acceleration.

The IL catalyzed IBI polymerization offers a facile route on the efficient fabrication of high molecular weight polymers under mild conditions of temperature and pressure. The IBI polymerization [11, 12] seems to be a powerful method for the generation of various polymer architectures. The use of ILs for these reactions which require solely catalytic amounts of IL improves the whole polymerization process enabling the creation of programmable initiator mixtures for a broad range of industrial and academic applications.

REFERENCES

[1] Abbott, A. P., Capper, G., Davies, D. L., Rasheed, R. K., Tambyrajah, G. *Green Chem.* 2002, 4, 24–26.

[2] Chiappe, C., Conte, V., Pieraccini, D. *Org. Lett.* 2001, 3, 1061–1063.

[3] Early, M. J., McCormac, P. B., Seddon, K. R. *Green Chem.* 1999, 1, 23–25.

[4] Lourenco, N. M. T., Afonso, C. A. M. *Tetrahedron* 2003, 59, 789–794.

[5] Allen, G. D., Buezzo, M. C., Davies, I. G., Villagran, C., Hardacre, C., Compton, R. C. *J. Phys. Chem. B* 2004, 108, 16322–16327.

[6] Conte, V., Floris, B., Galloni, P., Silvagni, A. *Pure Appl. Chem.* 2005, 77, 1575–1581.

[7] Lancaster, N. L., Welton, T. *J. Org. Chem.* 2004, 69, 5986–5992.

[8] Brinichi, L., Germani, R., Savelli, G. *Tetrahedron Lett.* 2003, 44, 2027–2029.

[9] Debreul, J. F., Bazureau, J. P. *Tetrahedron Lett.* 2000, 41, 7351–7355.

[10] Wheeler, C., West, K. N., Liotta, C. L., Eckert, C. A. *Chem. Commun.* 2001, 10, 887–888.

[11] Polenz, I., Schmidt, F. G., Spange, S. *J. Polym. Sci. Part A: Polym. Chem.* 2012, 50, 3324–3331.

[12] Polenz, I., Spange, S., Hase, K., Schmidt, F. G., *Patent Application*, Isocyanat/Organobase-vermittelte Polymerisation von (Meth-)Acrylaten, *2009E00284 DE*, 2009.

[13] Polenz, I., Schmidt, F. G., Spange, S. *J. Polym. Sci. Part A: Polym. Chem.* 2013, 51, 687–695.

[14] Polenz, I., Spange, S., Lungwitz, R., Schmidt, F. G., *Patent Application*, Aktivierung der Basen/Isocyanat-initiierten Polymerisation durch Ionic Liquids, *2010E00455 DE*, 2010.

[15] Ikegami, T., Hirai, H. *J. Polym. Sci. Part A: Polym. Chem.* 1970, 8, 195–208.

[16] Madruga, E. L., San Roman, J. *J. Polym. Sci. Part A: Polym. Chem.* 1981, 19, 1101–1108.

[17] Kargin, V. A., Kabanov, V. A., Zubov, P. V. *Vysokomol. Soedin.* 1960, 2, 765–769.

[18] Gaylord, N. G., Takahashi, A. *J. Polym. Sci. Part B: Polym. Phys.* 1968, 6, 743–748.

[19] Bamford, C. H., Brumby, S., Wayne, R. P. *Nature* 1966, 209, 292–294.

[20] Decken, A., Eisnor, C. R., Gossage, R. A., Jackson, S. M. *Inorg. Chim. Acta* 2006, 359, 1743–1745.

[21] Evans, D. A., Peterson, G. S., Johnson, J. S., Barnes, D. M., Campos, K. R., Woerpel, K. *J. Org. Chem.* 1998, 63, 4541–4544.

[22] Trofimenko, S. *Chem. Rev.* 1972, 72, 497–509.

[23] Köddmann, T., Wertz, C., Heintz, A., Ludwig, R. *Angew. Chem.* 2006, 118, 3780–3785.

[24] A. Stark, *Top. Curr., Chem., Ionic Liquids*, ed. B. Kirchner, Springer-Verlag Berlin, Heidelberg, 2009, 290, 58.

[25] R. Bini, O. Bortolini, C. Chiappe, D. Pieraccini, T. Siciliano, *J. Phys. Chem. B* 2007, 111, 598–604.

[26] Lungwitz, R., Spange, S. *New J. Chem.* 2008, 32, 392–394.

[27] Wasserscheid, P., Welton, T. *Ionic Liquids in Synthesis*, 2nd ed.; Wiley-VCH: Weinheim, 2007.

[28] Matyjaszewski, K., Davis, T. B. *Handbook of Radical Polymerization*; Wiley Interscience: Hoboken, 2002.

[29] Odian, G. *Principles of Polymerization*, 4th ed.; Wiley Interscience: Hoboken, 2004.

[30] Immergut, E. H., Brandrup, J. *Polymer Handbook*, John Wiley and Sons Inc., 4th Ed., 1999.

[31] Schmidt-Naake, G., Woecht, I., Schmalfuss, A., Glueck, T. *Macromol. Symp.* 2009, 275–276, 204–218.

[32] Zhang, H., Hong, K., Mays, J. W. *Macromolecules* 2002, 35, 5738–5741.

[33] Thurecht, K. J., Gooden, P. N., Goel, S., Tuck, C., Licence, P., Irvine, D. J. *Macromolecules* 2008, 41, 2814–2820.

[34] Puttick, S., Davis, A. L., Butler, K., Lambert, L., El Harfi, J., Irvine, D. J., Whittaker, A. K., Thurecht, K. J., Licence, P. *Chem. Sci.* 2011, 2, 1810–1816.

In: Ionic Liquids: Synthesis, Characterization and Applications ISBN: 978-1-63117-716-3
Editor: Anson Brooks

Chapter 4

THE ELECTROCHEMISTRY OF GOLD DEPOSITION FROM A DICYANAMIDE IONIC LIQUID

A. I. de Sá[1,*], S. Eugénio[2], C. M. Rangel[1] and R. Vilar[2]
[1]Laboratório Nacional de Energia e Geologia, Lisboa, Portugal
[2]Instituto Superior Técnico, Universidade de Lisboa, Lisboa, Portugal

ABSTRACT

Ionic liquids (ILs) have demonstrated an enormous potential in the electrodeposition of various materials including metals, alloys and oxides, providing opportunities to overcome the limitations of water-based solvents by extending the working electrochemical window and increasing current efficiencies. Also of relevance is the capability to tailor the redox chemistry and to control the nucleation characteristics, promoting nano-sized crystalline structures in the deposited layers. In addition, reduced health and environmental impacts are possible due to the low volatility of ILs, while maintaining suitable properties such as high chemical and thermal stability, paramount to a variety of industries to which electrodeposition is essential and that face stringent environmental regulations.

In this work electrochemical aspects of electrodeposition of gold using ILs are examined. Recent research has suggested a number of energy related applications for gold in batteries, fuel cells and hydrogen fuel processing, which include not only the need for coatings on less noble substrates but also the incorporation of gold as an electroactive deposit. The use of ionic liquids to engineer / develop new gold structures is a valid trend to investigate.

Gold deposition from a solution of 1-butyl-1-methylpyrrolidinium dicyanamide (BMP-DCA) containing $HAuCl_4$ and water is reported. The study includes the electrochemical characterization of the solvent, emphasizing admissible water contents and conditions for the deposition of thin films under laboratory atmospheric conditions. The electrochemistry of the reduction process itself is characterized by cyclic voltammetry as irreversible and under diffusion control. Diffusion coefficients of 1.15 x 10^{-7} cm^2 s^{-1} (293 K) and 2.60 x 10^{-7} cm^2 s^{-1} (333 K) for the Au^{3+} ionic species are estimated.

* Corresponding author: A. I. de Sá. Laboratório Nacional de Energia e Geologia, Paço do Lumiar 22, 1649 038 Lisboa, Portugal. E-mail: ana.sa@lneg.pt.

Electrodeposition conditions for gold nanocrystalline layers on copper, nickel and glassy carbon substrates are discussed and its relation with microstructure is established.

This is followed by the performance characterization of selected deposits in a phosphate buffer solution (pH 7.3) where well-defined voltammetric peaks associated with the oxygen chemisorption and reduction is found in all the substrates. Finally, the electrochemical behavior and stability of the gold deposited surfaces are fully addressed.

1. INTRODUCTION

Ionic liquids (ILs) have high chemical and thermal stability, good conductivity and a large electrochemical window [1, 2], which are crucial characteristics of solvents used for electrodeposition [3]. They have also extremely low vapor pressure [4] which can reduce enormously health impacts which strengthen even more the interest in using these liquids in electrodeposition and other practical applications.

Reactive metals and their alloys are among the materials than have been under research for electrodeposition using ionic liquids [5, 6]. However, these liquids are prone to water absorption [7, 8] and that can significantly modify both the characteristics of the IL itself and the electrodeposited material by favoring an oxide formation. Hence, normally it is necessary to work in extreme dry and inert conditions, since the absence of water and oxygen are essential conditions to maintain an extended electrochemical window and attain the reduction potentials of active metals. Electrodeposition of noble and common metals has also been investigated focusing on deposit morphology and structure and on the use of alternative electrolytes with less additives and environmental impact [5, 6].

The easy combination of anions and cations to form different liquids allows "solvent design" to a required function or process, as so often referred [6, 9]. Based in this concept, the ionic liquid chosen for the study of gold electrodeposition was the 1-butyl-1-methylpyrrolidinium dicyanamide (BMP-DCA) because of its low viscosity compared to other ionic liquids, favouring a more efficient mass transport and a higher electrical solution conductivity. One more advantage of using a dicyanamide anion ionic liquid is the ability demonstrated by this ion to dissolve many metal compounds, including metal chlorides, due to its donor ligand property [10, 11]. For example, deposition of aluminium, manganese, nickel, zinc, tin and copper using dicyanamide ionic liquids have already been demonstrated by Deng et al., [11]. The pyrrolidinium cation has been referred as contributing to the formation of nanoscrystalline structures [12, 13], which might exhibit improved catalytic properties due to higher roughness, particular morphology or porosity. Such structures are presently under investigation in new renewable energy devices [14] for ethanol oxidation or oxygen reduction reactions [15, 16]. Last, but not least the BMP-DCA ionic liquid viscosity (~50 cP [2]) and conductivity (12 mS cm^{-1}) for water content of 283 ppm [17] at room temperature, were also decisive parameters for this selection.

Previous studies reported by the authors have shown that the reduction of chloroauric complex $[AuCl_4]^-$ to Au^0 in 1-butyl-1methylpyrrolidinium dicyanamide (BMP-DCA) is a two-step process, where gold ionic species are reduced from Au^{3+} to Au^+ and from Au^+ to Au^0 [18, 19]. This two step reduction process of gold is similar to what was found by other authors using the same gold ionic species in different ionic liquids, including a mixture of aluminium chloride in 1-ethyl-3-methylimidazolium chloride ($AlCl_3$/ EMICl), 1-butyl-3-

methylimidazolium bis{(trifluoromethyl)sulfonyl} imide (BMI-TFSI) or 1-butyl-1-methyl-imidazolium tetrafluoroborate (BMI-BF_4) [20-23]. Gold films can be deposited under potentiostatic control from solutions of $[AuCl_4]^-$ in BMP-DCA on nickel and copper substrates under laboratory atmospheric conditions [18, 19].

For this last case, and due to the strong dissolution of copper in this liquid, gold layers contain a small percentage of copper.

Moreover, this effect is responsible for the electroless deposition of gold on copper by a galvanic effect [17, 19].

In this work, gold electrodeposition from a solution of $HAuCl_4.3H_2O$ in 1-butyl-1-methylpyrrolidinium dicyanamide is examined concerning the kinetics of the process and the effect of water content.

Gold layers deposited under potentiostatic control in different experimental conditions were tested in a phosphate buffer solution (pH 7.3) in order to evaluate their electrochemical behavior using cyclic voltammetry.

2. EXPERIMENTAL

The electrolytes were prepared by dissolving 0.02 M $HAuCl_4.3H_2O$ (ACROS, for synthesis) in the ionic liquid BMP-DCA (Merck for synthesis, purity > 98 %, < 0.1 wt. % of halides and < 1 wt. % H_2O). The amount of water initially present in the ionic liquid was determined by Karl Fischer titration using a Metrohm model 756KF coulometer and corresponds to 283 ppm (~ 0.03 wt. %). The conductivities of the ionic liquid in as received condition and with different concentrations of water, up to ~15 wt. %, were measured with a Profilab (LF197) conductivity meter.

The electrochemical experiments were performed with an EGandG PAR, model 173 potentiostat controlled by a PC with "Corrware" software. A 20 cm^3 cell with a three-electrode configuration was used for the experiments. Cyclic voltammetry was performed using glassy carbon as working electrode as working electrode (0.071 cm^2). The counter electrode was a platinum foil (~ 1.6 cm^2) and as reference was used a platinum wire. The potential of platinum wire in the electrolyte is stable, so it is considered a quasi-reference electrode (QRE), to which all potentials are referred.

Before the experiments, the glassy carbon electrode was polished with 0.25 μm diamond paste and the platinum electrodes were activated by dipping in HNO_3 (65 %) during one minute. Substrates used in the electrodeposition process were mechanically polished with 1000 grit SiC paper and activated in 1:1 HNO_3 solution for 15 s for copper and 30 s for nickel respectively. After electrodeposition, the specimens were rinsed with distilled water and dried with an air blast at room temperature.

Selected specimens were also tested by cyclic voltammetry in a phosphate buffer solution (pH 7.3) saturated with oxygen at room temperature to characterize the gold surfaces.

The films were characterized by scanning electron microscopy (SEM) and energy dispersive X-ray spectrometry (EDX) using a JEOL model 7001F field-emission gun scanning electron microscope and an Oxford X-ray EDS spectrometer.

The acceleration voltage was 15 keV.

3. Results

3.1. The Effect of Water Content in the Ionic Liquid

The BMP-DCA ionic liquid continuously absorbs water when exposed to the laboratory atmosphere. The water uptake is slow. After two hours of exposure to the atmosphere, the water content is below 1 wt. % [19] and increases to 15 wt. % (7.8 M) after three days [19, 24]. Figure 1 shows the voltamograms obtained for a glassy carbon electrode in BMP-DCA ionic liquid where the water content was increased from 1 wt. % to 15 wt. %.

The voltammograms show a cathodic peak associated to the reduction of the H_3O^+ ions from water (1 wt. %) located at -1.1 V, at 293 K. It is also evident that an increase in water content results in a peak current density increase. Even though the electrochemical window of BMP-DCA IL is affected by the presence of water, the deposition of gold is well within the exhibited potential stability limits.

Electrolytes of high conductivity are required for industrial electrodeposition processes in order to achieve high deposition-rates. This is easily obtained with aqueous electrolytes, but it is more dificult with organic electrolytes including ionic liquids.

Figure 2 shows the variation of the conductivity with temperature of the BMP-DCA ionic liquid in as received condition (0.03 wt. %) and with water content up to 15 wt. %. The data show a linear correlation between conductivity and temperature for all the tested IL-water solutions. At room temperature (293-298 K), the conductivity increases from 12 mS cm^{-1} at 0.03 wt. % H_2O to 24 mS cm^{-1}at 15 wt. % H_2O. At 353 K the same variation in water concentration raises the condutivity from 46 to 96 mS cm^{-1}, which is expected to increase electrodeposition rates.

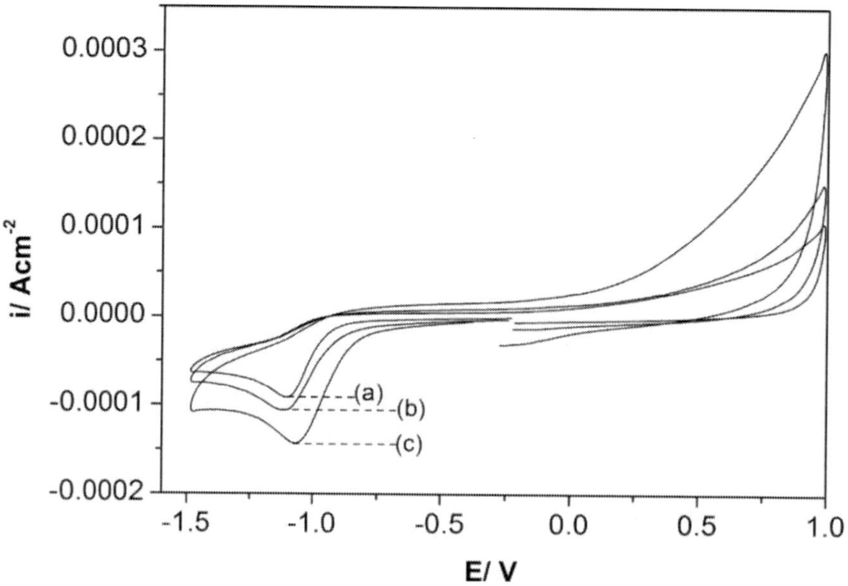

Figure 1. Cyclic voltammograms obtained at glassy carbon electrode in BMP-DCA with 1 (a) 5 (b) and 15 wt. % H_2O (c) at 293 K and 0.1 V s^{-1}.

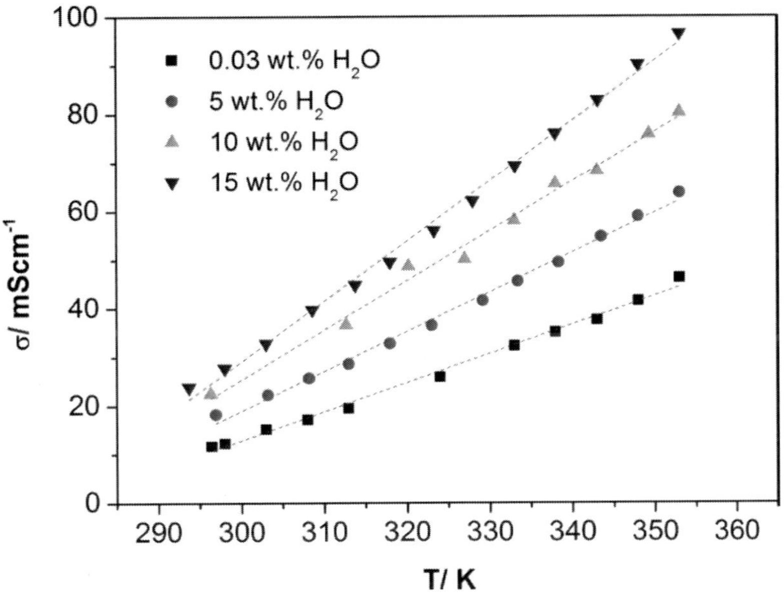

Figure 2. Conductivities of the BMP-DCA ionic liquid in as received condition and with increasing water content up to 15 wt %.

3.2. Kinetic Studies

Cyclic voltammetry was performed to study in detail the kinetics of the deposition process using the potential sweep-rate effect. The voltammograms were carried out from 0.10 to 0.01 Vs^{-1} on a glassy carbon electrode at 293 and 333 K in a solution of 0.02 M $HAuCl_4$. $3H_20$ in BMP-DCA. The deposition, as already referred, follows a two step reduction process which is confirmed in Figure 3 by the presence of two cathodic peaks well defined at both tested temperatures. The first peak (I_C) corresponds to the reduction of $AuCl_4^-$ to $AuCl_2^-$ and the second peak (II_C) is associated with the reduction of $AuCl_2^-$ to Au^0.

After each voltammogram the electrode was polished and clean to reproduce the same initial surface condition.

The current density of the cathodic peaks I_C and II_C vary linearly with the sweep-rate (v) with a slope close to 0.5 indicating that the deposition is diffusion controlled (Figure 4) [25].

Despite the presence of a small anodic peak detected at the lowest sweep-rate of 0.010 and 0.020 Vs^{-1}, which is possibly associated with some gold oxidation, there are no other peaks in the anodic side of the voltammogram showing that the reduction is an irreversible process. For these processes, the peak current density (i_p) should vary with the sweep-rate according to the equation [25, 26]:

$$i_p = 0.496\, n\, F^{3/2}\, C\, D^{1/2} \left(\frac{\alpha\, n_\alpha\, v}{RT}\right)^{1/2} \quad (1)$$

where n is the number of electrons transferred, F the Faraday constant (96485 C mol^{-1}), C the bulk concentration in the ionic species (mol cm^{-3}), D the diffusion coefficient ($cm^2\ s^{-1}$), α the charge coefficient, n_α the number of electrons transferred in the rate determining step, R the ideal gas constant (8.314 J mol^{-1} K^{-1}) and T the temperature (K).

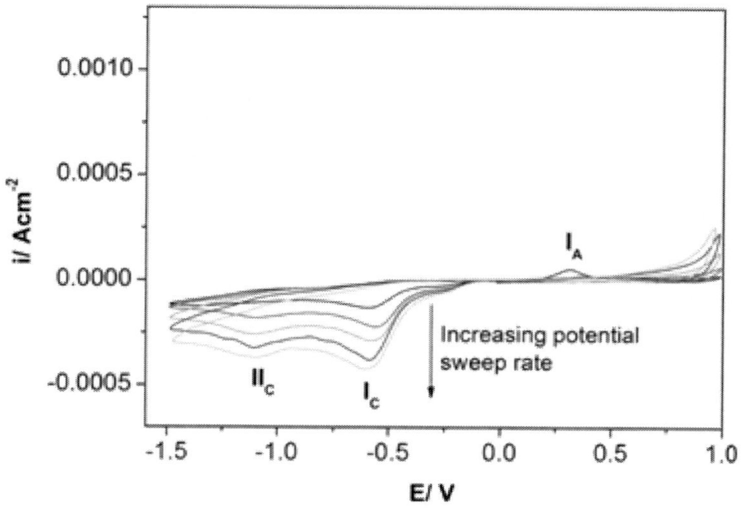

a

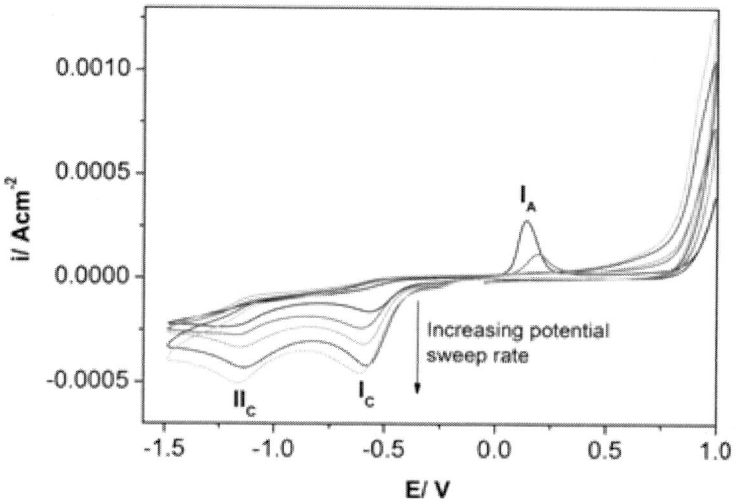

b

Figure 3. Cyclic voltammograms obtained at glassy carbon electrode in a solution of BMP-DCA +0.02 M $HAuCl_4.3H_2O$, at different potential sweep-rate, from 0.1 to 0.01 Vs^{-1}, at 293 K (a) and 333 K (b).

Furthermore, the peak potential (Ep) and the half-peak potential ($E_{p/2}$) should also be correlated according to the equation [25, 26]:

$$|E_p - E_{p/2}| = \frac{1.857RT}{F\alpha n_\alpha} \quad (2)$$

As peak(I) corresponds to the reduction of gold ionic species, from Au^{3+} to Au^{+}, the correspondent potential and current density can be used to determine the diffusion coefficient of the Au^{3+} species in the solution.

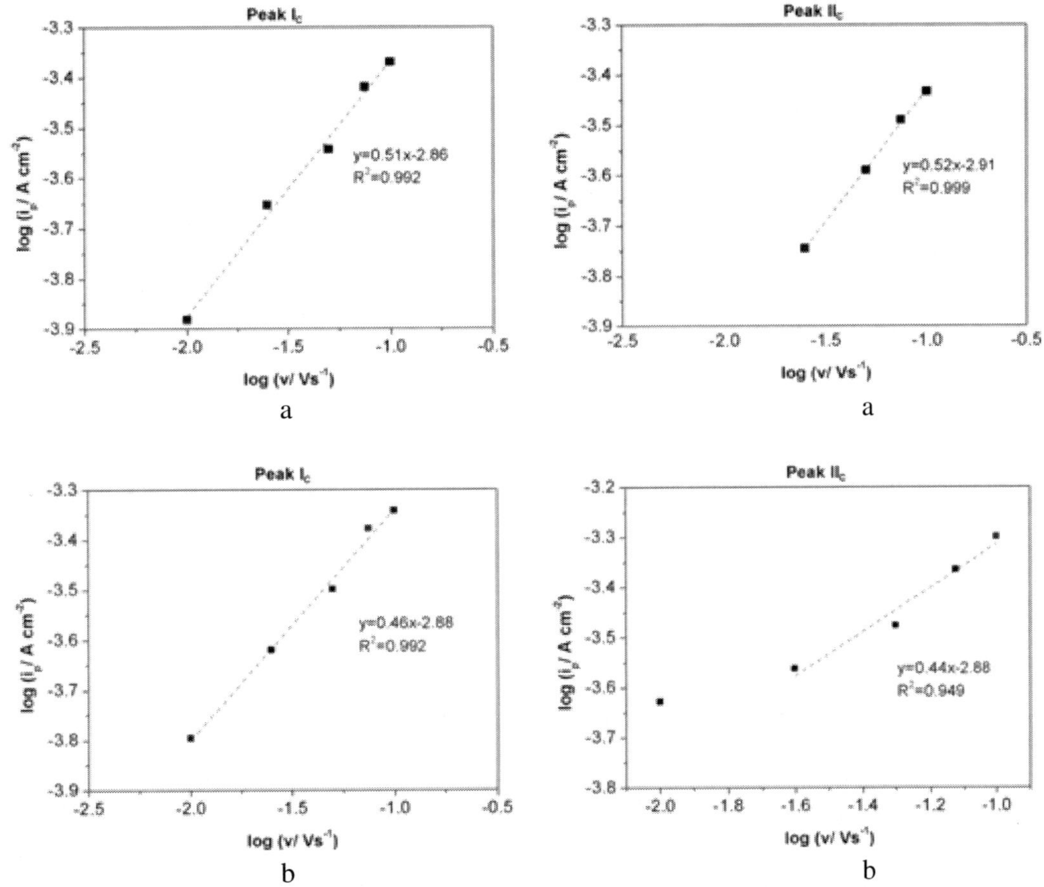

Figure 4. Variation of the cathodic peak current density (ip) with the potential sweep-rate (v) for peaks I_C and II_C (data from the voltammograms in Figure 3 at 293 K (a) and 333 K (b) in a log-log scale.

Therefore, using the data from the voltammogram obtained at 0.05 Vs^{-1} the coefficient, αn_α, was found to be 0.48 at 293 K and 0.56 at 333 K, corresponding to a α value of ~0.26, which is in the range of values found in the literature [27, 28].

These values were introduced in the equation 1 to calculate the diffusion coefficients, which are 1.15 x 10^{-7} cm^2 s^{-1} at 293 K and 2.60 x 10^{-7} cm^2 s^{-1} at 333 K, respectively.

The reduction potential of the H_3O^+ ions is close to the reduction of Au^+ ions to Au^o, meaning that the former reduction will be a concurrent process in gold deposition in this electrolyte. The sweep rate effect was also carried out for the solutions of BMP-DCA + x wt. % H_2O (x= 5 and 15) to estimate the interference of this process in the final step of gold deposition. The data have revealed that this process is also diffusion limited (Figure 5).

3.3. Gold Films Morphology

In previously reported work by the authors [17, 18], it was shown that nanocrystalline layers of gold are obtained at 333 and 353 K under potentiostatic control (-1 V) on nickel and copper substrates.

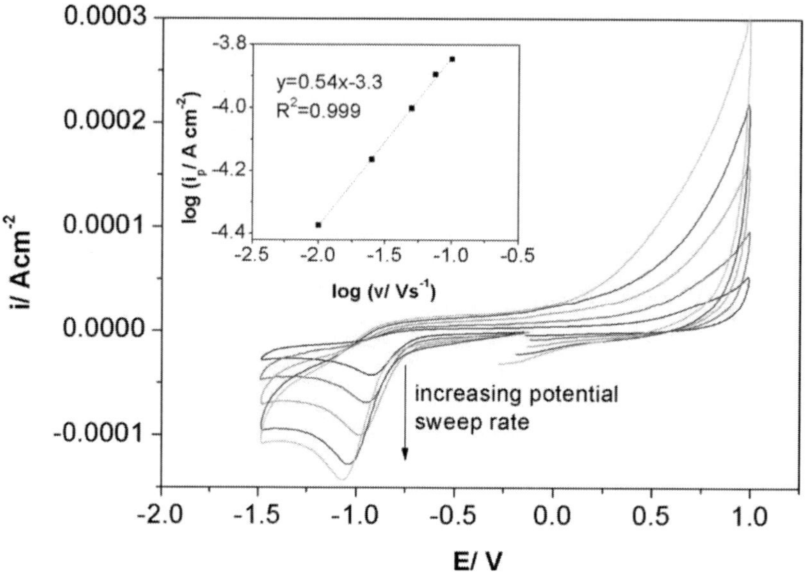

Figure 5. Cyclic voltammograms obtained at glassy carbon electrode in BMP-DCA + 15 wt % H_2O solution at different potential sweep-rate, from 0.1 to 0.01 Vs^{-1}. Inset shows the log-log variation of the peak current density with the potential sweep-rate.

The deposits were crystalline with a crystallite size in the range of 6-25 nm. The crystallites were aggregated in clusters with a typical dimension between 50 and 250 nm.

It was also found that temperature had a strong influence on the morphology of the films. Films deposited at 353 K showed a dendritic morphology. This effect did not change with the concentration of the gold salt or with the substrate used [18, 19].

Herein the effect of the electrodeposition time is illustrated. SEM observations of electrodeposits at different deposition times put in evidence, particularly at 353 K, that after 30 s the surface nucleation process is inhomogeneous occurring preferentially at scratches and other surface defects promoted as a result of the chemical activation of the surface performed prior to electrodeposition (Figure 6a and inset). After 1500 s the surface is almost completely covered with the gold layer, even though SEM observation revealed some bare areas (Figure 6b and 6c).

As shown in Figure 7, the compactness of the deposited layers was not improved when the electrolyte containing 15 wt. % of water was used.

Although, an increase of the conductivity is expected, similarly to what was found for the solvent (Figure 1), an increase in the rate of proton reduction is thought to interfere with the second step of gold ion reduction (Au^+ to Au^o).

Gold deposition was also performed for a deposition time of 4500 s, in the same experimental conditions. The thickness of the deposit was found to grow from 211 nm to 572 nm, for deposition times of 1500 and 4500 s, respectively (Figure 8). The thickness was estimated from SEM micrographs taken with the specimens tilted at 70º.

Results indicate an average constant deposition rate of ~ 0.13 nm s^{-1}, however a complete coverage of the surface was not achieved.

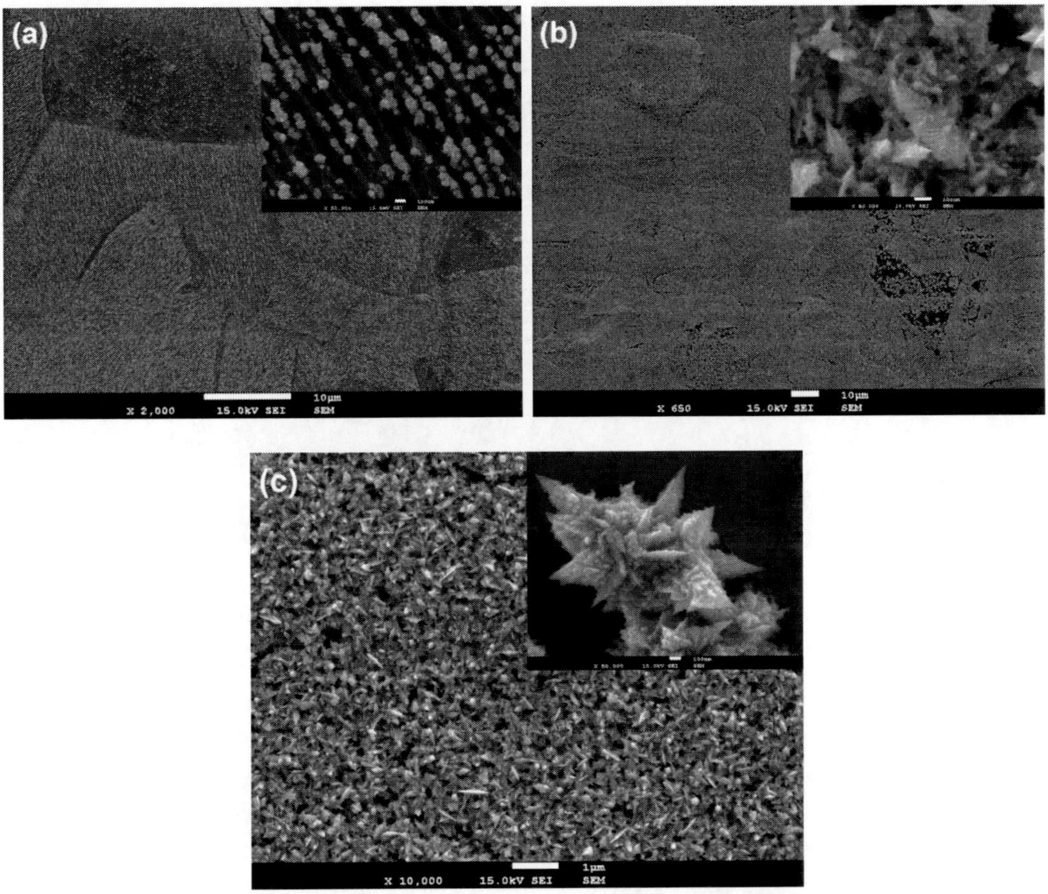

Figure 6. SEM micrographs of gold deposited at -1 V in a solution of BMP-DCA + 0.02 M $HAuCl_4.3H_2O$ at 353 K for 30 s (a) and 1500 s (b,c) on nickel. Insets show details of the specimens at high magnification.

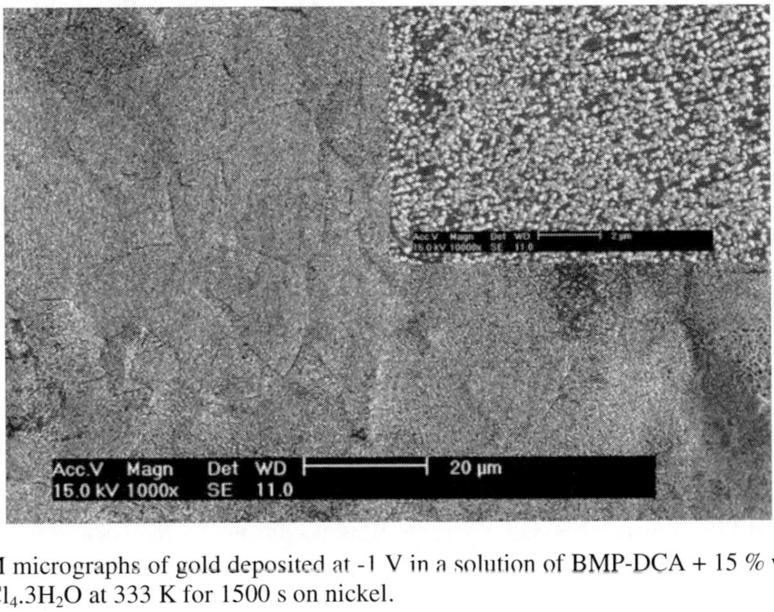

Figure 7. SEM micrographs of gold deposited at -1 V in a solution of BMP-DCA + 15 % wt H_2O + 0.02 M $HAuCl_4.3H_2O$ at 333 K for 1500 s on nickel.

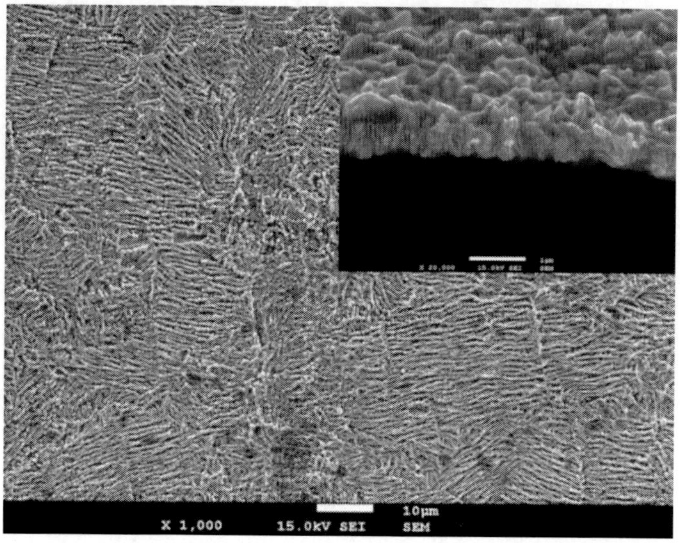

Figure 8. SEM micrographs of gold deposited at -1 V in in a solution of BMP-DCA+ 0.020 M $HAuCl_4$. $3H_2O$ at 333 K for 4500 s on copper (inset is a micrograph with the specimen tilted at 70°).

3.4. Electrochemical Behaviour of Gold Deposited Surfaces

The electrochemical characterization of gold deposited layers was also performed by cyclic voltammetry in a phosphate buffer solution (pH 7) saturated with oxygen.

This technique and solution were chosen because oxygen is chemically absorbed on gold surfaces during the anodic potential scan forming an oxygen monolayer with a ratio of gold-oxygen atoms of 1:1. This monolayer is then reduced during the cathodic potential scan forming a well defined peak with a corresponding charge of 400 $\mu C\ cm^{-2}$ for a polycrystalline gold flat area [29, 30].

Based on this value, the surface roughness factor (Rf) can be calculated from the ratio of the $Q_{AuO}/400\ \mu C\ cm^{-2}$ and used as a measure of the microscopic surface area for comparative purposes [29, 30].

Gold layers were also deposited at 333 and 353 K on glassy carbon, under potentiostatic control (-1 V) for 1500 s.

The glassy carbon surface is completely inert in the electrodeposition electrolyte and in the buffer solution, which is advantageous to obtain a voltammetric response only related to the deposited layer. The voltammograms show that the typical peaks associated to oxygen adsorption and reduction are well defined and located at the expected potentials (Figure 9). The stability of the electrodeposited surfaces was assessed by performing up to 50 cycles at 50 mV s^{-1}. From the 50th sweep the roughness factor is calculated, corresponding to 1.8 and 3.7 respectively for 333 and 353 K. As expected, the gold surface formed at 353 K has a higher roughness factor due to the dendritic morphology.

Gold layers deposited under potentiostatic control at -1 V for 1500 s on nickel and copper at 353 K were also tested in the same buffer solution (Figure 10). The voltammograms obtained for gold deposited on nickel are similar to those obtained for gold deposited on glassy carbon, since nickel is inert in this buffer solution (Figure 10a).

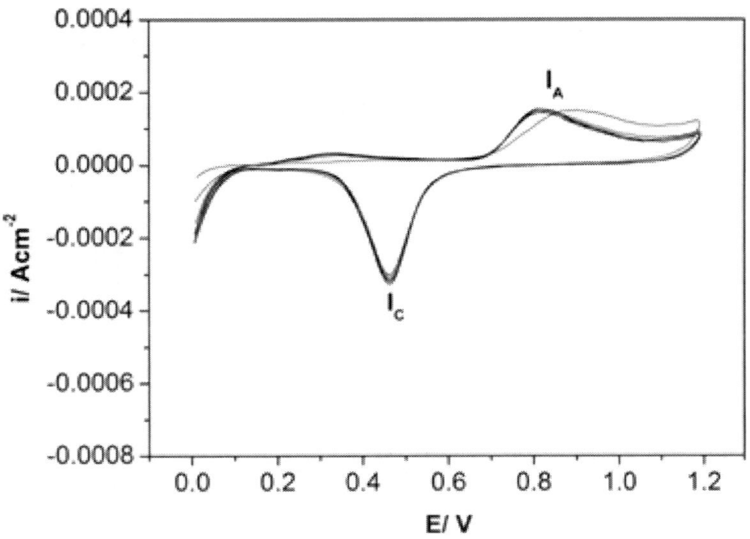

a

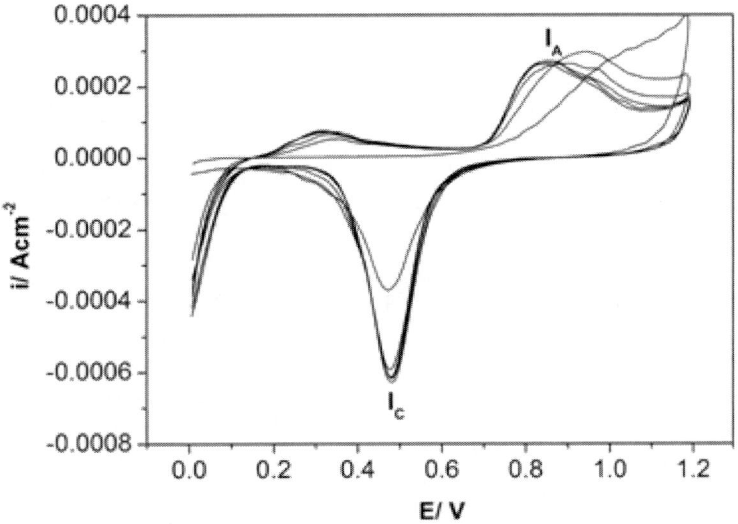

b

Figure 9. Cyclic voltammograms in a phosphate buffer solution (pH 7.3) and O_2 saturated at 50 mV s^{-1} obtained at gold deposited under potentiostatic control of -1 V for 1500 s at 333 K and 353 K on glassy carbon substrate (1st, n x 10th sweep ($1 \leq n \leq 5$).

For this surface the roughness factor is 11 which is associated not only to the dendritic morphology but also to the increase of the substrate area promoted by the strong etching with HNO_3 1:1 solution.

The cyclic voltammograms obtained for gold deposited on copper show a different behaviour, since two anodic peaks and the correspondent cathodic peaks were identified (Figure 10b). Peaks IIa and Ic are located in the same potential range than those obtained in Figure 9 for gold on glassy carbon and on nickel substrates, and are associated to oxygen adsorption and reduction in phosphate buffer solutions.

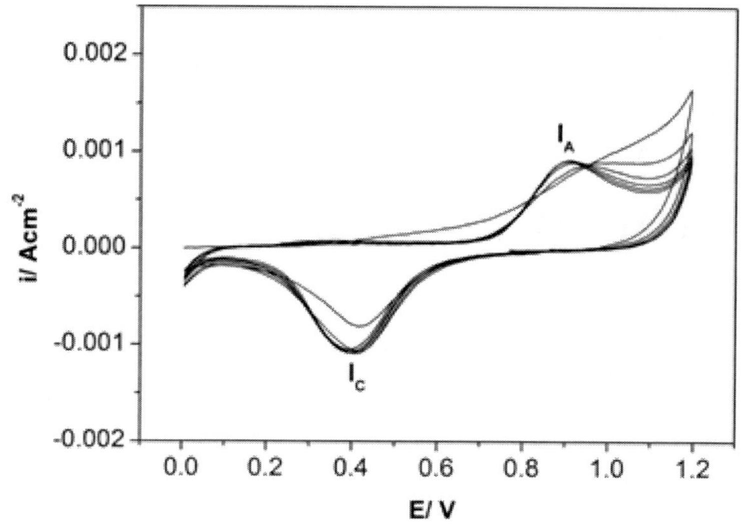

a

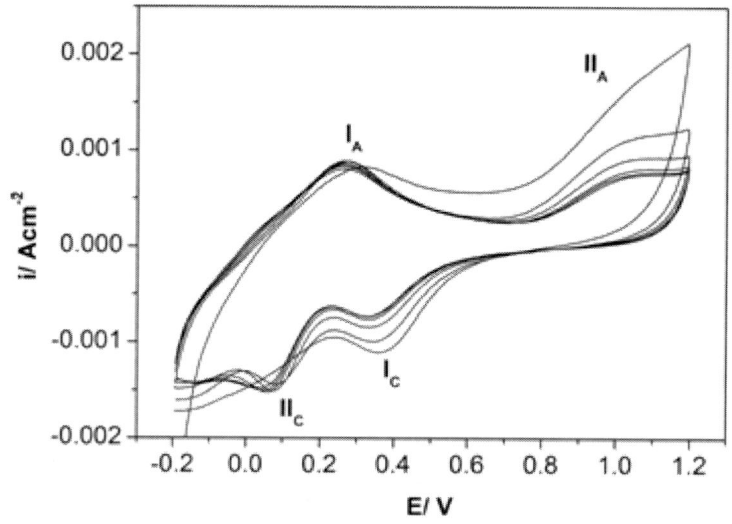

b

Figure 10. Cyclic voltammograms in a phosphate buffer solution (pH 7.3) and O_2 saturated at 50 mV s^{-1} at gold deposited under potentiostatic control of -1 V for 1500 s at 353 K on nickel (a) and on copper (b) (1st, n x 10th sweep ($1 \leq n \leq 5$).

Peaks Ia and IIc correspond to the anodic oxidation and reduction of copper respectively. Copper from the substrate may contribute to this effect as well as copper co-deposited with gold during the electrodeposition. Voltammograms appear quite stable after 50 cycles.

4. Discussion

The kinetics studies show that the electrodeposition of Au from $HAuCl_4.3H_2O$ + BMP-DCA solution is an irreversible process under diffusion control.

The diffusion coefficient for the $[AuCl_4]^-$ ionic species in this electrolyte, calculated from the voltammetric data, are 1.15 x 10^{-7} cm^2 s^{-1} at 293 K and 2.60 x 10^{-7} cm^2 s^{-1} at 333 K. This values are within the range usually found for the diffusion coefficients of metal ions in ionic liquids (10^{-8} to 10^{-6} cm^2 s^{-1}) [27, 28, 31, 32] but lower than those found in aqueous solutions (usually ~10^{-5} cm^2 s^{-1}), so lower deposition rates were expected when using IL as electrolyte.

SEM observation of the deposited layers obtained with 15 wt. % of water do not show complete coverage of the substrate in spite of the increase of conductivity (Figure 2), showing that water reduction as parallel reaction is also increased and consequently the electrodeposition is not improved. From the electrochemical characterization by voltammetry of the deposited layers in the phosphate buffer solution, it is found that gold deposited on inert materials, as nickel or glassy carbon, reveal the typical peak associated to oxygen chemisorption during the anodic potential scan followed by the correspondent reduction peak during the cathodic scan in the voltammetric studies.

Despite the good results, the adherence of the gold deposited layers to these inert substrates is low, since they were easily detached with adhesive tape. On the other hand, films deposited on copper show better adherence and only by scratching with a spatula it was possible to remove them. This fact is related with the initial dissolution and codeposition of copper with gold in this electrolyte, promoting a bond layer. To improve adherence of the gold deposits in nickel, a surface treatment with an etching ionic liquid electrolyte might be efficient in improving adherence. For example, preliminary experiments performed using the ionic liquid BMI-BF_4 seem to improve the adherence of gold deposited on nickel.

CONCLUSION

- The electrodeposition of gold from the electrolyte of 0.02M $HAuCl_4.3H_2O$ in BMP-DCA under laboratory atmospheric conditions is an irreversible process under diffusion control. The diffusion coefficients, for the Au^{3+} ionic species, calculated from the voltammetric data, are 1.15 x 10^{-7} cm^2 s^{-1} at 293 K and 2.60 x 10^{-7} cm^2 s^{-1} at 333 K (< 1 wt. % H_2O).
- The increase in water content up to 15 wt. % increases the conductivity of the electrolyte. However, the proton reduction rate also increases and interferes with the second step of the reduction process, from Au^+ to Au^0, and consequently, the compactness of the deposited layers did not improved.
- An average deposition rate for gold of 0.13 nm/s was estimated for films formed on copper at -1 V (333K). However, SEM observation evidenced that coverage is not uniform even after extending deposition time to 4500 s.
- Gold layers electrodeposited on inert surfaces, such as nickel or glassy carbon, show the typical peaks associated to oxygen chemisorption as it was shown by the voltammetry performed in the phosphate buffer solution. The surfaces deposited at 353 K show higher roughness for the same geometric area due to the dendritic morphology.

Acknowledgments

This work was financed by Fundação para a Ciência e Tecnologia (FCT) under contract nº PTM/CTM68847/2006.

References

[1] M. Galinski, A. Lewandowski, I. Stepniak, *Electrochim. Acta* 51 (2006) 5567-5580.

[2] D. R. MacFarlane, J. Golding, S. Forsyth, M. Forsyth, G. B. Deacon, *Chem. Commun.* (2001) 1430-1431.

[3] M. Schlesinger, M. Paunovic (Eds.), *Modern Electroplating*, John Wiley and Sons, New York, 2000.

[4] M. Bier, S. Dietrich, *Molecular Physics*, 108 (2010) 211–214.

[5] F. Endres, D. R. MacFarlane, A. P. Abbott (Eds.), *Electrodeposition of Metals from Ionic Liquids*, Wiley-VCH, 2008.

[6] A. P. Abbot, G. Frisch, K. S. Ryder, *Annu. Rev. Mater. Res.* 43 (2013) 335-358.

[7] F. Di Francesco, N. Calisi, M. Creatini, B. Melai, P. Salvo, *Green. Chem.* 13 (2011) 1712-1717.

[8] A. M. Mahony, D. S. Silvester, L. Aldous, C. Hardcre, R. G. Compton, *J. Chem. Eng. Data* 53 (2008) 2884-2891.

[9] M. J. Earle, K. R. Seddon, *Pure Appl. Chem.* 72 (2000) 1391-1398.

[10] T.-I. Leong, I.-W. Sun, M.-J. Deng, C.-M. Wu, P.-Y. Chen, *J. Electrochem. Soc.* 155 (2008) F55-F60.

[11] M.-J. Deng, P.-Y. Chen, T.-I. Leong, I.-W. Sun, J.-K. Chang, W.-T. Tsai, *Electrochem. Commum.* 10 (2008) 213-216.

[12] S. Z. El Abedin, E. M. Moustafa, R. Hempelmann, H. Natter, F. Endres, *Chem. Phys. Chem.* 7 (2006) 1535-1543.

[13] S. Z. El Abedin, M. Pölleth, S. A. Meiss, J. Janek, F. Endres, *Green. Chem.* 9 (2007) 549-553.

[14] D. Cameron, R. Holliday, D. Thompson, *Journal of Power Sources* 118 (2003) 298–303.

[15] R. B. Lima, H. Varella, *Gold Bulletin* 41 (2008) 15-22.

[16] A. Sarapuu, M. Nurmik, H. Mändar, A. Rosental, T. Laaksonen, K. Kontturi, D. J. Schiffrin, K. Tammeveksi, *J. Electrocanal. Chem.* 612 (2008) 78-86.

[17] A. I. de Sá, S. Eugénio, S. Quaresma, C. M. Rangel, R. Vilar, *Ciência e Tecnologia dos Materiais,* 23 (2001) 53-58.

[18] A. I. de Sá, S. Eugénio, S. Quaresma, C. M. Rangel, R. Vilar, *Thin Solid Film* 519 (2011) 6278-6282.

[19] A. I. de Sá, S. Eugénio, S. Quaresma, C. M. Rangel, R. Vilar, *Surf. and Coat. Technology* 223 (2013) 645-651.

[20] X.-H. Xu, C. L. Hussey, *J. Electrochem. Soc.* 139 (1992) 3103-3108.

[21] L. Aldous, D. S. Silvester, C. Villagrán, W. R. Pitner, R. C. Compton, M. C. Lagunas, C. Hardacre, *New J. Chem.* 30 (2006) 1576-1583.

[22] L. Aldous, D. S. Silvester, W. R. Pitner, R. G. Compton, M. C. Lagunas, C. Hardacre, *J. Phys. Chem.* C 111 (2007) 8496-8503.
[23] T. Oyama, T. Okajima, T. Ohsaka, *J. Electrochem. Soc*. 154 (2007) D322- D327.
[24] M. C. Tsai, D. X. Zhuang, P. Y. Chen, *Electrochim. Acta* 55 (2010) 1019-1027.
[25] X. R. Greef, R. Peat, L. M. Peter, D. Pletcher, J. Robinson, *Instrumental Methods in Electrochemistry,* Ellis Horwood, New York, 1990.
[26] A. J. Bard, L. R. Faulkner, *Electrochemical Methods: Fundamentals and Applications*, 2nd ed. John Wiley and Sons, New York, 2001.
[27] S. Eugénio, C. M. Rangel, R. Vilar, S. Quaresma, *Electrochim. Acta*, 56 (2011) 10347-10352.
[28] C. J. Rao, K. A. Venkatesan, K. Nagarajan, T. G. Srinivasan, P. R. V. Rao, *Electrochim. Acta* 54 (2009) 4718-4725.
[29] S. Trasatti, O. A. Petrii, *Pure Appl. Chem*. 63 (1991) 711-734.
[30] C. Hoogvliet, M. Dijksma, B. Kamp, W. P. Bennekon, *Anal. Chem*. 72 (2000) 2016-2021.
[31] M.-J. Deng, P.-Y. Chen, T.-I. Leong, I.-W. Sun, J.-K. Chang, W.-Ta Tsai, *Electrochem. Acta* 53 (2008) 5812-5818.
[32] A. P. Abbott, G. Capper, D. L. Davies, and R. K. Rasheed, *Chem. Eur. J.* 10 (2004) 3769-3774.

In: Ionic Liquids: Synthesis, Characterization and Applications ISBN: 978-1-63117-716-3
Editor: Anson Brooks

Chapter 5

EXPLORATION OF ELECTROLYTES FOR ZN-ANODE RECHARGEABLE BATTERIES: ROOM TEMPERATURE IONIC LIQUIDS AS MAJOR OR SUPPORTING COMPONENTS

M. Xu[1,2], D. G. Ivey[1], E. Dy[1,2], W. Qu[2] and Z. Xie[2]

[1]Department of Chemical and Materials Engineering,
University of Alberta, Edmonton, Alberta, Canada T6G 2V4

[2]Energy, Mining and Environment, National Research Council,
Vancouver, British Columbia, Canada

ABSTRACT

For Zn-anode rechargeable batteries, there are a number of shortcomings associated with using traditional KOH aqueous electrolytes. These include drying-out of the electrolyte due to water evaporation and dendrite formation at the anode, which severely impair battery performance (e.g., cycle life and capacity) and limit their application. It is, therefore, critical to either modify conventional KOH aqueous electrolytes or explore alternative electrolytes to eliminate these bottlenecks to the development of a feasible Zn-anode rechargeable battery system. Room temperature ionic liquids (RTILs) in recent years have been increasingly recognized as potential electrolytes or electrolyte components for rechargeable batteries. Compared with alkaline electrolytes, a simple electrolyte system composed of an RTIL as the sole component faces the challenge of improving its low conductivity before it can be practically applied in a battery. In this chapter, water and/or dimethyl sulfoxide modified RTILs (composed of pyrrolidinium or imidazolium cations and bis(trifluoromethanesulfonyl)imide or dicyanamide anions), with improved conductivity and Zn redox kinetics, are investigated as potential electrolyte systems for Zn-anode rechargeable batteries. In addition, with the capability of modifying metal deposit morphology, RTILs are found to be beneficial for depressing Zn dendrite formation in KOH aqueous electrolytes. An electrolyte, composed of 9M KOH + 5 wt% ZnO with a hydrophilic RTIL (0.5 wt% 1-ethyl-3-methylimidazolium dicyanamide), appears to be a promising electrolyte system for Zn-anode rechargeable batteries.

1. INTRODUCTION

Zn-anode rechargeable batteries, characterized by low cost, abundant supply and widespread availability of Zn raw materials and environmentally benign nature, are vital energy storage devices to electrify road/marine transportation, as well as stationary utilities. The inherent drawback of a conventional Zn-anode rechargeable battery with an alkaline aqueous electrolyte, however, is its limited cycle life. The short lifetime, which is due in part to Zn electrode failure, is typified by four problems:

1) Dendrite formation. The uneven deposition of Zn at active sites, where discharged Zn species gather, results in the formation of Zn dendrites. Dendrites can grow and then penetrate the separator, reach the counter electrode and cause eventual short circuiting during charging [1].
2) Shape change. This is associated with the redistribution of active materials over the Zn electrode surface on cycling. The cause is preferential dissolution of solid Zn from the electrode edges during discharging, followed by preferential redeposition at the electrode center with more compact deposits forming during charging [2, 3]. This problem leads to an eventual decrease in the effective surface area and limited cycle life.
3) Passivation. The precipitation of ZnO can occur during Zn discharge, giving rise to the formation of a passivation layer on the surface of the electrode. This problem is ascribed to the limited mass transfer of soluble species (OH^- and $Zn(OH)_4^{2-}$) to/from the dissolving Zn anode [4]. Decreased zinc electrode utilization (which depends on the capability of the electrode to remain active during discharge) and capacity loss are the undesirable consequences of passivation.
4) Self-discharge (hydrogen evolution). This refers to Zn corrosion along with hydrogen evolution during the charging period. Self-discharge not only impairs the cycle life of Zn-anode batteries, but also leads to a decline in battery capacity due to the corrosion of active Zn materials.

During the past half century, the most common ways to solve these four problems has been to modify the Zn electrode and the electrolyte, as well as to choose a desirable and robust separator on the basis of conventional alkaline aqueous electrolytes. Of these solutions, the incorporation of additives in the electrode or electrolyte (either inorganic [5-7] or organic [8]) has proven to be an effective way to concurrently minimize some of the problems. In addition, the combination of a microporous separator and a multi-layer separator can solve mixing problems, with the former reducing shape change and passivation and the latter limiting dendrite propagation [9]. Nevertheless, the current methods cannot completely eliminate the problems, particularly dendrite formation. Alternative methods are still needed to achieve a breakthrough for improving cycle life performance for Zn-anode batteries. Replacement or modification of traditional aqueous electrolytes with novel electrolyte systems appears to be a promising way to bridge the current research gap. Room temperature ionic liquids (RTILs) (which will be addressed in the next section), with their desirable properties of low or no volatility, low melting temperatures, high thermal stability, low flammabilities and wide electrochemical windows (EW), have received growing interest for application as advanced and safe electrolytes/electrolyte components for rechargeable batteries.

This chapter focuses on the research progress made in our group in adding diluent(s) to RTILs to modify both Zn redox kinetics and Zn deposit morphology, with the benefit of eliminating Zn dendrite formation. Four kinds of ionic liquids (with pyrrolidinium or imidazolium cations and bis(trifluoromethanesulfonyl)imide or dicyanamide anions) with H_2O and/or dimethyl sulfoxide (DMSO) as diluents are discussed and compared as potential electrolyte systems for Zn-anode rechargeable batteries. In addition, RTILs as additives to traditional alkaline aqueous electrolytes are also investigated, with the purpose of developing novel electrolyte systems that can harness the benefits from both RTILs (Zn morphology control) and aqueous electrolytes (rapid Zn/Zn^{2+} kinetics).

2. Room Temperature Ionic Liquids (RTILs) with Diluents

RTILs, i.e., salts with low melting temperatures around or below room temperature, composed exclusively of cations and anions, are novel solvents that have attracted great interest globally. They exhibit unique characteristics that make them promising electrolyte candidates for rechargeable batteries. First of all, the volatility disadvantage inherent to aqueous electrolytes is not a problem for RTILs. A battery using an ionic liquid as its electrolyte may function significantly longer because drying out due to water evaporation is avoided. Secondly, high thermal stability and low flammability make RTILs safe alternatives in batteries. Currently, RTILs have been broadly adopted as electrolytes in lithium ion batteries to alleviate their safety issues [10]. Thirdly, being completely ionic is a special feature that distinguishes RTILs from aqueous electrolytes. Originating from the dissolution of salts in a water solvent, aqueous electrolytes consist of solvated ions, their charged or neutral combinations and solvent molecules. In contrast, RTILs are salts that are melted (liquified) by providing heat to the system to overcome the salt lattice energy and are only composed of ions and their combinations, without any molecular solvent [11]. This ionic nature guarantees a wide electrochemical window (EW) when RTILs are used in batteries.

Different choices and combinations of cations and anions give rise to the diversity of RTILs. Among the various RTILs, aprotic RTILs (without dissociable H^+ and unable to donate hydrogen) have attracted attention as electrolytes for energy storage devices. Compared with conventional aqueous electrolytes, aprotic RTILs have the intrinsic ability to avoid hydrogen evolution and improve current efficiency [12-14], which makes them fascinating electrolytes for rechargeable batteries with potentially longer cycle lives. Aprotic RTILs also provide the benefit of modifying the deposit morphology of various metals [15, 16] and show promise for preventing Zn dendrite formation [17]. In this section, four kinds of aprotic RTILs, 1-butyl-1-methylpyrrolidinium bis(trifluoromethanesulfonyl)imide (BMP-TFSI), 1-methyl-1-propylpyrrolidinium bis(trifluoromethanesulfonyl)imide (MPP-TFSI), 1-ethyl-3-methylimidazolium bis(trifluoromethanesulfonyl)imide (EMI-TFSI) and 1-ethyl-3-methylimidazolium dicyanamide (EMI-DCA), are discussed as potential electrolytes for Zn-anode rechargeable batteries. The structures of the studied RTILs, together with their physicochemical properties, are presented in Table I.

It is clear from Table I that all the studied RTILs suffer a critical shortcoming, i.e., relatively low conductivity (3.4-25 mS•cm^{-1} vs. 540 mS•cm^{-1} for a 29.4 wt% KOH aqueous solution [18, 19]), which could severely restrict their use in energy storage applications. Henceforth, it is crucial to improve the conductivity of RTILs before their merits can be fully utilized as electrolytes. One approach to alleviate the low conductivity of RTILs is to incorporate other solvents as diluents to reduce the viscosity of the electrolyte. Among the potential diluents, H_2O is the most green and economically affordable diluent. DMSO is also a potential additive for RTILs. Its low vapour pressure helps maintain low electrolyte volatility [20]. In addition, DMSO, as a polar solvent, can facilitate the solvation of Zn^{2+} ions [21] and stabilize oxygen reduction reaction intermediates (superoxide radicals) [22]. The effects of H_2O and/or DMSO on the electrochemical properties of the Zn anode and Zn deposit morphology in the studied RTILs are the focus of this section. Suitable diluent concentrations for the studied RTILs are also suggested.

2.1. Water Added RTILs with Bis(trifluoromethanesulfonyl)imide Anions

BMP-TFSI, MPP-TFSI and EMI-TFSI, with bis(trifluoromethanesulfonyl)imide (TFSI$^-$) anions, have wide EWs (about 4.5-6 V as shown in Table I) and are favored for both electrochemical deposition and Li-based batteries [23-26].

Table I. Structure and physicochemical properties of the studied RTILs

Electrolyte	Structure	Melting temperature (°C)	Electrochemical window (V)	Conductivity (mS•cm^{-1})
BMP-TFSI	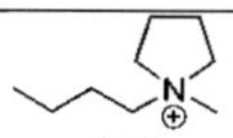 $(CF_3SO_2)_2N^{\ominus}$	-18.1	6.3	3.4
MPP-TFSI	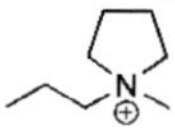 $(CF_3SO_2)_2N^{\ominus}$	+9.0	6.2	3.9
EMI-TFSI	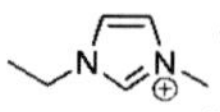 $(CF_3SO_2)_2N^{\ominus}$	-13.5	4.5	8.6
EMI-DCA	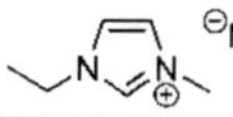 $^{\ominus}N(CN)_2$	-21.0	3.5	25

Table II. Effect of water on the electrochemical windows (EW) and conductivities of BMP-TFSI, MPP-TFSI and EMI-TFSI

	BMP-TFSI		MPP-TFSI		EMI-TFSI	
Water concentration (wt%)	EW (V)	Conductivity (mS•cm^{-1})	EW (V)	Conductivity (mS•cm^{-1})	EW (V)	Conductivity (mS•cm^{-1})
0.0	6.3	3.4	6.2	3.9	4.5	8.6
1.0	4.3	4.3	4.0	5.2	3.3	10.9
2.0	3.6	4.5	3.4	5.7	3.1	12.9
2.5	3.0	4.5	3.1	5.8	3.0	13.3

The addition of H_2O leads to an increase in conductivity, but results in a decrease in the EW at the same time. Finding the proper balance between the loss in EW and the improvement in conductivity for the three RTILs with H_2O addition is the main concern for developing workable electrolyte systems. Table II compares the EW and conductivity of the three RTILs before and after adding different amounts of H_2O. The EWs of these RTILs, especially for the two with pyrrolidinium cations, are sensitive to H_2O addition and EW decreases with increasing H_2O content. An approximately 45% reduction in the EW occurs when the H_2O amount exceeds 2 wt%. Conversely, the conductivity reaches its peak value with H_2O additions in the range of 2-2.5 wt%. When the amount of H_2O exceeds 2.5 wt%, the conductivity data are not stable and reliable, since the solubility limit for H_2O in the RTILs is exceeded. Compared with the conductivity of a 29.4 wt% KOH aqueous electrolyte, which is about 540 mS•cm^{-1} [18, 19], the conductivity of the ILs with added H_2O is still 1-2 orders of magnitude lower. Nevertheless, the RTIL electrolytes have conductivities comparable to or higher than electrolytes using lithium salts as solutes for state-of-the-art Li-based batteries (2-12 mS•cm^{-1}) [27].

2.1.1 State of Water in RTILs with TFSI⁻ Anions

Improving the conductivity of RTILs by incorporating H_2O as a diluent has been confirmed by many studies. Different mechanisms have been proposed to explain the enhanced conductivity of H_2O added RTILs. It has been suggested that H_2O addition increases the diffusion constant. The energies of interaction between H_2O and the ions of 1-butyl-3-methylimidazolium bromide are smaller than the ion-ion interactions, which leads to an increase in conductivity [28]. Another investigation on 1-butyl-3-methylimidazolium tetrafluoroborate [29] demonstrated that the presence of H_2O molecules reduces the electrostatic attraction between the cation and anion, resulting in lowered viscosity and increased conductivity for this IL. A recent study on the measurement of the frequency-dependent shear viscosity of mixtures of H_2O with two ionic liquids, composed of imidazolium cations and tetrafluoroborate anions, pointed out that H_2O reduces the shear viscosity of mixtures through a decrease in the structural relaxation time [30].

The state of H_2O in RTILs can serve as a key to understanding the change in physicochemical properties, including the viscosity and conductivity. By using Fourier transform infrared spectroscopy (FTIR) and differential scanning calorimetry (DSC) tests, our group [31] studied the interaction between H_2O and three RTILs with the same $TFSI^-$ anion (BMP-TFSI, MPP-TFSI and EMI-TFSI). Water exists as either bound water or both bound and free water in all the RTILs studied, depending on the H_2O concentration.

1) Bound water

DSC studies show that when H_2O is added in amounts less than 2 wt% H_2O, the melting peaks M for the bulk RTILs shift toward or remain at 0 °C (Figure 1); this is attributed to a bound water effect. Correspondingly, FTIR spectra show two distinct bands, located at ~3630 cm^{-1} and ~3560 cm^{-1}, respectively (Figure 2). These are attributed to the asymmetric v_3 and symmetric v_1 stretching modes of H_2O bound to the $TFSI^-$ anion. In addition, according to the studies in [32, 33], it is suggested that one H_2O molecule interacts with two $TFSI^-$ anions by H-bonding since the wave number difference at maximum absorbance for these two bands is around 70-80 cm^{-1}.

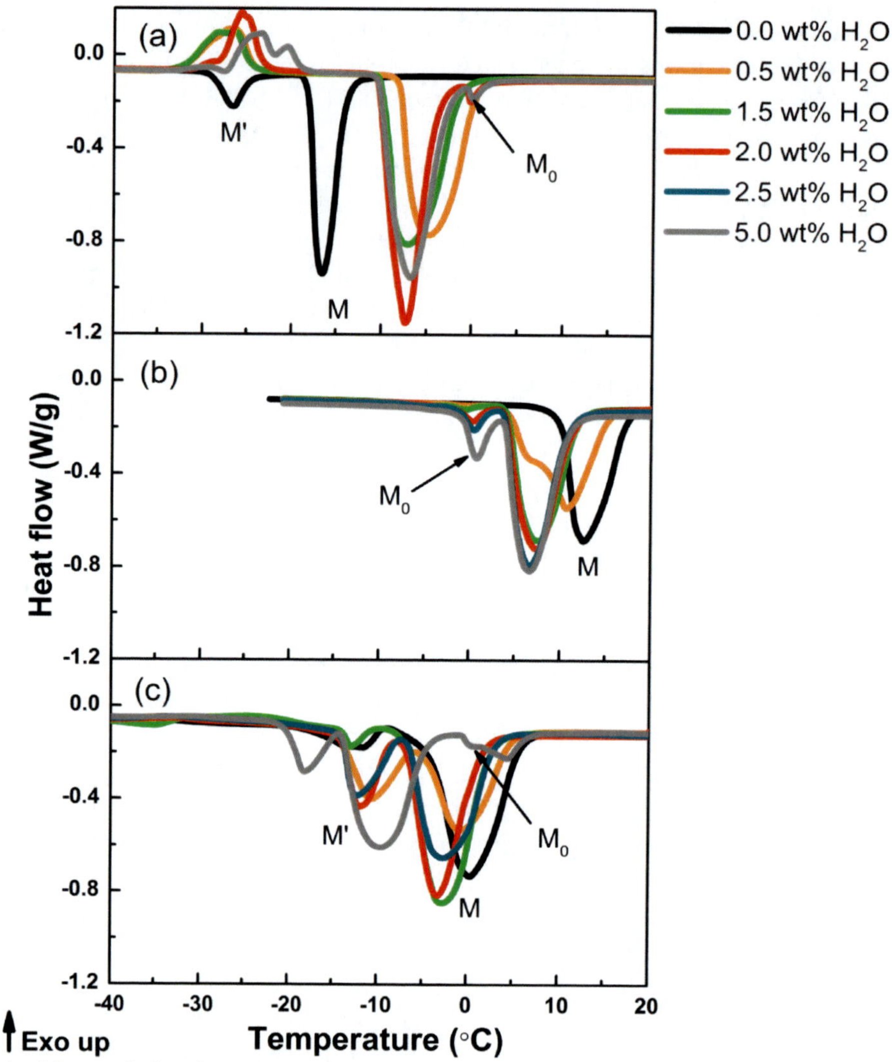

Figure 1. DSC traces for each RTIL with different amounts of water additions: (a) BMP-TFSI, (b) MPP-TFSI and (c) EMI-TFSI (M, M' and M_0 denote melting peaks) [31].

2) Bound and free water

When H_2O amounts exceed 2-2.5 wt%, in addition to the bulk RTIL melting peaks M and M' (M' for both BMP-TFSI and EMI-TFSI represents melting of a metastable liquid crystal), additional small melting peaks M_0, with melting temperatures near 0 °C, are detected in the DSC curves (Figure 1). These peaks are related to the melting of free water, which is not linked to the RTILs. Corroborating results are found in the FTIR spectra in Figure 2, where a broader band (~3500 cm^{-1} to ~3350 cm^{-1}) is visible at high H_2O concentrations in addition to the two distinct bands mentioned above, suggesting the presence of free water aggregates.

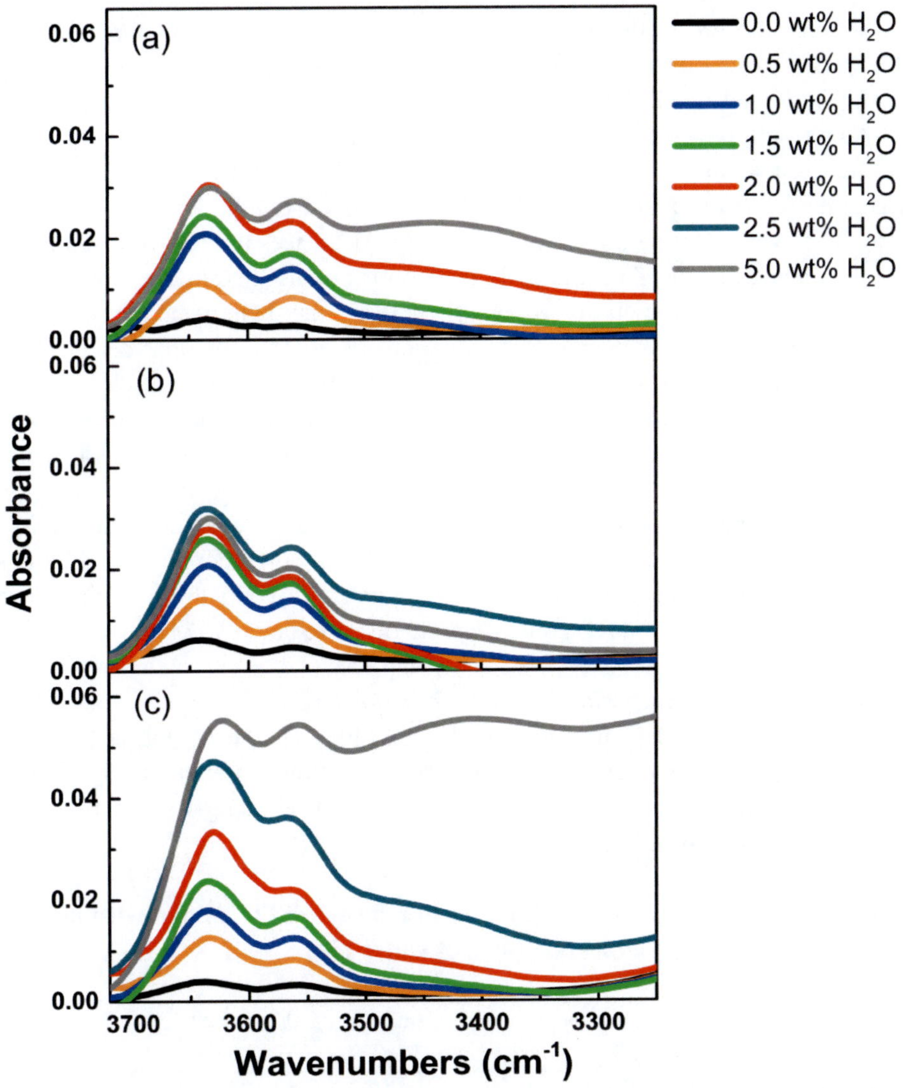

Reprinted with permission from *Journal of The Electrochemical Society*, 161 (1) A128-A136 (2014). Copyright 2014, The Electrochemical Society.

Figure 2. FTIR spectra for each RTIL with different amounts of water additions: (a) BMP-TFSI, (b) MPP-TFSI and (c) EMI-TFSI [31].

It is suggested that the enhanced conductivity that occurs by adding 2-2.5 wt% H_2O to BMP-TFSI, MPP-TFSI and EMI-TFSI is mainly due to the portion of bound water. The smaller water-ion interactions, i.e., through H-bonding, interfere with the strong ion-ion interaction of each RTIL, resulting in decreased viscosity and increased conductivity.

2.1.2. Effect of Water on Kinetic Properties of Zn/Zn^{2+} Redox Reactions in RTILs with $TFSI^-$ Anions

Adding H_2O not only improves the conductivity of BMP-TFSI, MPP-TFSI and EMI-TFSI, but also kinetically facilitates the Zn/Zn^{2+} redox reaction in the three RTILs. An increase in H_2O addition leads to an increase in exchange current density j_0 (reaction rate) for the Zn/Zn^{2+} redox reaction and an increase in diffusion coefficient D_0 (mass transfer) for the Zn species (Table III). As a consequence, a smaller surface overpotential (η_s) required to activate the redox reaction and a lower concentration overpotential (η_T) are achieved. Moreover, the improved conductivity with H_2O addition minimizes the ohmic drop in the electrolyte. As a result, the total overpotential η (where $\eta = \eta_s + \eta_T$ + ohmic drop) for driving the Zn/Zn^{2+} redox reaction, as indicated by peak potential separation ΔE_p between the Zn reduction and Zn oxidation reactions (Table III), is reduced for the three RTILs as H_2O is added (up to 2-2.5 wt%). On the other hand, since H_2O decreases the EW for each RTIL, it is important to guarantee that the Zn/Zn^{2+} redox reaction is not affected in the shortened EW. In the three RTILs, cyclic voltammetry (CV) experiments (Figure 3) indicate that when the H_2O content is kept below 2-2.5 wt%, the Zn/Zn^{2+} redox reactions are unaffected. When the amount of H_2O exceeds 2-2.5 wt%, in addition to bound water, water can also exist as free water, which can decompose and interfere with the Zn redox reaction during the CV scan.

In order to understand the mechanism for the improved Zn/Zn^{2+} redox kinetics with H_2O, gas-phase density functional theory (DFT) calculations on cluster models composed of BMP^+, $TFSI^-$, H_2O and Zn^{2+} species were applied to explore the interactions among the ions. Table IV shows the various gas-phase reactions, together with the corresponding binding energies among the ions present in the electrolyte. The interactions among the ions present in the electrolyte can be classified into three groups and are discussed below [31].

1) H_2O molecules interacting with Zn^{2+} ions (reactions a-d)

The binding energy of H_2O to Zn^{2+}, in the absence of other species, is somewhere between 4.38 and 2.30 eV depending on the level of hydration of the central Zn cation. The results are similar to those obtained by Pavlov et al., [34]. Binding for the first two H_2O molecules is stronger than for subsequent ones.

Table III. Kinetic properties for Zn/Zn^{2+} redox in each RTIL added with different concentrations of water [31]

	Water concentration (wt%)	0.0	1.0	2.0	2.5
	BMP-TFSI	0.90	0.52	0.57	/
ΔE_p (V)	MPP-TFSI	0.75	0.58	0.48	0.46
	EMI-TFSI	0.97	0.56	0.55	0.48
D_0	BMP-TFSI	0.22	0.38	0.76	/
($x10^{-6}$ $cm^2 \cdot s^{-1}$)	MPP-TFSI	0.18	0.30	0.46	0.43
	EMI-TFSI	0.29	0.31	0.42	0.58
j_0	BMP-TFSI	0.018	0.017	0.056	/
($mA \cdot cm^{-2}$)	MPP-TFSI	0.018	0.026	0.049	0.040
	EMI-TFSI	0.021	0.055	0.079	0.13

ΔE_p: Peak potential separation; D_0: Diffusion coefficient of Zn species; j_0: Exchange current density.
Reprinted with permission from *Journal of The Electrochemical Society*, **161** (1) A128-A136 (2014).

Table IV. Gas-phase binding energies between BMP^+, $TFSI^-$ and Zn^{2+} and H_2O species determined by B3LYP/6-311G(df,p) DFT calculations [31]

Scheme	Reactants		Product	Binding Energy, eV
a	$Zn^{2+} + H_2O$	→	$Zn(H_2O)^{2+}$	4.38
b	$Zn(H_2O)^{2+} + H_2O$	→	$Zn(H_2O)_2^{2+}$	3.72
c	$Zn^{2+} + 6\ H_2O$	→	$Zn(H_2O)_6^{2+}$	13.84 (2.30/H_2O molecule)
d	$Zn^{2+} + 6\ H_2O$	→	Zn $(H_2O)_6^{2+}$, PCM*	21.61 (3.60/H_2O molecule)
e	$BMP^+ + H_2O$	→	BMP^+-H_2O	0.20
f	$TFSI^- + H_2O$	→	$TFSI^-$-H_2O	0.33
g	$BMP^+ + TFSI^-$	→	BMP^+-$TFSI^-$	2.92
h	$Zn^{2+} + TFSI^-$	→	$Zn(TFSI)^+$	15.94
i	$Zn(TFSI)^+ + TFSI^-$	→	$Zn(TFSI)_2$	6.96
j	$Zn(TFSI)_2 + H_2O$	→	$Zn(TFSI)_2(H_2O)$	0.46
k	$Zn(TFSI)_2(H_2O) + H_2O$	→	$Zn(TFSI)_2(H_2O)_2$	0.43
l	$Zn(TFSI)_2(H_2O)_2 + H_2O$	→	$Zn(TFSI)_2(H_2O)_2 \cdots OH_2$	0.36
m	$Zn(TFSI)(H_2O)_2^+ + TFSI^-$	→	$Zn(TFSI)_2(H_2O)_2$	4.71

* PCM: Polarizable continuum model is used to approximate the effect of additional layers of water molecules surrounding the Zn $(H_2O)_6^{2+}$ cluster.

2) H_2O molecules interacting with ionic liquid species (reactions e-f)

These binding energies are the weakest, since ion-dipole interactions are involved rather than chemical bonding. The partially negative oxygen atom of H_2O faces BMP^+, while the partially positive hydrogen atoms of H_2O face $TFSI^-$.

3) $TFSI^-$ anions interacting with other ions in the absence of H_2O (reactions g-i)

The interaction of Zn^{2+} with $TFSI^-$ is the strongest because of Zn's large 2^+ charge. This is followed by the interaction strength of $Zn(TFSI)^+$ with $TFSI^-$, where the central Zn cation has already been partially neutralized. The relative weakness of the BMP^+-$TFSI^-$ interaction (reaction g) can be attributed to the distribution of the cationic charge over a large surface.

It can be concluded that H_2O will preferentially bind to Zn^{2+} when added to a mixture of $Zn(TFSI)_2$ in BMP-TFSI, since the interaction of H_2O with $Zn(TFSI)_2$ (reaction j-l) is more energetically favored than that with BMP^+ (reaction e) or $TFSI^-$ (reaction f). Furthermore, H_2O weakens the ionic interaction between Zn^{2+} and $TFSI^-$. By comparing reaction m to reaction i, it is clear that adding two molecules of H_2O reduces the $Zn(TFSI)^{+} \cdots TFSI^-$ binding energy considerably from 6.96 to 4.71 eV. This could facilitate the dissolution of Zn^{2+} in BMP-TFSI, which in turn can increase the concentration of Zn^{2+} involved in the redox reaction and explains the improved overall Zn redox reaction rates with H_2O addition in the RTIL.

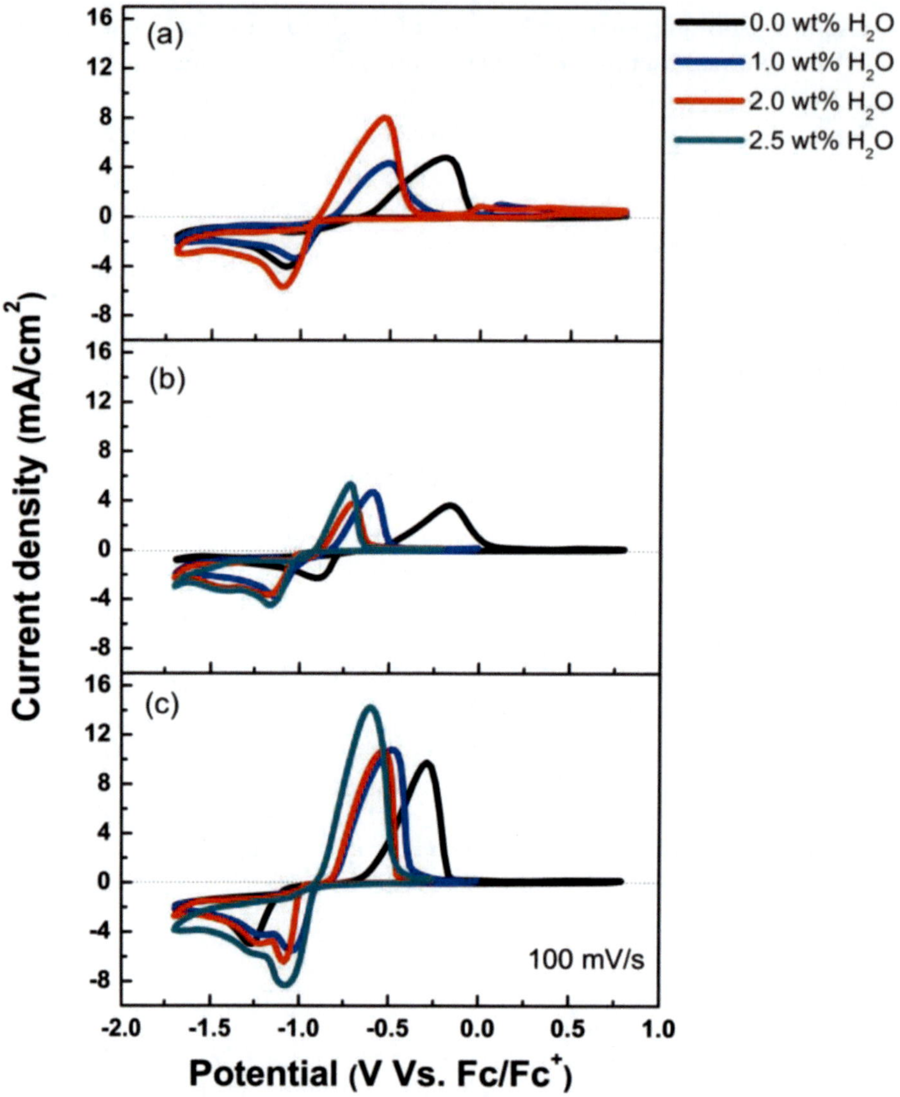

Figure 3. Cyclic voltammetry plots for Zn/Zn^{2+} redox in: (a) BMP-TFSI, (b) MPP-TFSI and (c) EMI-TFSI [31].

2.1.3. Morphology of Electrodeposited Zn in RTILs with TFSI⁻ Anions with/without Water

The morphology of Zn electrodeposited from BMP-TFSI, MPP-TFSI and EMI-TFSI is of value for study to determine whether dendrite formation occurs. Zinc was electrodeposited at a constant potential (the peak cathodic potential based on the CV curves for Zn/Zn^{2+} redox in Figure (3)) from a 0.1 mol•L^{-1} Zn^{2+} solution produced by dissolving Zn-TFSI in each RTIL and the RTIL-H_2O mixtures. No Zn dendrites are observed from the three RTILs in the absence of H_2O (Figure 4 (a-c)). The morphology and size of the Zn deposits are strongly dependent on the RTIL cations. In RTILs with pyrrolidinium cations (Figure 4 (a-b)), clustered, nanoscale, needle-like Zn deposits are generated. Coarser, faceted Zn (micron size range) is formed from the RTIL with imidazolium cations (Figure 4 (c)). The difference in Zn deposit morphology and size is most likely related to the IL solvation layers formed at the substrate/IL interface, which affect the electrodeposition process. As revealed in a study done on electrodeposition of Al from RTILs [35], ILs with pyrrolidinium cations are about 4 times more strongly adsorbed on the Au substrate than from ILs with imidazolium cations. The improved adsorption hinders underpotential deposition of Al and facilitates the electrodeposition of nanocrystalline Al instead of microcrystalline Al.

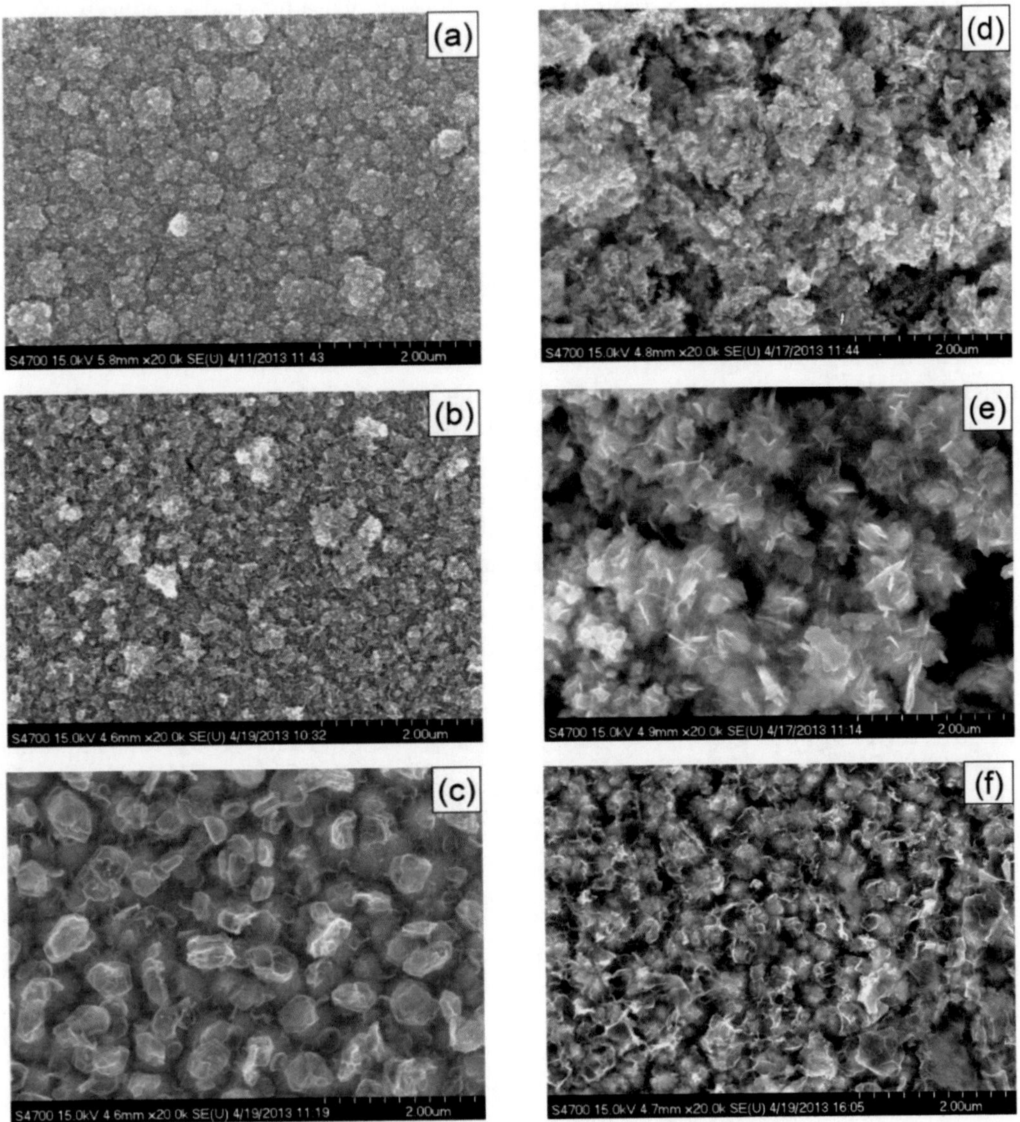

Reprinted with permission from *Journal of The Electrochemical Society*, 161 (1) A128-A136 (2014). Copyright 2014, The Electrochemical Society.

Figure 4. SEM secondary electron (SE) images of Zn deposited on Pt coated Si substrates prepared by potentiostatic deposition from: (a) BMP-TFSI, (b) MPP-TFSI, (c) EMI-TFSI, (d) BMP-TFSI with 2.0 wt% H_2O, (e) MPP-TFSI with 2.0 wt% H_2O and (f) EMI-TFSI with 2.5 wt% H_2O [31].

Water addition to RTILs affects both deposit morphology and adhesion of the deposited film on the substrate. The influence of H_2O varies for different types of RTILs. In the two RTILs with pyrrolidinium cations, i.e., BMP-TFSI and MPP-TFSI, adding H_2O increases the mass transfer rate for the Zn species and, consequently, facilitates deposit growth as seen in Figure 4 (d-e). Unlike BMP-TFSI and MPP-TFSI, Zn deposits from EMI-TFSI with H_2O addition show negligible change in size compared with H_2O-free EMI-TFSI, but the morphology changes from faceted to granular in shape (Figure 4 (f)). The adhesive properties of the deposited films seems to correlate with the size of the Zn deposits. The larger scale Zn deposits obtained in the BMP-TFSI and MPP-TFSI RTILs with H_2O easily delaminate from the substrate. The poor adhesion of the Zn deposit indicates that a reduced amount of the Zn material would be involved in subsequent electrochemical reactions, leading to decreased capacity upon cycling. Good adhesion of Zn deposits to the substrate is maintained for EMI-TFSI with H_2O additions. Zinc electrodeposited from RTILs with imidazolium cations appears to be less affected by H_2O additions than Zn obtained from RTILs with pyrrolidinium cations.

2.2. Water and/or DMSO Added RTILs with Dicyanamide Anions

Compared with other candidates in the RTIL family, especially those with $TFSI^-$ anions, 1-ethyl-3-methylimmidazolium dicyanamide (EMI-DCA) stands out with its relatively high conductivity (~25 $mS{\bullet}cm^{-1}$). Additionally, electrodeposition of non-dendritic and compact Zn films was achieved in EMI-DCA [36], which is preferred for Zn-anode batteries to avoid short-circuit issues.

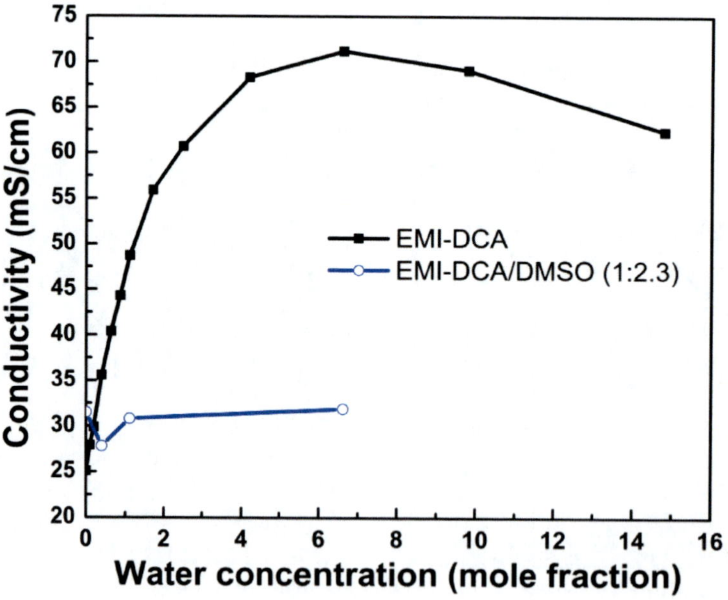

Figure 5. Conductivity of EMI-DCA with addition of H_2O and/or DMSO. (EMI-DCA/DMSO (1:2.3) is the mole fraction of EMI-DCA to DMSO.)

With the aim of modifying/improving Zn redox kinetics, H_2O and/or DMSO were added to EMI-DCA as diluents to optimize the electrochemical system with respect to Zn/Zn^{2+} electrochemical behavior. As seen in Figure 5, the electrolyte conductivity for EMI-DCA with H_2O and/or DMSO additions is improved. A mole ratio of EMI-DCA to H_2O of 1:6.6 and EMI-DCA to DMSO of 1:2.3 correspond to the highest measured electrolyte conductivities of ~70 $mS{\bullet}cm^{-1}$ and ~30 $mS{\bullet}cm^{-1}$, respectively. EMI-DCA with H_2O at a mole ratio of 1:1.1 has a conductivity of ~50 $mS{\bullet}cm^{-1}$ (twice the EMI-DCA value).

2.2.1. Mechanisms for Zn/Zn^{2+} Redox Reactions in EMI-DCA with H_2O and DMSO

As with the RTILs containing $TFSI^-$ anions discussed in Section 2.1, H_2O plays a key role in improving the Zn/Zn^{2+} redox kinetics in EMI-DCA, including enhancing the diffusion rate and the exchange current density, as well as decreasing the overpotential for redox reactions (peak potential separation). Unlike hydrophobic RTILs with $TFSI^-$ anions, where H_2O improves Zn redox reaction rates by increasing the concentration of Zn^{2+} involved in the redox reaction, hydrophilic EMI-DCA is miscible with H_2O and relatively large H_2O additions alter the Zn/Zn^{2+} redox reaction mechanism in EMI-DCA and change the Zn/Zn^{2+} redox reaction rate. As discussed in our recent work [37], for electrolytes based on EMI-DCA with no H_2O, complex ions of $[DCA]^-$ anions and Zn^{2+} ions, i.e., $Zn(DCA)_x^{(x-2)-}$, are reduced to Zn in two steps with one electron received per step. The proposed electrode reactions during reduction are as follows:

$$Zn(DCA)_x^{(x-2)-} + e^- \leftrightarrow Zn(DCA)_x^{(x-1)-} \qquad (1)$$

$$Zn(DCA)_x^{(x-1)-} + e^- \leftrightarrow Zn + x(DCA)^- \qquad (2)$$

In electrolytes with H_2O added, $Zn(DCA)_x^{(x-2)-}$ ions are not the only complex ions, since Zn^{2+} ions also likely coordinate with H_2O to form $Zn(H_2O)_y^{2+}$ complex ions. $Zn(H_2O)_y^{2+}$ ions compete with $Zn(DCA)_x^{(x-2)-}$ ions for involvement in the reduction reactions, facilitating the occurrence of Reactions (3) and (4) in addition to Reactions (1) and (2).

$$Zn(H_2O)_y^{2+} + e^- \leftrightarrow Zn(H_2O)_y^{+} \qquad (3)$$

$$Zn(H_2O)_y^{+} + e^- \leftrightarrow Zn + y\,H_2O \qquad (4)$$

The dominant reactions are determined by the H_2O content and can be predicted from the shift in peak reduction potentials from the CV curves for Zn/Zn^{2+} redox reactions, as shown in Figure 6. In EMI-DCA with no H_2O (Figure 6(a) and (d)), potentials for the first and second reduction steps range from -1.3 to -1.35 V and -1.5 to -1.55 V, respectively. For electrolytes EMI-DCA + H_2O (1:1.1) and EMI-DCA + H_2O + DMSO (1:1.1:2.3), with relatively small amounts of H_2O (Figure 6(b) and (e)), $Zn(DCA)_x^{(x-2)-}$ remains as the dominant complex ion involved in the electrode reactions. The peak reduction potentials are similar to those observed for EMI-DCA with no H_2O. Nevertheless, after adding a large amount of H_2O to EMI-DCA (Figure 6(c) and (f)), the resultant peak potentials for Zn^{2+} reduction in EMI-DCA + H_2O (1:6.6) and EMI-DCA + H_2O + DMSO (1:6.6:2.3) ranged from -1.2 to -1.25 V and -1.35 to -1.45 V, respectively.

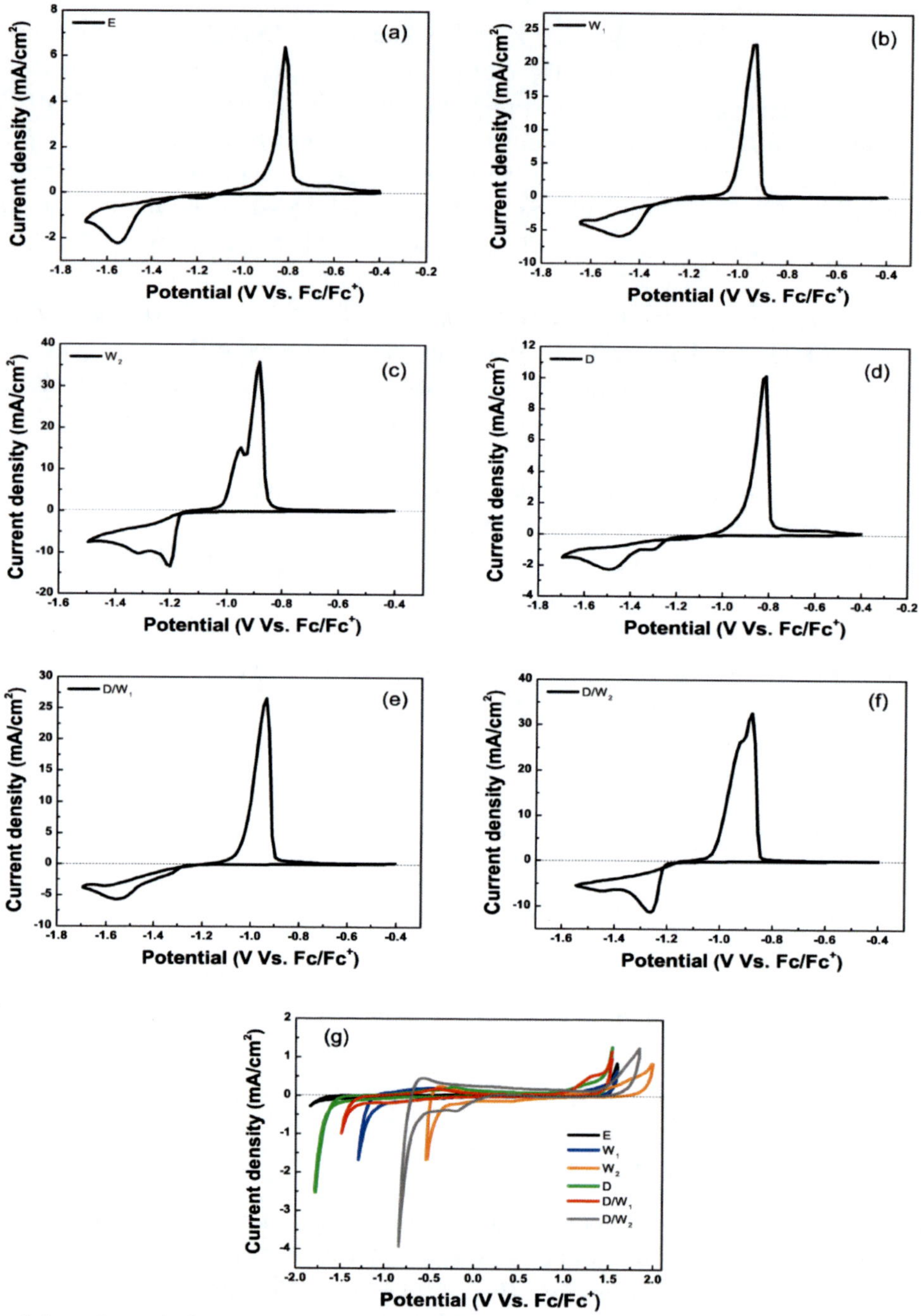

Figure 6. Cyclic voltammetry plots at 50 mV•s^{-1} for Zn/Zn^{2+} redox in EMI-DCA based electrolytes: (a) E: EMI-DCA; (b) W$_1$: EMI-DCA + H_2O (1:1.1); (c) W$_2$: EMI-DCA + H_2O (1:6.6); (d) D: EMI-DCA + DMSO (1:2.3); (e) D/W$_1$: EMI-DCA + H_2O + DMSO (1:1.1:2.3); (f) D/W$_2$: EMI-DCA + H_2O + DMSO (1:6.6:2.3) and (g) electrochemical window for each electrolyte in the absence of Zn^{2+} ions [37].

Table V. Electrochemical behavior of Zn/Zn^{2+} redox in EMI-DCA based electrolytes [37]

Electrolyte	D_0 [a] ($x10^{-6}$ cm^2•s^{-1})	j_0 [b] (mA•cm^{-2})	ΔE_p [c] (V)	Q_a/Q_c [d] at CV cycle 2
E	0.10	0.0016	0.73	0.84
W_1	0.90	0.014	0.58	0.62
W_2	6.00	0.12	0.36	0.59
D	0.30	0.0089	0.67	0.89
D/W_1	0.80	0.032	0.57	0.91
D/W_2	8.80	0.13	0.41	0.84

[a] Diffusion coefficient; [b] Exchange current density; [c] Peak potential separation; [d] Ratio of Q_a to Q_c (Q_c is the charge integrated for Zn reduction and Q_a is the charge integrated for Zn oxidation) (E: EMI-DCA; W_1: EMI-DCA + H_2O (1:1.1); W_2: EMI-DCA + H_2O (1:6.6); D: EMI-DCA + DMSO (1:2.3); D/W_1: EMI-DCA + H_2O + DMSO (1:1.1:2.3); D/W_2: EMI-DCA + H_2O + DMSO (1:6.6:2.3))

These reduction potentials are more positive compared with that observed for EMI-DCA with no H_2O and the effect is attributed to the transition from Reactions (1) and (2) to Reactions (3) and (4) for Zn/Zn^{2+} redox.

Compared with H_2O, DMSO seems to have no obvious effect on the kinetic behavior for Zn/Zn^{2+} redox reactions. Nevertheless, DMSO is important from the standpoint of good Zn/Zn^{2+} redox reversibility. As displayed in Table V, adding only H_2O to EMI-DCA impairs Zn/Zn^{2+}redox reversibility, which is evaluated from the ratio of Q_a/Q_c. Q_c is the charge integrated for Zn reduction and Q_a is the charge integrated for Zn oxidation; the closer Q_a/Q_c is to 1, the better the reversibility. Increased H_2O content leads to poorer reversibility. On the other hand, the drawbacks brought about by H_2O addition can be compensated somewhat by adding DMSO as a supplementary diluent. The addition of mixed diluents, e.g., H_2O and DMSO, maintains good Zn/Zn^{2+} redox reversibility.

It is speculated that the Zn/Zn^{2+} redox reversibility is strongly associated with the stability, i.e., the EW, of the electrolytes. As shown in Figure 6(g), for electrolytes based on EMI-DCA plus H_2O, the EW is severely reduced as H_2O content is increased, which is most likely caused by water electrolysis. DMSO not only has a negligible effect on the EW of EMI-DCA, but also seems to stabilize H_2O molecules (probably through H-bonding), which is good for inhibiting H_2O decomposition and maintaining Zn/Zn^{2+} redox reversibility in the electrolytes.

2.2.2. Morphology of Electrodeposited Zn in EMI-DCA with H_2O and/or DMSO Addition

Electrodeposition was done at a constant potential (the peak cathodic potential based on the CV curves for Zn/ Zn^{2+} redox in Figure (6)). Non-dendritic Zn deposits are obtained for all electrolytes based on EMI-DCA containing 0.1 mol•L^{-1} Zn^{2+} ions, although the Zn morphology varies with electrolyte composition.

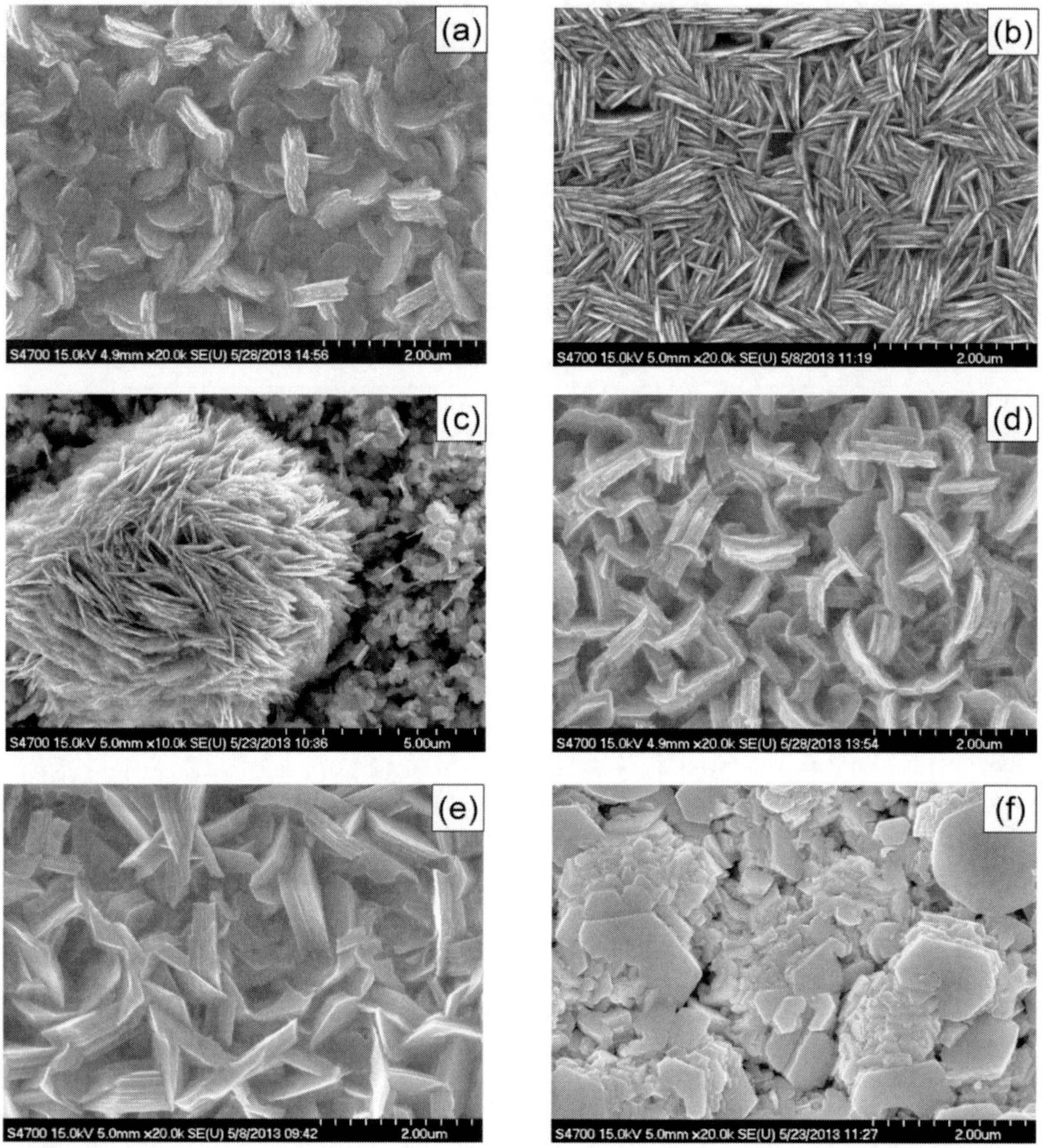

Figure 7. SEM SE images of Zn deposited on Pt coated Si substrates from EMI-DCA with addition of H_2O and/or DMSO: (a) E: EMI-DCA; (b) W_1: EMI-DCA + H_2O (1:1.1); (c) W_2: EMI-DCA + H_2O (1:6.6); (d) D: EMI-DCA + DMSO (1:2.3); (e) D/W_1: EMI-DCA + H_2O + DMSO (1:1.1:2.3) and (f) D/W_2: EMI-DCA + H_2O + DMSO (1:6.6:2.3)).

Uniform and compact Zn films with plate-like morphology are obtained from EMI-DCA with no H_2O and a small amount of H_2O addition (EMI-DCA + H_2O (1:1.1)), as evident in Figure 7(a-b). In EMI-DCA containing DMSO (EMI-DCA + DMSO (1:2.3) and EMI-DCA + H_2O + DMSO (1:1.1:2.3)), plate-like Zn deposits grow in different orientations, so that the Zn films appear to be more porous (Figure 7(d-e)). This effect is desirable for battery electrode applications, since power density is related to electrode surface area. In EMI-DCA diluted with large amounts of H_2O (Figure 7(c) and (f)), the Zn morphology is quite different. Non-uniform and less compact Zn deposits, composed of flower-like Zn nodules surrounded by dispersed Zn particles, are observed for EMI-DCA + H_2O (1:6.6) (Figure 7(c)). Faceted Zn particles are the dominant morphology for deposits from EMI-DCA + H_2O + DMSO (1:6.6:2.3), where micro-voids adjacent to Zn facet clusters are present in Figure 7(f). These results indicate that nucleation and growth of Zn is greatly influenced by the electrolyte composition. Water and/or DMSO additions to EMI-DCA likely modify the Zn morphology by changing the diffusion rate of Zn ions and the IL/substrate interface. In EMI-DCA, the Zn deposits are all in the micron size range; this is also the case for EMI-TFSI. Deposit size is, therefore, mainly determined by the type of RTIL cations, while the morphology and the growth orientation of deposits are closely related to the RTIL anions.

2.3. Promising RTILs-based Electrolytes for Zn-anode Rechargeable Batteries

For Zn-anode rechargeable battery applications, any candidate RTIL-based electrolyte should exhibit high conductivity and, equally important, balanced and good electrochemical behavior for Zn/Zn^{2+} redox reactions. Moreover, the deposition of non-dendritic Zn is preferred so as to avoid short-circuiting during recharging. Among the four studied RTILs, BMP-TFSI, MPP-TFSI, EMI-TFSI and EMI-DCA, the first three with $TFSI^-$ anions show superior performance to EMI-DCA in terms of Zn/Zn^{2+} redox kinetics, including diffusion coefficients and exchange current densities. However, the conductivity for EMI-DCA is one order of magnitude higher than the other three RTILs. In addition, EMI-DCA containing both H_2O and DMSO at a mole ratio of EMI-DCA:H_2O:DMSO = 1:1.1:2.3 shows considerable promise. It has high conductivity and reaction kinetics (compare the data in Tables III and V), good Zn/Zn^{2+} redox reversibility and desirable deposit morphology characterized by compact and porous Zn films. As such, an electrolyte system composed of EMI-DCA, H_2O and DMSO in the appropriate amounts is the most promising RTIL for Zn-anode rechargeable batteries among the studied RTILs-based electrolytes.

The present study also demonstrates the Zn/Zn^{2+} redox reactions exhibit sluggish kinetics in RTILs-based electrolytes (about two orders of magnitude lower) when compared with Zn/Zn^{2+} redox in traditional alkaline aqueous electrolytes. As such, it may be advantageous to combine the benefit of Zn morphology control using RTILs with the benefit of high Zn/Zn^{2+} redox kinetics in alkaline aqueous electrolytes by using RTILs as additives to alkaline aqueous electrolytes. This novel electrolyte system approach for Zn-anode rechargeable batteries is currently under investigation; preliminary results are discussed in the next section.

3. Conventional Alkaline Electrolytes with Ionic Liquids As Additives

As revealed above, the three RTILs with $TFSI^-$ anions and EMI-DCA show the capability of modifying Zn deposit morphology. Two RTILs, i.e., EMI-TFSI and EMI-DCA with the same EMI^+ cation, were selected as additives to modify alkaline electrolytes composed of 9M KOH + 5 wt% ZnO with the purpose of eliminating Zn dendrite formation. It is well known that the electrolyte composition and deposition conditions are critical factors in controlling the morphology of Zn deposits. Two aspects covered in this section include: 1) a comparison between EMI-TFSI and EMI-DCA additions in terms of suppressing Zn dendrite formation in alkaline electrolytes and 2) the variation of Zn deposit morphology with RTIL concentration and the applied current density.

3.1. Comparison of EMI-TFSI and EMI-DCA As Additives in Alkaline Electrolytes

The first issue of concern when adding RTILs to an alkaline aqueous electrolyte is solubility. Compared with EMI-TFSI, which is hydrophobic in nature, hydrophilic EMI-DCA is more soluble in aqueous electrolytes. The solubility limit for EMI-DCA in an electrolyte containing 9M KOH + 5 wt% ZnO is about 1 wt% and the mixed electrolyte turns from colorless to light yellow (the color of EMI-DCA) once the concentration exceeds 1-2 wt% EMI-DCA. Only about 0.1 wt% EMI-TFSI can be completely dissolved into an electrolyte composed of 9M KOH + 5 wt% ZnO; insoluble EMI-TFSI droplets are visible at higher concentrations.

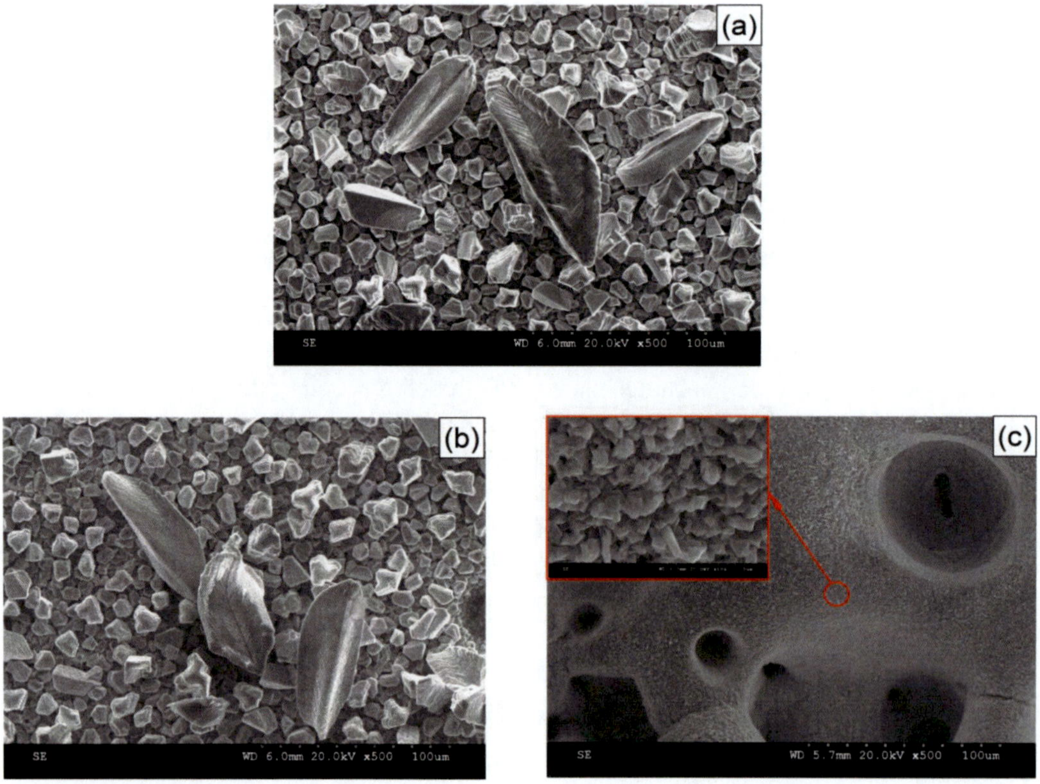

Figure 8. SEM SE images of Zn electrodeposited on Pt coated Si substrates at a constant potential (-0.15 V vs. Zn wire) from: (a) 9M KOH + 5 wt% ZnO; (b) 9M KOH + 5 wt% ZnO + 0.1 wt% EMI-TFSI and (c) 9M KOH + 5 wt% ZnO + 1 wt% EMI-DCA.

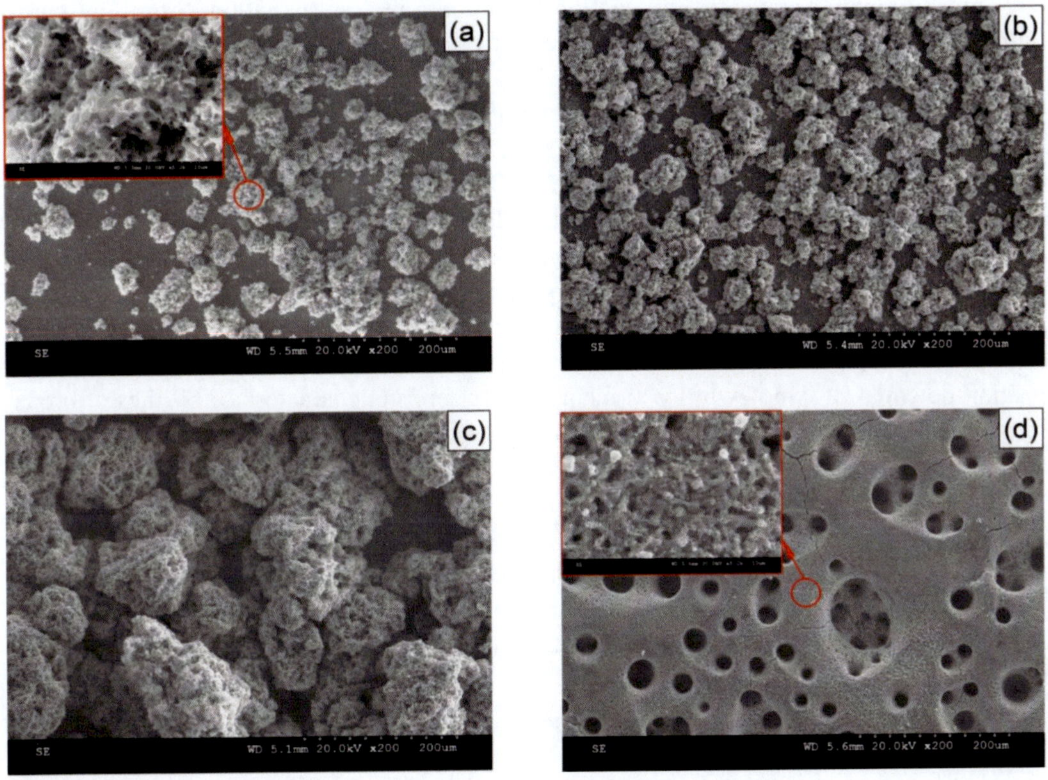

Figure 9. SEM SE images of Zn electrodeposited on Pt coated Si substrates at a constant current density (-20 mA•cm^{-2}) from: (a) 9M KOH + 5 wt% ZnO; (b) 9M KOH + 5 wt% ZnO + 0.01 wt% EMI-DCA; (c) 9M KOH + 5 wt% ZnO + 0.1 wt% EMI-DCA and (d) 9M KOH + 5 wt% ZnO + 1 wt% EMI-DCA.

Zinc dendrites (Figure 8(a)) are clearly observed in deposits from the 9M KOH + 5 wt% ZnO electrolyte with no additives, under constant potential deposition conditions (-0.15 V vs. Zn). The addition of 0.1 wt% EMI-TFSI does not depress the formation of Zn dendrites (Figure 8 (b)). On the other hand, 1 wt% addition of EMI-DCA significantly changes the deposit morphology (Figure 8 (c)). Granular Zn deposits predominate and no dendrites were found in the deposit. EMI-DCA is clearly more effective at suppressing Zn dendrite formation than EMI-TFSI. The next section is, therefore, focused on EMI-DCA as the additive to 9M KOH + 5 wt% ZnO electrolytes.

3.2. Zn Deposit Morphology As a Function of EMI-DCA Concentration and Current Density

The morphology of Zn electrodeposited under constant current density conditions (-20 mA•cm^{-2}) from an electrolyte containing 9M KOH + 5 wt% ZnO with different concentrations (0.01, 0.1 and 1 wt%) of EMI-DCA is shown in Figure 9. Zinc deposits obtained from the 9M KOH + 5 wt% ZnO electrolyte, with no EMI-DCA added, are composed of fibrous clusters dispersed on a continuous Zn layer as seen in Figure 9 (a). It is clear that no Zn dendrites are present, unlike the deposits obtained at a constant potential of -0.15 V vs. Zn (Figure 8 (a)). The potential-current density profiles obtained during electrodeposition may provide an explanation for the morphology variation. From Figure 10 (a), it can be seen that at an applied current density close to -20 mA•cm^{-2}, the corresponding Zn deposition potential in 9M KOH + 5 wt% ZnO is far lower than -0.15 V, i.e., approximately -0.035 V or less. Under these conditions, Zn deposition is not under diffusion control and dendrite initiation is not favored. For Zn deposition from the 9M KOH + 5 wt% ZnO electrolyte, with 0.01 and 0.1 wt% EMI-DCA additions (Figure 10 (b) and (c)), the relationship between deposition potentials and applied current densities is similar to that with no EMI-DCA. Compared with the Zn morphology displayed in Figure 9 (a), there is no significant change in Zn deposit morphology after adding 0.01 and 0.1 wt% EMI-DCA, respectively (Figure 9 (b) and (c)). Only the surface density of the fibrous clusters increases with increasing amount of EMI-DCA added, which may be ascribed to the slight increase in deposition potential. It is noteworthy to mention that deposits characterized by these fibrous clusters show poor adhesion properties. With the addition of 1 wt% EMI-DCA, the Zn deposit is much finer, changing from a fibrous structure to a granular texture that is adherent on the substrate (Figure 9 (d)). As shown in Figure 10 (d), the deposition potential is around -0.1 V vs. Zn in the 9M KOH + 5 wt% ZnO + 1 wt% EMI-DCA electrolyte. As expected, the Zn morphology in Figure 9 (d) is similar to that deposited at a constant potential of -0.15 V vs. Zn observed in Figure 8 (c). It is, therefore, apparent that a critical amount of EMI-DCA, i.e., more than 0.1 wt%, is required to modify the Zn deposit morphology obtained from a 9M KOH + 5 wt% ZnO electrolyte.

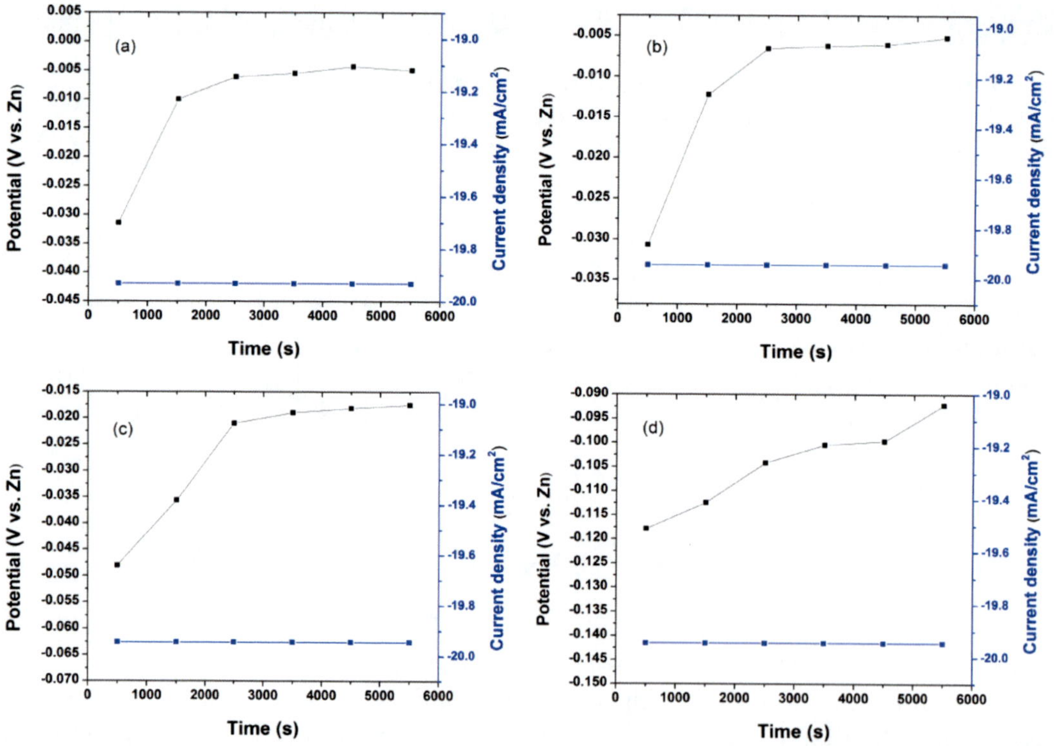

Figure 10. Potential-current density profiles for Zn electrodeposition at a constant current density (-20 mA•cm^{-2}) from: (a) 9M KOH + 5 wt% ZnO; (b) 9M KOH + 5 wt% ZnO + 0.01 wt% EMI-DCA; (c) 9M KOH + 5 wt% ZnO + 0.1 wt% EMI-DCA and (d) 9M KOH + 5 wt% ZnO + 1 wt% EMI-DCA.

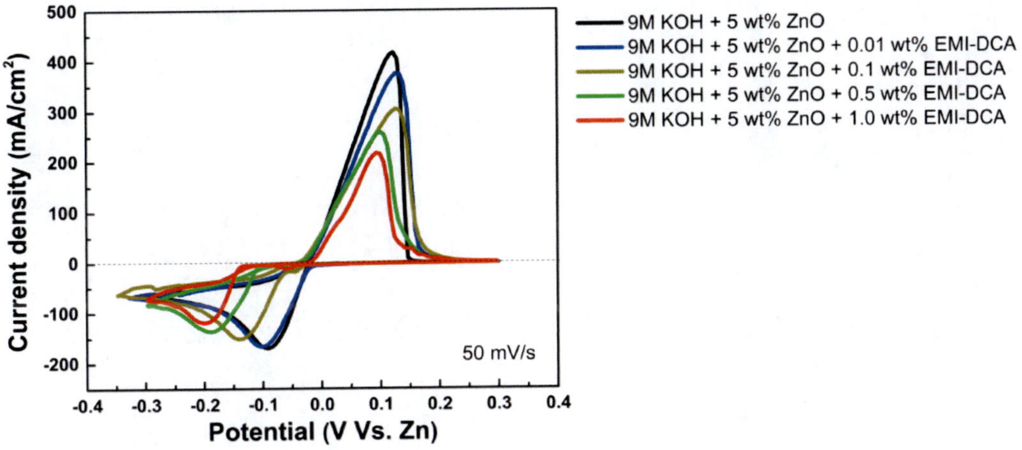

Figure 11. Cyclic voltammetry plots at 50 mV•s^{-1} for Zn/Zn^{2+} redox in 9M KOH + 5 wt% ZnO electrolytes with the addition of different concentrations EMI-DCA.

One adverse effect of adding EMI-DCA to 9M KOH + 5 wt% ZnO is the weakened Zn kinetic performance, as indicated by the decreased current density and increased potential separation between the anodic and cathodic peaks for Zn/Zn^{2+} redox (CV results in Figure 11). In order to maintain good Zn kinetic performance and to achieve desirable Zn deposit morphology at the same time, a concentration of 0.5 wt% (within 0.1-1 wt%) of EMI-DCA was added to the 9M KOH + 5 wt% ZnO electrolyte. Zinc deposits were obtained from 9M KOH + 5 wt% ZnO electrolytes, with and without 0.5 wt% EMI-DCA addition, at applied current densities ranging from -40 to -100 mA•cm^{-2}; the results are shown in Figure 12. The formation of dendritic Zn is preferred at high current densities, while the addition of 0.5 wt% EMI-DCA postponed the initiation of Zn dendrites. As seen in Figure 12 (a) and (c), dendritic formation of Zn deposits occurs in the 9M KOH + 5 wt% ZnO electrolyte when the magnitude of the current density increases from -40 to -60 mA•cm^{-2}. A further increase in the cathodic current density promotes the growth of Zn dendrites (Figure 12 (e) and (g)). The formation of Zn dendrites does not take place in the 9M KOH + 5 wt% ZnO + 0.5 wt% EMI-DCA electrolyte until the applied current density reaches -100 mA•cm^{-2} (Figure 12 (h)). Moreover, it is interesting that at a current density of -80 mA•cm^{-2}, a desirable Zn deposit morphology characterized by homogeneous granular clusters is obtained (Figure 12 (f)) without the presence of cracks (Figure 12 (b)) or dendrites (Figure 12 (h)).

The current section demonstrates that a RTIL can be added to an alkaline aqueous electrolyte to beneficially modify the Zn deposit morphology. It is worth mentioning that there is a threshold concentration for the added RTIL to eliminate the formation of Zn dendrites. As such, hydrophobic RTILs may be of no benefit in modifying Zn deposit morphology because of their low solubility (<0.1 wt%) in aqueous electrolytes. Hydrophilic RTILs appear to be promising additives to alkaline aqueous electrolytes in terms of developing novel electrolyte systems for Zn-anode rechargeable batteries.

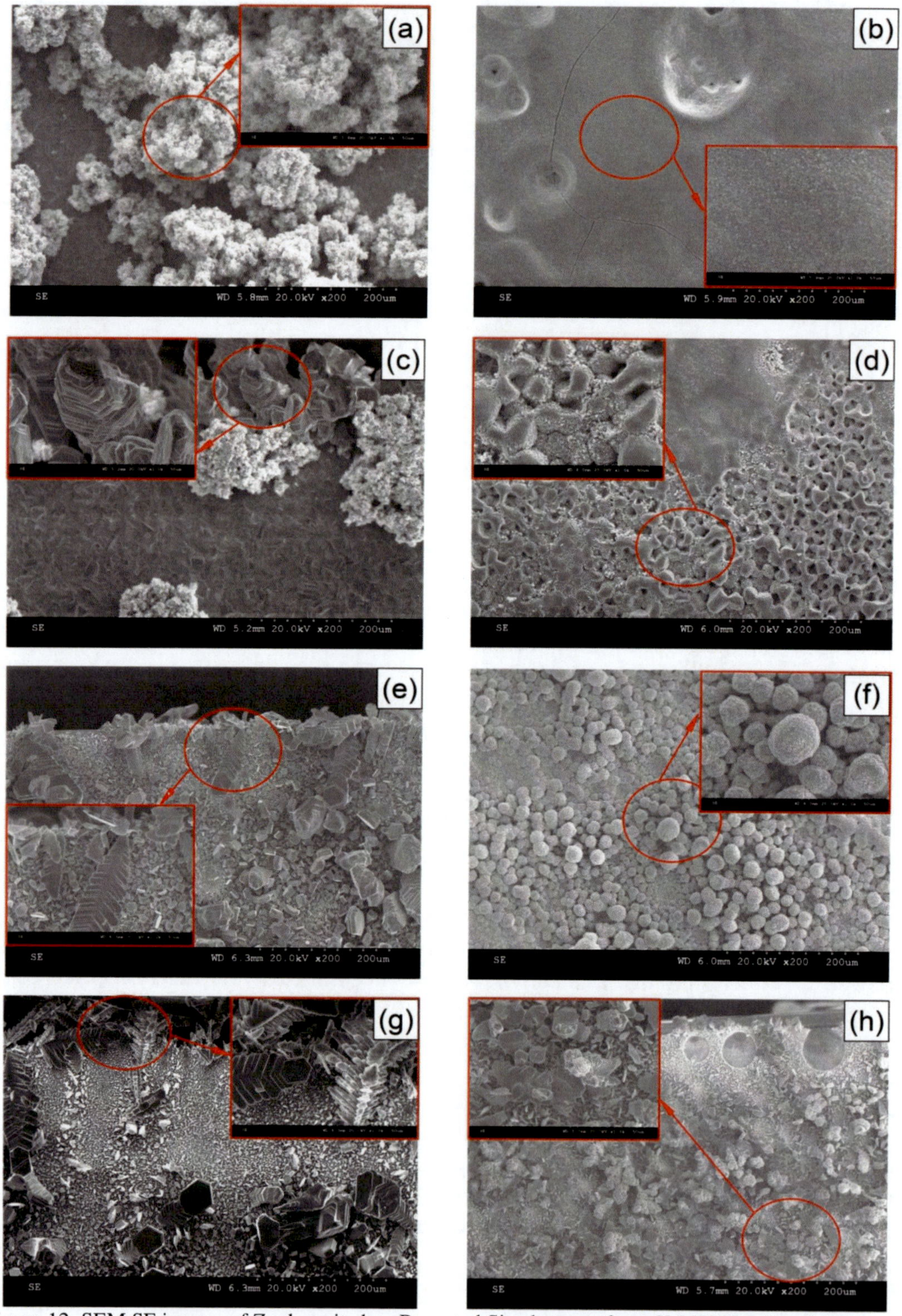

Figure 12. SEM SE images of Zn deposited on Pt coated Si substrates from 9M KOH + 5 wt% ZnO (left column) and 9M KOH + 5 wt% ZnO + 0.5 wt% EMI-DCA (right column) at different current densities: (a) (b) -40 mA•cm^{-2}; (c) (d) -60 mA•cm^{-2}; (e) (f) -80 mA•cm^{-2}; (g) (h) -100 mA•cm^{-2}.

Conclusion

This chapter described the fundamentals and recent progress in developing room temperature ionic liquids (RTILs) as electrolytes or additives to alkaline aqueous electrolytes applied for Zn-anode rechargeable batteries. The following main points were highlighted:

1) For RTILs-based electrolytes, it is necessary to incorporate diluents into the RTILs to improve electrolyte conductivity and Zn/Zn^{2+} redox kinetic performance. Four kinds of RTILs, i.e., BMP-TFSI, MPP-TFSI, EMI-TFSI and EMI-DCA, modified with H_2O and/or DMSO, were investigated. Among the studied RTILs, an electrolyte system composed of EMI-DCA with the addition of both H_2O and DMSO exhibited the best performance with respect to electrolyte conductivity, electrochemical properties for Zn/Zn^{2+} redox and Zn deposit morphology.
2) For conventional alkaline aqueous electrolytes, Zn dendrite formation during electrodeposition can be effectively eliminated by adding the appropriate RTIL as an electrolyte additive. In particular, hydrophilic RTILs are better relative to hydrophobic RTILs in terms of obtaining desirable Zn morphologies and preventing dendritic Zn formation. These results give insights into developing novel alkaline aqueous electrolytes, which are deliberately modified with hydrophilic RTILs, for Zn-anode rechargeable batteries.

In spite of the current progress, many aspects regarding Zn electrodeposition from mixed electrolytes composed of aqueous KOH and RTILs are not fully understood. Additional investigation is required to elucidate the mechanisms for morphology control induced by adding RTILs.

Acknowledgments

The authors acknowledge funding support from the Transportation Development Centre of Transport Canada, the International Zinc Association (IZA) and the Natural Sciences and Engineering Research Council (NSERC) of Canada. The authors would also like to thank Xiao-Zi Yuan and Ken Tsay (NRC) for their assistance during conductivity measurements.

References

[1] K. Bass, P. J. Mitchell, G. D. Wilcox, J. Smith, *Journal of Power Sources*, 35 (1991) 333-351.
[2] D. A. J. Rand, *Journal of Power Sources*, 4 (1979) 101-143.
[3] W. G. Sunu, D. N. Bennion, *Journal of The Electrochemical Society*, 127 (1980) 2007-2016.
[4] E. D. Farmer, A. H. Webb, *Journal of Applied Electrochemistry*, 2 (1972) 123-136.
[5] R. F. Thornton, E. J. Carlson, *Journal of The Electrochemical Society*, 127 (1980) 1448-1452.

[6] Y. Sato, M. Kanda, H. Niki, M. Ueno, K. Murata, T. Shirogami, T. Takamura, *Journal of Power Sources*, 9 (1983) 147-159.
[7] J. McBreen, E. Gannon, *Journal of Power Sources*, 15 (1985) 169-177.
[8] Y. Sharma, M. Aziz, J. Yusof, K. Kordesch, *Journal of Power Sources*, 94 (2001) 129-131.
[9] S. Martirosyan, *Journal of Power Sources*, 172 (2007) 984-987.
[10] G.T. Kim, S.S. Jeong, M. Joost, E. Rocca, M. Winter, S. Passerini, A. Balducci, *Journal of Power Sources*, 196 (2011) 2187-2194.
[11] M. Galinski, A. Lewandowski, I. Stepniak, *Electrochimica Acta*, 51 (2006) 5567-5580.
[12] N. Tachikawa, N. Serizawa, Y. Katayama, T. Miura, *Electrochimica Acta*, 53 (2008) 6530-6534.
[13] M. J. Deng, P. Y. Chen, I. W. Sun, *Electrochimica Acta*, 53 (2007) 1931-1938.
[14] P. Y. Chen, M. J. Deng, D. X. Zhuang, *Electrochimica Acta*, 54 (2009) 6935-6940.
[15] Y. L. Zhu, Y. Kozuma, Y. Katayama, T. Miura, *Electrochimica Acta*, 54 (2009) 7502-7506.
[16] T. I. Leong, Y. T. Hsieh, I. W. Sun, *Electrochimica Acta*, 56 (2011) 3941-3946.
[17] Z. Liu, S. Z. E. Abedin, F. Endres, *Electrochimica Acta*, 89 (2013) 635-643.
[18] Y. Yoshida, O. Baba, C. Larriba, G. Saito, *The Journal of Physical Chemistry B*, 111 (2007) 12204-12210.
[19] P. Hapiot, C. Lagrost, *Chemical Reviews*, 108 (2008) 2238-2264.
[20] X. Qian, B. Han, Y. Liu, H. Yan, R. Liu, *Journal of Solution Chemistry*, 24 (1995) 1183-1189.
[21] I. Persson, *Acta Chemica Scandinavica A*, 36 (1982) 7-13.
[22] M. J. Trahan, S. Mukerjee, E. J. Plichta, M. A. Hendrickson, K. M. Abraham, *Journal of The Electrochemical Society*, 160 (2013) A259-A267.
[23] S. Zein El Abedin, E. M. Moustafa, R. Hempelmann, H. Natter, F. Endres, *Chemphyschem*, 7 (2006) 1535-1543.
[24] A. I. Bhatt, A. S. Best, J. Huang, A. F. Hollenkamp, *Journal of The Electrochemical Society*, 157 (2010) A66-A74.
[25] T. Kuboki, T. Okuyama, T. Ohsaki, N. Takami, *Journal of Power Sources*, 146 (2005) 766-769.
[26] G. B. Appetecchi, G. T. Kim, M. Montanino, F. Alessandrini, S. Passerini, *Journal of Power Sources*, (2011).
[27] K. Xu, *Chemical Reviews*, 104 (2004) 4303-4418.
[28] P. N. Tshibangu, S. N. Ndwandwe, E.D. Dikio, *International Journal of Electrochemical Science*, 6 (2011) 2201 - 2213
[29] K. R. Seddon, A. Stark, M. J. Torres, *Pure and Applied Chemistry*, 72 (2000) 2275-2287.
[30] T. Yamaguchi, K. Mikawa, S. Koda, *Bulletin of the Chemical Society of Japan*, 85 (2012) 701-705.
[31] M. Xu, D. G. Ivey, W. Qu, Z. Xie, E. Dy, X. Z. Yuan, *Journal of The Electrochemical Society*, 161 (2014) A128-A136.
[32] L. Cammarata, S. G. Kazarian, P. A. Salter, T. Welton, Physical Chemistry Chemical Physics, 3 (2001) 5192-5200.
[33] J. R. Scherer, Advances in Infrared and Raman Spectroscopy, in: R. J. H. Clark, R. E. Hester (Eds.), vol. 5, Heyden, London, 1978.

[34] M. Pavlov, P. E. M. Siegbahn, M. Sandström, *The Journal of Physical Chemistry A,* 102 (1998) 219-228.

[35] F. Endres, O. Hofft, N. Borisenko, L. H. Gasparotto, A. Prowald, R. Al-Salman, T. Carstens, R. Atkin, A. Bund, S. Zein El Abedin, Physical Chemistry Chemical Physics, 12 (2010) 1724-1732.

[36] T. J. Simons, A. A. J. Torriero, P. C. Howlett, D. R. MacFarlane, M. Forsyth, *Electrochemistry Communications*, 18 (2012) 119-122.

[37] M. Xu, D. G. Ivey, W. Qu, Z. Xie, *Journal of Power Sources*, 252 (2014) 327-332.

In: Ionic Liquids: Synthesis, Characterization and Applications ISBN: 978-1-63117-716-3
Editor: Anson Brooks

Chapter 6

Ionic Liquids: Promising Solvents and Catalysts for Conversion of Lignocellulose Biomass into Chemicals and Biomaterials

***Huiling Li*[1], *Wenju Wang*[1], *Shuaiyang Wang*[1], *Weiqing Kong*[1], *Junli Ren*[1,*] *and Runcang Sun*[1,2,]**

[1]State Key Laboratory of Pulp and Paper Engineering,
South China University of Technology, Guangzhou, China
[2]Institute of Biomass Chemistry and Utilization,
Beijing Forestry University, Beijing, China

Abstract

Due to the global challenge of the face of depleting fossil carbon resources and growing concerns about environmental issues, the increasing focus has paid on the conversion of the renewable lignocelluloses biomass into biomaterials, chemicals and energy to replace fossil-based products. Recently, ionic liquids (ILs) as desirable green solvents and reaction media have attracted increased interest in the lignocellulose biomass conversion because of their non-volatile, non-flammable, non-corrosive and thermally stable properties. In this chapter, the pretreatment and fractionation of lignocellulosic biomass with ILs as green solvents are summarized up, which are promising and challenging processes in a lignocellulosic biomass biorefining industry. The fractioned main components such as cellulose, hemicelluloses and lignin can efficiently converted into biopolymers and biomaterials with functional properties in ILs as non-derivatizing solvents. Main components derivatives with high DS and high yields could be synthesized by esterification, etherification, oxidation and grafting copolymerization. Moreover, another important focus is discussed on the synthesis of chemicals with low M_w from lignocellulosic biomass in ILs media. The important platform chemicals such as furfural, 5-hydroxymethyl furfural, levulinic acid, 5-ethoxy-methyl-2-furfural, and lignin-depolymerized compounds can be obtained from main components of lignocellulosic biomass using ILs as green reaction media or high efficient catalyst. The difficulties encountered in the process of lignocelluosic biomass conversion

* Correpsonding authors: renjunli@scut.edu.cn.

using ILs are critically discussed. ILs display the great potential for application on a commercial scale due to their broad structural diversity, high dissolution ability, great thermal stability and recyclability.

1. INTRODUCTION

Lignocellulosic biomass is widely available than fossil fuels and serves as the renewable resource for the manufacture of bio-based materials, fuels, energy products, and chemicals considering the uncertain availability of natural energy resources and world wide environmental issues. [1, 2] Since the increasing concerns of environmental problems as well as the demands for healthy food and alternative medicines, novel approaches for the production of high-value added chemicals as a promising alternative of petroleum-based products and clean, safe bio-based materials from lignocellulosic biomass have received immense attentions in recent years. Producing high-value added chemicals and bio-based materials from lignocellulosic biomass is of strategic significance for an environmentally sustainable society.

Lignocellulosic biomass is an abundant resource on the earth and can be generated through photosynthesis worldwide. A large fraction of lignocellulosic biomass was regarded as waste in our normal life and only 3.0% is used by human consumption (food and non-food), leading to an enormous waste of resources. [3] So the conversion of lignocelluloses biomass into bioenergy, chemicals and biomaterials with special function have attracted increasing interests during the last few decades. Great efforts are therefore being made to exploit the potential application of this renewable material for different industries.

Lignocellulosic biomass is composed of three primary components: cellulose, hemicelluloses and lignin. The structures of main components are illustrated in Figure 1.

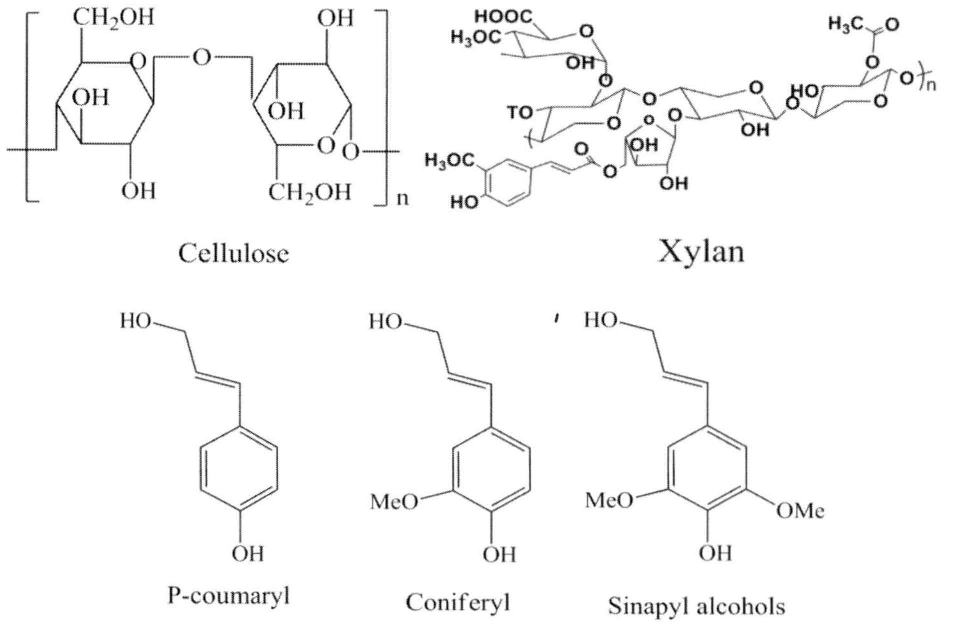

Figure 1. Structures of main components in lignocellulosic biomass.

Lignin is an irregular three-dimensional macromolecule resulted from random dehydrogenation polymerization of phenyl propane building units, together with cellulose and hemicelluloses and provides the structural strength for plants. [4, 5] Cellulose is a crystalline linear polymer of β-D-glucose linked by the uniform 1, 4-glycosidic bond. It is synthesized by protein complexes embedded and traveling in plasma membranes of cells and excreted as a fibrillar from consisting of 24-36 chains into extracellular matrices where the cellulose fibrils are deposited and mixed with other polysaccharides forming cell walls in plants. [6] Hemicelluloses are the second most abundant biopolymer to cellulose in lignocellulosic biomass comprising roughly one-third of most plant materials. [7] Different from cellulose, hemicelluloses represent a type of hetero-polysaccharides with complex structures containing glucose, xylose, mannose, galactose, arabinose, fucose, glucuronic acid, and galacturonic acid in various amounts according to the particular plant species. [8] The hemicellulosic components have the great difference in content and structural features based on the plant species and the present cell walls, such as wheat straw 32.0%, barley straw 32.0%, oat straw 27.0%, rye straw 31.0%, rice straw 25.0%, sunflower husk 23.0%, sugarcane rind 22.0%, and corn cobs 37.0%. [9, 10] In hardwoods, the principal hemicelluloses are an *O*-acetyl-4-*O*-methylglucuronoxylans with amounts between 15% and 30% depending on the species, except for small portions of glucomannan (2–5%). The main hemicelluloses of softwoods are an *O*-acetyl-galactoglucomannan with a content range of 10%-25%. Here the arabino-4-*O*-methyl-glucuronoxylan exits in minor quantities of around 5-10%. Gramineae such as cereal straws, the main hemicelluloses is arabinoxylans (AX), which are similar to harwoods xylan but have the higher amount of L-arabinose.

Recently, ionic liquids (ILs) have received growing attentions as promising green solvents or catalysts for the production of bio-based materials and bioenergy from lignocellulosic biomass. [11-14] ILs, which are also referred to as room temperature ILs, room temperature molten salts (RTILs), organic ionic liquids, etc., are the ionic system that takes on a liquid state at room temperature or slightly warmer. ILs are molten salts comprised entirely of organic cations and inorganic or organic anions with low melting points. [11] Anions and cations of typical ionic liquids as solvents and catalysts for the manufacture of bio-based products from lignocellulosic biomass are listed in Table 1.

Table 1. Anions and cations of typical ionic liquids

Cations	Anions
$[EMIM]^+$	$Cl^-/Br^-/[BF_4]^-/[DEP]^-/Ac^-/[CF_3SO_3]^-$
$[BMIM]^+$	$Cl^-/Ac^-/Br^-/[BF_4]^-$
$[HMIM]^+$	Cl^- /$[HSO_4]^-$
$[C_2OHMIM]$	$[BF_4]^-$
$[C_4SO_3HMIM]^+$	$[CH_3SO_3]^-$
$[AMIM]^+/[CMIM]^+/[ABIM]^+/[OMIM]^+$	Cl^-
$[NMP]^+/[DMA]^+$	$[CH_3SO_3]^-$
$[C_4H_6N_2(CH_2)_3SO_3H]_{3-n}^+$	$[H_nPW_{12}O_{40}]^-$
$[TetraEG(mim)_2]^+$	$[OMs]_2^-$
$[ASCBI]^+$	$[CF_3SO_3]^-$

In terms of reducing environmental pollution, ILs are very popular nowadays because of their unique properties including high-boiling, negligible vapour pressures, high thermal stabilities and the ability to interact and disrupt the recalcitrant lignocellulose structure. [12, 13] The most promising ILs in the dissolution of lignocelluloses are imidazolium-based chloride salts: the cations are able to solvate the aromatic character of lignin by means of π–π interaction, whereas the chloride counteranion is usually the most effective in disrupting the extensive inter- and intramolecular hydrogen bonds pattern of cellulose microfibrils. [14, 15] Therefore, utilizing ILs as green solvents or catalysts is expected to be an ideal process for the production of high-value added chemicals and bio-based materials.

In this present paper, a brief review on the pretreatment and fractionation of lignocellulosic biomass in ILs and the conversion of main components from lignocellulosic biomass into bio-based materials and high-value added chemicals in ILs system is presented. Attentions are paid to the derivative reactions of major components from lignocellulosic biomass to product bio-based materials. Thermal-chemical approaches for the conversion of lignocellulosic biomass into high-value chemicals are also be importantly discussed.

2. The Pretreatment and Fractionation of Lignocellulose in Ionic Liquids

Lignocellulose is the most abundant non-edible biomass and can be broken down into cellulose, hemicelluloses, and lignin. Many high value-added chemicals can be obtained from lignocellulose by the biorefinery processes (Figure 2).

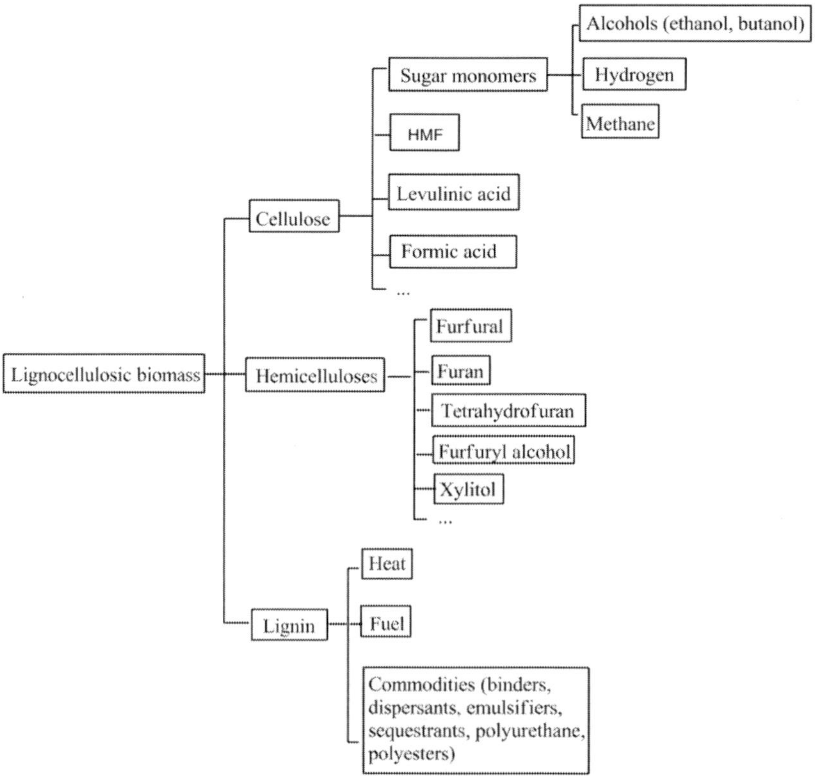

Figure 2. High value-added chemicals obtained from lignocelluloses.

Cellulose, as the most abundant component of the lignocellulose, is mostly hydrolyzed to sugar monomers and subsequently converted into alcohols (ethanol, butanol), hydrogen or methane by fermentation processes, on biochemical platforms. Apart from biofuel and energy production, cellulose can be acid-catalyzed to produce valuable products, such as hydroxymethylfurfural (HMF), levulinic and formic acids. [16] Meanwhile, the applications for hemicelluloses in chemical and pharmaceutical fields have been increasingly emphasized in recent years. A broad range of chemicals such as furfural, furan, tetrahydrofuran, furfuryl alcohol and xylitol can be obtained from hemicellulose. [2] Moreover, lignin is a complex three-dimensional macromolecule composed mainly of three hydroxycinnamyl-type alcohols: coniferyl, sinapyl, and *p*-coumaryl alcohol. It is required not only for the production of heat or fuel but also for the manufacturing of several commodities, such as binders, dispersants, emulsifiers and sequestrants, polyurethane and polyesters. [17] Moreover, fractionated components in lignocellulose can also be used as raw materials to prepare the functional materials in fields of food, water treatment, packaging, papermaking and pharmaceuticals industry as well as in agriculture field. Therefore, the lignocellulose biomass can be widely used as a substitute of fossil-based feedstock.

However, one of the major limitations of the biorefinery process is the complex structure of lignocellulose, which is arduous to realize the fractionation of those three compositions from biomass raw materials and subsequently bioconversion or chemical transformations. Thus, fractionation of cellulose, hemicelluloses and lignin from lignocellulose biomass is the first step to the conversion of lignocellulosic biomass into higher value-added chemicals and biomaterials.

Currently, many processes of biomass pretreatment and fractionation were developed to efficiently overcome the recalcitrance of lignocellulose and reduce the cost of biorefinery processes. Various attempts in the pretreatment and fractionation have been performed, such as liquid hot water [18, 19], dilute sulfuric acid [20-22], alkaline pretreatment [23, 24], ammonia fiber expansion (AFEX) [25, 26], ammonia recycle percolation (ARP) [18, 27] and ILs. [28, 29] Typically, pretreatment aims to convert native lignocellulosic biomass, in which cellulose is recalcitrant to hydrolysis, into a form where hydrolysis is effectively achieved. While fractionation, a newer class of pretreatment, aims to not only improve hydrolysis, but also separate the core constituents (cellulose, hemicelluloses, and lignin) of lignocellulosic biomass, so that each individual component may be more readily processed or functionalized. [30] Among numerous methods for pretreatment and fractionation, the use of ionic liquid has been proposed as solvents in an environment-friendly pretreatment of lignocellulose biomass since ILs are capable of breaking down the extensive hydrogen-bonding network in the polysaccharides and promoting its solubilization. [31]

1-Ethyl-3-methylimidazole acetate [EMIM]Ac is regarded as one of the most suitable ILs for the pre-treatment of lignocellulosic biomass because it presents good solubility properties for lignocellulosic material and also achieves substrate delignification and the decrystallization of cellulose. [32, 33] An efficiency fractionation method to pretreat wheat straw using ionic liquid [EMIM]Ac as a pretreatment agent was developed to fractionate cellulose, hemicellulose and lignin from lignocellulosic biomass. [29] When pretreated with [EMIM]Ac, a suite of feedstocks, such as softwoods [34], hardwoods [35, 36], and agricultural wastes [37, 38] were transferred to biomass that is amenable to enzymatic saccharification and downstream fermentation. [39, 40] Additionally, 1-butyl-3-methylimidazolium chloride ([BMIM]Cl) was employed under microwave irradiation and/or pressure to dissolve lignocellulosic materials. [41] Comparatively, the ionic liquid 1-allyl-3-methylimidazolium chloride ([AMIM]Cl), which was proved to be a good solvent for lignocellulosic biomass [42], was used to directly dissolve the ball milling bamboo, and the dissolved fractions were then regenerated by consecutive alkaline treatments to obtain lignin, hemicelluloses and cellulose-rich fractions. In fact, the fractionation of main components from lignocellulosic biomass after the complete dissolution in ILs is influenced by the different solubility in organic solvent and water under different pH media. Figure 3 depicts the pretreatment of biomass in ILs and the further fractionation of main components.

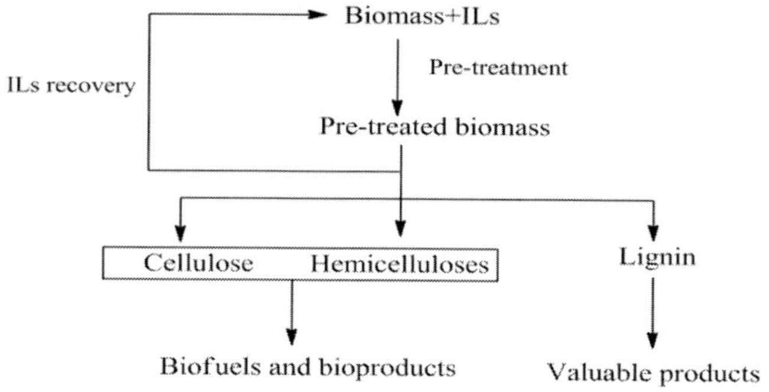

Figure 3. Pretreatment of biomass in ILs.

Pretreatment of lignocellulose with ILs offered advantages over conventional methods: (1) altering physicochemical properties of the biomass macromolecular components, such as a reduction of lignin content [32] and cellulose crystallinity [43]; (2) extracting a specific macromolecular component, such as isolation of lignin [44] and cellulose [46]; (3) performing different fractionation approaches after biomass dissolution in ILs systems. [47, 48, 49] The use of ILs as green solvents for biomass fractionation exhibits a great potential to maximize the recovery of biomass and simultaneously to add value to the fractionated components within the biorefinery concept [50].

3. The Synthesis of Biomaterials from Fractionated Cellulose, Hemicelluloses and Lignin in Ionic Liquids

Cellulose and hemicelluloses belong to polysaccharides, which contain exclusively hydroxyl groups, the modification reactions preferably occur at primary OH groups. Furthermore, there are more active OH groups on lignin. The presence of the active OH groups provides additional active sites for the derivative reaction of cellulose, hemicelluloses and lignin. Modification can create novel opportunities to exploit the various valuable properties of polymers for previously unconceived applications. Recently, the reaction of cellulose, hemicelluloses and lignin from lignocellulosic biomass in different systems has received more attention for the preparation of biopolymers and materials, while their modification reactions in ILs have gotten, comparatively, little attention. The modification reaction of cellulose, hemicelluloses and lignin such as esterification, etherification, oxidation and graft polymerization is summarized below.

3.1. Crosslinking and Graft Copolymerization in Ionic Liquids

Cellulose is one of the most important polysaccharides. Due to its many unique properties, cellulose has a wide variety of current and potential future application. Cellulose-based materials have been used for more than 150 years as food, paper production, biomaterials and pharmaceuticals in industries. The crosslinking reactions happened on the combinations cellulose with crosslinking agents such as methanal, urea, polyprotic acid, etc. The reactions resulted in the decrease of strength, degree of swelling of cellulose. The mechanical properties (deformability and swelling capacity) of cellulose could be changed by graft copolymerization with free radicals and ionic compounds. In recent years, many researchers have paid extensive attentions to the utilization of ionic liquids in the chemical reaction of cellulose due to their higher hydrophilicity, thermal stability, higher reaction rate, and non-flammability.

Ionic liquid can be used as non-derivatizing solvents to dissolve cellulose and then cellulose can be regenerated from ionic liquid to change its some properties and shape. Ionic liquid-regenerated cellulose beads as solid support matrices for papain immobilization were produced. [51] Regenerated cellulose beads were modified with silane. Papain was immobilized on the beads using two different methods including glutaraldehyde and covalent cross-linking method. Glutaraldehyde cross-linking method is more suitable for amino-modified according to the enzyme activity and immobilization rate. Kunchornsup and Sirivat [52] investigated the effects of crosslinking ratio and aging time on the properties of physical and chemical properties of cellulose gels via ionic liquid [BMIM]Cl with glutaraldehyde acting as the crosslinking agent. Results showed that the crosslinking reaction conversion could increase with the increase of the crosslinking ratios and aging time. Moreover, dissolution and homogeneous graft copolymerization of cellulose were performed in an ionic liquid [AMIM]Cl with 2-hydroxyethyl methacrylate. [53] Results showed that [AMIM]Cl dissolved cellulose directly by destroying intermolecular and intramolecular hydrogen bonds and the crystalline form of cellulose was transformed from cellulose I to cellulose II after regeneration from [AMIM]Cl. The rate of grafting reached 77.3%. Poly (L-lactide) was successfully grafted onto cellulose, dissolved in an ionic liquid [AMIM]Cl. The results indicated the grafting rate of the polymer reached 4.44. These materials are biocompatible and biodegradable, and can be dissolved in common organic solvents and water, which provides the basis for its application in drug release systems for specific drug delivery. [54] Cellulose-gaft-poly(acrylamide) and poly (*N*-isopropylacrylamide) have been prepared by radical polymerization, induced by persulfate initiation or γ-ray irradiation in an ionic liquid [BMIM]Cl. [55, 56] ILs might be useful for grafting via atom transfer radical polymerization (ATRP), when preparation of the cellulose macro-initiator and subsequent polymerization can be combined in a one-pot synthesis without previous isolation and purification of the intermediate compound. Cellulose-based macro-initiators, synthesized by homogeneous acylation in ILs, have been applied for subsequent preparation of methacrylate- and styrene graft-copolymers via ATRP in *N,N*-dimethylformamide (DMF) or 1,4-dioxane. [57-60] Therefore, ionic liquids have proved to be efficient solvents for the homogeneous chemical modification of cellulose for improving its some properties to widen the range of applications.

Hemicellulosic materials are abundant and relatively under-utilized polymeric materials present in lignocellulosic materials because of their more complex structure. Zheng et al., [61] have proved that it was efficient to modify the surrounding matrix of hemicelluloses in ionic liquid prior to the enzymatic treatment of the hemicelluloses. Graft copolymerization of hemicelluloses and other polymers was paid great attention to synthesize biomaterials in recent years due to their abundance and biocompatibility. Hemicelluloses were grafted by functional monomers by copolymerization to form the intelligent hydrogels in drug delivery application and waste treatment application. But the graft copolymerization of hemicelluloses occurred in water system. [62-66] Up to now, few studies have been reported about the graft copolymerization of hemicelluloses in ionic liquids. Some researchers have focused on the fractionation of cellulose, hemicelluloses and lignin from lignocelluloses biomass in ionic liquids. [67] Yang et al., [68] investigated the structural characterization of lignin, hemicelluloses and celluloses fractionated by ILs followed by alkaline extraction from bamboo. [AMIM]Cl dissolution and regeneration was confirmed to decrease the total yield of fractionation. Moreover, the results illustrated that during [AMIM]Cl treatment lignin and hemicelluloses were slightly degraded and the side-chains of hemicelluloses were partially cleaved. These indicated that the fractionation of lignocellulose in ILs needs relatively high temperature to entirely dissolve the main components, resulting in the degradation of hemicelluloses and lignin. The grafting copolymerization of hemicelluloses in ILs media will meet the challenge of high temperature of hemicelluloses dissolution, degradation and the isolation and purification of resulting products, which limits the development of grafting copolymerization of hemicelluloses in ILs media.

Lignin is a three-dimensional amorphous polymer based on methoxylated phenylpropanoid units, and it is crosslinking to different polysaccharides with covalent bond. [69] In the production of ethanol from cellulose, the presence of lignin is a barrier to cellulose hydrolysis by cellulase by fermentation. Usually, there are three approaches to fractionate biomass into useful chemicals. One promising procedure is dissolution of biomass using ionic liquid to extract lignin, and then re-precipitate the cellulosic residue. The lignin extraction was investigated by using a purpose designed ionic liquid 1-butyl-3-methylimidazolium acesulfamate [BMIM]Ace at room temperature and open atmosphere. [70] It was synthesized from the sugar acesulfamate potassium, and with good property of greater thermal stability, increased viscosities, and inability to dissolve crystalline cellulose. In addition, it is of low cost and nontoxic, and is expected to interact with lignin owing to its aromatic character of anion. The lignin could be obtained by using acetone as the reconstitution solvent (ILs/acetone 1:3, v/v). The lignin and the reconstitution cellulose-rich material that have been achieved in one dissolution/reconstitution cycle by fresh ILs are approximately 11.5% and 96.5%, respectively. It was found that the presence of some organic solvents mix with ILs can increase the extraction efficiency. The regenerated ionic liquid showed good retention of structure and properties. The regenerated lignin fractions exhibited large molecular weights (M_w, 3700-7810 g mol^{-1}). This designed ionic liquid will have the potential application as the green and efficient solvents for the modification reaction of lignin.

3.2. Oxidation in Ionic Liquids

Chemical modification of cellulose could introduce functional groups into the macromolecules to substitute the free hydroxyl groups in the heterogeneous phase or homogeneous phase, which continues to play a dominant role in improving the overall utilization of cellulosic polymers. However, the intrinsic lack of solubility of native cellulose in water and most organic solvent systems constitutes a major obstacle for cellulose homogeneous modification. [71] As desirable green solvents and reaction media for a wide range of processes, ILs have received significant attention. [72] It was testified that ionic liquid could efficiently dissolve cellulose. [73] Hydrophilic ionic liquids (ILs) are promising solvents for cellulose, which can be used for the formation and conversion of cellulose fibers. [74] In most cases, the side reactions often happen and become predominant during the oxidation reactions of celluloses in the organic solvents. However, the side reactions with the reducing end of cellulose can be completely avoided by applying 2-alkylsubstituted ILs. Ionic liquids with 1-alkyl-3-methyl-imidazolium cations react at C-2 with the reducing ends of cellulose forming a carbon-carbon bond, and are thus no inert cellulose solvents. The reaction is catalyzed by bases, such as commonly present impurities in ILs. [75] This is of importance for the processing of cellulose and its derivatives for use in medical and biological applications where even minor impurities might induce adverse effects. Up to date, the processing of cellulose by its dissolution in ionic liquids is still an entirely new industrial application. [76] The use of ionic liquids as a solvent system enables fibre manufacture in a manner has a positive effect in a direct dissolution process. Ionic liquids are almost entirely recovered (>99.5%) after use in the process and can be reused.

Hemicelluloses are amorphous polymers with heterogeneous branches. The separation of hemicelluloses from wood pulp was studied by means of ionic liquid/co-solvent systems. [77] It was a selective and quantitative separation of a hemicellulose-rich pulp into a hemicellulose fraction, without any depolymerization or yield losses as occurs during conventional dissolving pulp manufacturing processes. Tamarind xyloglucan was oxidized by reaction with sodium hypochlorite in the presence of 2, 2, 6, 6-tetramethyl-1-piperidinyloxy free radical (TEMPO). [78] Galactose residues and non-xylosylated glucose residues were thus converted into galacturonic and glucuronic acid residues, respectively, producing an anionic polysaccharide. Acid hydrolysis of oxidised xyloglucan yielded two aldobiouronic acids, deduced to be β-D-GalpA-(1-2)-D-Xyl and β-D-GlcpA-(1-4)-D-Glc. Anionic xyloglucan had a decreased ability to hydrogen-bond to cellulose and to complex with iodine. It was almost totally resistant to digestion by cellulase [endo-(1-4)-β-glucanase] and did not serve as a donor substrate for xyloglucan endotransglucosylase activity. Anionic derivatives of xyloglucan, acting as acceptor but not donor substrates, may be valuable tools for exploring the biological roles of ionic liquid modified xyloglucan endotransglucosylase/hydrolases (protein) in the integration versus the re-structuring of xyloglucan in the plant cell wall [77].

Oxidation plays an important role in surface modification of lignin as hydroxyl groups are abundant. When hydroxyl groups are oxidized, carbonyl groups are formed, which are more susceptible to reactions as compared to hydroxyl groups. In addition, it increases the hydrophilicty of lignin as well. The different oxidants are used to react with lignin, such as oxygen, hydrogen peroxide, chlorine dioxide, etc. [79] However, there are few attempts on the oxidation reactions of lignin in ionic liquids media. Ambrose et al., [80] studied recyclable ionic liquid tagged co(salen) catalysts for the oxidation of lignin model compounds using air or pure oxygen as the source of oxygen. The results showed good performance in the oxidative degradation reaction of lignin compared with the traditional solvent of NaOH aqueous solution. Moreover, the mixture of product and IL phase was easily separated and had a good reusability. Zakzeski and co-workers [81, 82] effectively performed oxidation of lignin using 1-ethyl-3-methylimidazolium diethyl phosphate ([EMIM][DEP]), cobalt catalyst and molecular oxygen. The benzyl and other alcohol functionalities (i.e., hydroxyl groups at the side-chains) were oxidized to form aldehydes/acids. Mild conditions were used for the oxidation process due to the special properties of IL, which includes high oxygen solubility. The abundance of diethyl phosphate, [DEP], anions from the IL also allows simple metal salts to give exceptionally high activity. Similar oxidation reaction was studied by Kumar, Jain and Chauhan [83] using water soluble iron (III) porphyrins and horseradish peroxidase (HRP) as catalyst, with hydrogen peroxide in IL. The non-coordinating nature and weak nucleophilicity of IL significantly enhanced the activity of water soluble iron(III) porphyrins as compared to aqueous solutions. The stability of HRP was also improved when IL was employed. Additionally, both catalysts can be recycled with appreciable activity for up to five runs.

3.3. Esterification in Ionic Liquids

The ionic liquids as green solvents have proved to be every efficient to synthesize esters of nature polymers such as cellulose and hemicelluloses. Too many papers published have reported the synthesis of cellulose esters in the different ionic liquids system. [84] Cellulose can react with carboxylic acid chlorides or anhydrides at an elevated temperature (80-100 °C) in ILs to form cellulose ester with high degree substitution (DS). Cellulose acetates with high DS of 3.0 was prepared within a short reaction time (0.5-8 h). [85-87] Bacterial cellulose with high degree of polymerization, high crystalliniy and comparably fine fibris could be easily esterified using [BMIM]Cl as the reaction medium. [88] Ionic liquid [AMIM]Cl was confirmed to have the high solubility of cellulose and tosylation and acylation of cellulose in the [AMIM]Cl medium under the mild reaction condition were preformed. [89] Cellulose was modified with bulky acid chlorides in [AMIM]Cl and the products with a DS rang of 0.9-1.4 showed some regioselectivity. [90] Long chain fatty acid esters of cellulose were prepared in ILs but not under completely homogeneous conditions. The acylation of cellulose with the fatty acid chloride lauroyl chloride led to cellulose laurates with DS from 0.34 to 1.54 in [BMIM]Cl. The reaction starts homogeneously and continues heterogeneously, because cellulose esters became increasingly hydrophobic. [91] Cellulose benzoates with different moieties at the aromatic ring have been prepared homogeneously in [AMIM]Cl by conversion of cellulose with benzoyl chlorides. [92] These materials with relatively high DS values between 1.0 and 3.0 might find use as chiral stationary phase in chromatography. Moreover, ionic liquids [AMIM]Cl and [BMIM]Cl have been applied as reaction media for the homogeneous preparation of dicarboxylic acid esters like cellulose phthalates and cellulose succinates. [93, 94] Cellulose dehydroabietates have been prepared by DMAP catalyzed esterification in 1-butyl-3-methylimidazolium bromide ([BMIM]Br). [95] Microwave irradiation can increase yields of products and reduce the reaction time to several minutes. Microwave assisted acetylation of cellulose in 1-allyl-3-butylimidazolium chloride ([ABIM]Cl) has been reported to yield a higher DS of 2.8 compared to the derivatization under conventional heating (DS = 2.2). [96, 97] After the esterification of cellulose, ILs could be recycled by treatment procedures and reused as solvents. Cellulose esters prepared are of great interest for the production of fibers, films, coatings, and additives used in many commercial applications.

Recently, the esterification of hemicelluloses in ILs has gotten ever increasing attentions because higher DS and the reduction of reaction time could be achieved in ionic liquids, compared with other reaction systems. Since Rogers first reported cellulose could be dissolved by ILs, [14] public attentions have focused on the use of ILs on the dissolving of hemicelluloses, lignin and lignocellulose because of their various advantages such as eco-friendliness,[38, 98] lower hydrophobicity and viscosity, enhanced electrochemical and thermal stability, and higher reaction rates. [46, 99,100] However, the applications of ILs on preparation of functional biopolymers and materials from hemicelluloses are very limited. ILs such as [BMIM]Cl was first reported being used in the acetylation of wheat straw hemicelluloses using iodine as a catalyst. [101] It was found that wheat straw hemicelluloses were completely soluble in ionic liquid [BMIM]Cl at 90 °C for 1.5 h, up to 2.6% (w/w) concentration, and the DS of acetylated hemicelluloses between 0.49 and 1.53 could be obtained. The succinoylated hemicelluloses was investigated by using [BMIM]Cl as the reaction medium without the presence of a catalyst to improve the reaction efficiency. A wide DS range between 0.61 and 1.80 could be available at 70-100 °C for 20-60 min. In comparison to the conventional solvents such as the aqueous alkaline system [102] and the DMF/LiCl system [103], a higher DS of 1.80 could be achieved at a shorter scale in [BMIM]Cl, which indicated that ILs created a completely homogenous system for the succinoylation of hemicelluloses. [104] Succinoylated hemicelluloses prepared in ionic liquid [BMIM]Cl without any catalyst, had good metal ions adsorption. [105] The homogeneous esterification of xylan-rich hemicelluloses with maleic anhydride in [BMIM]Cl using LiOH as a catalyst was reported [106]. The novel biopolymers with carboxyl groups and carbon-carbon double bonds can be used as materials for the further copolymerization with monomers at unsaturated double bonds, such as acrylamide, acrylonitrile and crylic acid to form grafted copolymers which could be applied in pharmaceutical, agriculture, food, and wastewater treatment. Hemicelluloses react with long chain anhydride and acyl chlorides to form the hydrophobic materials, which have been widely researched in the DMF/LiCl system and DMSO. [99, 100] Few studies on using ILs as reaction media have been reported. Lauroylated hemicelluloses with DS between 0.43 and 1.82 were synthesized in ionic liquid [BMIM]Cl. [107] Lauroylated hemicelluloses with a DS of 1.82 could be achieved at 90 °C for 60 min. The results indicated that homogeneous modification was successfully conducted and highly substituted hemicelluloses esters were obtained.

Carboxymehyl lignin from organosolv lignin with monochloroacetic acid was prepared. [108] One of the uses for this carboxymethyl lignin produced is as a stabilizing agent in aqueous ceramic suspension. Zhu et al., prepared polylactide-lignin hybrids with M_w of ca. 5.3×10^4 (DPI 2.56) and 77% yield. [109] The tensile strength was approximately doubled upon blending lignin with PLLA. So this reaction endows lignin-based materials with high tensile strength.

3.4. Etherification in Ionic Liquids

Compared with the number of publications describing esterification of cellulose in ILs, studies on the preparation of cellulose ethers in these reaction media are scarce. Cellulose ethers such as carboxymethyl cellulose, hydroxyethyl- and hydroxypropyl celluloses are important and employed in different commercial applications, e.g., as additive in paint, cement, and household products. The synthesis of carboxymethyl cellulose was performed in ionic liquid [BMIM]Cl. [110] Only small DS values below 0.5 could be obtained.

Carboxyethylation and propylation of cellulose has also been studied in [BMIM]Cl and in [AMIM]Cl, but the reactions yielded relatively low DS values that ranged from 0.01 to 1.53. [111] Water-soluble hydroxyalkyl cellulose with a molar degree of substitution of up to 2.79 was prepared under completely homogeneous reaction conditions in various ionic liquids without addition of inorganic bases. [112] Results showed that [EMIM]Ac was the most suitable reaction medium because the acetate anion catalyzes ring opening of the oxiranes and the reaction with the polysaccharide. Tritylation of cellulose is a well-known reaction that produces regioselectively of C-6 substituted cellulose derivatives leaving positions C-2 and C-3 available for subsequent modifications. Tritylation of cellulose in the ionic liquid [BMIM]Cl medium was investigated with triphenylmethyl chloride (tritylchloride, TrCl) in the presence of pyridine. [113] Selectively protected 2,6-di-O-(4-methoxytrityl) cellulose was obtained in one reaction step with a DS of close to 2 in [AMIM]Cl. [114] Trimethylsilylation of cellulose in different 1,3-dialkylimidazolium ionic liquids (IL) with hexamethyldisilazane (HMDS) as a silylating agent was performed. [115] Trimethylsilyl (TMSi) cellulose with a DS greater than 1 was insoluble in the IL. The maximum DS obtained depends on the nature of the anion. Trimethylsilyl cellulose with high DS up to 2.89 could be efficiently synthesized with HMDS in [EMIM]Ac [116].

Up to now, compared to the reports on cellulose ethers prepared in ILs, no studies have been investigated on the etherification of hemicelluloses and lignin in ILs. The etherification will endow biopolymers with some special properties such as better solubility, antimicrobial ability, charge groups, viscosity, and film forming ability. These features provide the extensive opportunity for their application.

3.5. Blending with Other Nature Biopolymer to Form Hydrogels in Ionic Liquids

Gels, are defined as three-dimensional polymer networks swollen by large amounts of solvent. Hydrogels are usually structures formed from natural or synthetic polymers, and contain large amounts of trapped water. [117] Various hydrogels have been fabricated by using hyaluronate, alginate, agarose, starch, gelatin, cellulose, chitosan, and their derivatives. Natural polysaccharides such as cellulose, starch, and chitin are the most abundant organic substances on earth and have received attention for use as resources of new green and sustainable materials because of their properties. [118] Hydrogels prepared from plant fiber and the relevant cellulose derivatives have many other potential applications, such as invisible eyes, wound dressing, super-absorbent and drug release systems because of their perfect hydrophily. [32-36] Up to now, a few reports have focused on the blending of natural polymer. A novel adsorbent, biodegradable collagen/cellulose hydrogel bead (CCHB) was prepared by dissolving and blending cellulose and collagen in [BMIM]Cl. [119] CCHB was first prepared using [BMIM]Cl, and the adsorption behavior for Cu(II) ions was investigated. This work revealed the potential application of CCHBs as a heavy-metal adsorbent due to their biodegradation, reusability, and low cost. Izawa et al., [120] reported a facile preparation of a manipulatable gel of cellulose with [BMIM]Cl. When a solution of cellulose in [BMIM]Cl (13.0 wt% (0.15: 1, w/w)) was left standing at room temperature for one week, a gel was gradually formed with exclusion of excess [BMIM]Cl from the gel matrix. This procedure has been further applied to cellulose/starch and cellulose/chitin systems as well as carrageenan to produce the corresponding gels with ionic liquids. These work described above will give us an indication that there is the probability of yielding more unique and higher performance gel systems from some polysaccharides using ionic liquids.

4. Important Platform Chemicals Derived from Cellulose, Hemicelluloses and Lignin in Ionic Liquids System

4.1. Chemicals Derived from Cellulose

4.1.1. 5-Hydroxymethylfurfural

5-hydroxymethylfurfural (HMF), which is formed from the dehydration of hexose, has been recognized as a significant biomass-based platform molecule (Figure 4). It can be used to synthesize a broad range of high-performance fuels such as DMF [121], 5-ethoxymethylfurfural (EMF) [122], ethyl levulinate (EL) [123] and linear alkanes [124] and high value-added chemicals such as 2,5-diformylfuran (DFF) [125], 2,5-furandicarboxylic acid (FDCA) [126] and levulinic acid (LA). [127] HMF is simultaneously an aromatic aldehyde, an aromatic alcohol and a furan ring system. Thus, HMF has been called as ''sleeping giant''.

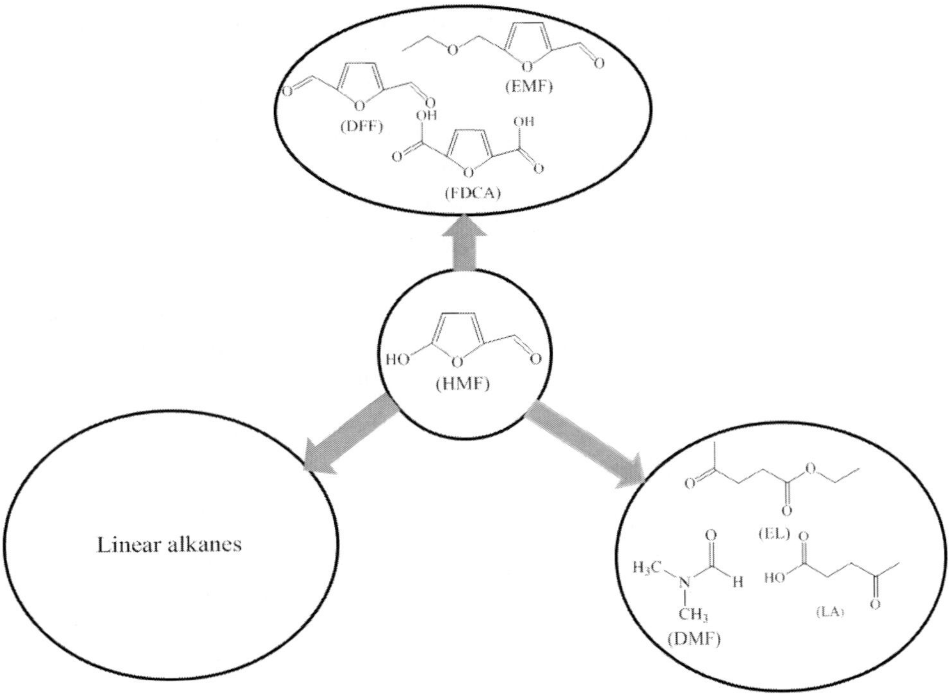

Figure 4. HMF-based applications.

In recent years, the synthesis of HMF from dehydration of biomass-based sugars has been widely studied in various systems including water [128], organic solvents [129, 130], mixture solvents [131], biphasic systems [132], and ionic liquids. [133, 134] Among these systems, ionic liquids have been shown to be an excellent reaction media for many chemical reactions in terms of their transcendent advantages. [135, 136] For example, some ILs can dissolve monosaccharides, polysaccharides, cellulose and even can dissolve lignocellulosic biomass completely, which have already discussed above. Therefore different ILs have been used as the reaction solvent for the dehydration of carbohydrates into HMF. Aside from functioning as a solvent, some ILs have also used as catalysts to promote the formation of HMF. A few ILs acted as both the solvent and catalyst. [137, 138]. Currently, the production of HMF is primarily started from carbohydrates such as fructose, glucose and cellulose. Some studies also used biomass materials as feedstock to produce HMF. Each of these will be discussed in the following sections respectively.

Studies involving the utilization of ionic liquid as reaction media for the dehydration of fructose into HMF have been extensively reported. Lansalot-Matras and Moreau found that 1-butyl 3-methylimidazolium tetrafluoroborate ([BMIM]BF_4) was suitable reaction media for the dehydration of D-fructose to HMF with up to 50% yield in the presence of Amberlyst-15 at 80 ^{o}C for around 3 h. When DMSO was added as a co-solvent in this reaction media, nearly 80% HMF yield was achieved from fructose within 24 h. It was proved that DMSO can prevent the formation of LA and humans. [139] Shi et al., developed a novel catalytic system that consists of [AMIM]Cl and DMF without any additional metal salts or acids to produce HMF and led to reasonable HMF yields (84.9%), which provides a simple, efficient and environmentally friendly way for the production of HMF. [140] A high HMF yield of 93.4% was achieved from fructose in [BMIM]Cl by using lignosulfonic acid (LS) as an acidic catalyst under mild conditions. The catalytic activity of the LS–ILs catalytic system after removing HMF decreased from 77.4 to 62.9% after five cycles. [134] $N_3P_3Cl_6$ as the catalyst and [BMIM]Cl as the solvent were employed for dehydration of fructose to HMF, and it was found that the $N_3P_3Cl_6$/[BMIM]Cl catalytic system was very effective for catalyzing fructose into HMF, yielding 92.8% of HMF in 20 min at the optimized conditions. [20] The conversion of fructose to HMF in [BMIM]Cl with acidic ion-exchange resins D001-cc as catalyst was investigated. At 75 ^{o}C for 20 min, a 93.0% yield of HMF from the dehydration of fructose was obtained. This system of D001-cc resin and [BMIM]Cl exhibited a constant activity with a 86.2% yield of HMF was obtained after seven recycles. [21] Ning et al., synthesized tin(IV) phosphonate (SnBPMA) and zirconium phosphonate (ZrBPMA) to catalyzed fructose into HMF with 1-ethyl-3-methylimidazolium bromide ([EMIM]Br) as solvent, and the yield of HMF could reach 86.5%. The SnBPMA and SnBPMA/[EMIM]Br catalytic system could be reused five times without considerable reduction in catalytic efficiency [22].

Taken together, acid catalyzed dehydration of fructose is the most convenient and efficient method for the synthesis of HMF [141], as excellent yields can be readily achieved. [142-144] However, it should be pointed out that fructose is not abundant in nature and its cost is very high, the two major problems limit the large-scale production of HMF. [145] In contrast, glucose, the most abundant and the cheapest monosaccharide, has been increasingly regarded as a preferred raw material for the preparation of HMF. [122] However, effective production of HMF from glucose is challenging because the isomerization of glucose into fructose is usually achieved in the reaction systems with powerful basicity while the dehydration of fructose into HMF demands an acidic environment.

In 2007, an unprecedented HMF yield of nearly 70% was firstly achieved from glucose in ionic liquid 1-ethyl-3-methylimidazolium chloride ([EMIM]Cl) using $CrCl_2$ as catalyst. [146]

Hu et al., used 1-ethyl-3-methylimidazolium tetrafluoroborate ([EMIM][BF_4]) with $SnCl_4$ as catalyst, for glucose conversion and obtained a 5-HMF yield of ca. 60% at 100 °C for 3 h. [147] Stahlberg et al., examined ionic liquids with lanthanide catalysts, and obtained a 5-HMF yield of 24%. [13] Chidambaram and Bell discovered that 12-molybdophosphoric acid in [EMIM]Cl/acetonitril or [BMIM]Cl/acetonitril can selectively convert glucose to HMF. [148] Zhang and Zhao studied the production of 5-HMF from glucose catalyzed by hydroxyapatite-supported chromium chloride in the ionic liquid [BMIM]Cl, to obtain a 5-HMF yield of 40%. [149] The conversion of glucose into HMF was performed in ionic liquid [BMIM]Cl using carbon-based solid acid as catalysts and 46.4% HMF yield was obtained at 160 °C for 15 min. [150] A method for converting glucose into 5-HMF using ionic liquid–water mixtures with a ZrO_2 catalyst was developed. A 5-HMF yield of 53% was obtained within 10 min at 200 °C in a 50:50 w/w% 1-hexyl-3-methyl imidazolium chloride ([HexylMIM]Cl)-water mixture in the presence of ZrO_2. [151]

Synthesis of HMF directly from cellulose is desirable for large-scale production of HMF because it can reduce the step of the production of glucose and the corresponding cost. However, cellulose is hardly soluble in conventional solvents due to the tight H-bond network and van der Waals interactions in cellulose. It was reported that some hydrophilic ionic liquids and some organic solvents (e.g., *N,N*-dimethylacetamide containing lithium chloride (DMA–LiCl)) are capable of dissolving cellulose well. [152,153] A pair of metal chlorides ($CuCl_2$ and $CrCl_2$) could effectively depolymerize cellulose in [EMIM]Cl, and a high HMF yield of about 55.4 % was obtained at 120 °C for 8 h. [154] Zhang and Zhao studied the conversion of lignocellulose into furans in the presence of $CrCl_3$ in ionic liquids. A HMF yield of 62% was obtained from cellulose in [BMIM]Cl. [155] Similarly, Wang et al., reported a HMF yield of 62.3% with $CrCl_3$/LiCl as catalysts in [BMIM]Cl. [153] Cho and his co-workers found that the joint use of $CrCl_2$ and $RuCl_3$ could result in 60% HMF yield in [BMIM]Cl. [156] Liu et al., investigated the conversion of cellulose into HMF catalyzed by $ZrCl_4$ in ionic liquids [BMIM]Cl. Under optimal conditions, a high HMF yield up to 51.4% was obtained from cellulose. [157] A new process for the efficient conversion of cellulose into HMF was developed by the use of $AlCl_3$ as the catalyst in DMSO-ionic liquid mixtures, achieving a high HMF yield of 54.9% from cellulose in the mixed solvent of DMSO-[BMIM]Cl (10 wt.%). [158]

Apart from fructose, glucose and cellulose, sucrose and inulin are always used as the feedstock for HMF production, which also reached desirable yield of HMF. [39, 40] Nevertheless, the direct utilization of the lignocellulosic biomass to produce chemicals is more meaningful and environmentally friendly than the use of carbohydrates. However, lignocellulosic biomass is relatively recalcitrant and heterogeneous, making its conversion typically inefficient and uneconomical, so lots of efforts have been devoted to conversion of raw biomass materials into chemicals, such as HMF in ionic liquids. Zhang and Zhao studied the conversion of pine wood and corn stalk into HMF in the presence of $CrCl_3$ in ionic liquids [BMIM]Cl, reaching 52% and 47% of HMF yield, respectively. [155] Cho and his co-workers investigated the conversion of lignocellulosic raw material (reed) in [EMIM]Cl using $CrCl_2$/$RuCl_3$ (4:1) at 120 °C for 2 h, reed was converted into HMF and furfural in 41% and 26% of yields, respectively, based on glucan and xylan contents. [156] A technology for direct conversion of citrus peel waste into HMF was developed in ionic liquid 1-octyl-3-methylimidazoliumchloride ([OMIM]Cl) using fluorinated metal catalysts. [159] The HMF yields from lignocellulosic biomass are relatively lower than cellulose and its monosugars due to the different structure.

The functional ILs have crossed the barrier of solvent and entered very successfully into the area of catalyst. Recently, different ILs have been used, along with catalysts, for the dehydration of carbohydrates into HMF. Bao et al., reported that the Lewis acidic 3-allyl-1-(4-sulfurylchloridebutyl)-imidazolium trifluoromethanesulfonate ([ASCBI][CF_3SO_3]) could effectively catalyze the dehydration of D-fructose under microwave irradiation. [160] On the other hand, a Brønsted acidic ionic liquid, 1-H-3-methyl imidazolium chloride ([HMIM]Cl), was also used as both solvent and catalyst for the dehydration of fructose and sucrose to give HMF in relatively high yield and selectivity. [137] Similarly, a series of Brønsted-acidic ionic liquids were synthesized, in which *N*-methyl-2-pyrroli-donium methyl sulfonate ($[NMP]^+[CH_3SO_3]^-$) showed very high catalytic activity and selectivity in the dehydration of D-fructose, and a 72.3% HMF yield and 87.2% selectivity were obtained at 90 °C after 2 h. [133] Dicationic room temperature ionic liquids (RTILs) were employed as catalysts to selective dehydration of fructose and sucrose into HMF. As a result, 92.3% of HMF yield was obtained from fructose in 40 min with one equivalent of $[TetraEG(mim)_2][OMs]_2$ at 120 °C. While, 67.2% of HMF was achieved from dehydration of sucrose at 120 °C in 150 min using two equivalents of triethylene glycol-bis(3-methyli-midazolium) dimesylate $[TetraEG(mim)_2][OMs]_2$. [40] Qu et al., used the room temperature ionic liquids (ILs), 1-hydroxyethyl-3-methylimidazolium tetrafluoroborate ($[C_2OHMIM]BF_4$) as a catalyst in the conversion of glucose to 5-HMF and the yield of 5-HMF reached as high as 67.3% after 1 h at 180 °C in DMSO as solvent. [161] An efficient process was brought forward by Hu and his co-workers for HMF production from carbohydrates using DMSO as the solvent and an acidic ionic liquid 1-carboxypropyl-3-methyl imidazoliumchloride ([CMIM]Cl) as the catalyst. HMF yields of 95.7% and 88.4% were obtained from fructose and inulin when reactions were carried out 120 °C after 120 min and 180 min, respectively. Tao et al., investigated the hydrolysis of cellulose in the presence of catalytic amounts of $FeCl_2$ (or $MnCl_2$) in ILs functionalized with SO_3H, simultaneously using methyl isobutyl ketone (MIBK) as the extracting solvent, and an HMF yield of 34% (or 37%) was achieved. [48, 49] The combination of $ZrOCl_2$ and [CMIM]Cl resulted in a 50.6% yield of HMF from glucose. [39] Ding and co-workers converted microcrystalline cellulose (MCC) to HMF using acidic ionic liquids (ILs) as the catalysts and metal salts as co-catalysts in the solvent of [EMIM]Ac). The results showed that $CuCl_2$ in 1-(4-sulfonic acid) butyl-3-methylimidazolium methyl sulfate ($[C_4SO_3HMIM][CH_3SO_3]$), was found to be an efficient catalyst for catalytic conversion of MCC to HMF, and 69.7% yield was obtained. [162] The direct transformation of cellulose and carbohydrates into HMF was investigated in the solvent [BMIM]Cl using dual-core sulfonic acid ionic liquids (ILs) as catalysts and metal salts as co-catalysts. HMF was directly afforded from cellulose in 66.5% yield using a combination of $[bi\text{-}C_3SO_3HMIM][CH_3SO_3]$ (IL-2) and $MnCl_2$ as catalysts. [163] One-pot production of HMF and EMF from different weed substrates was reported with ionic liquid (ILs) and silica supported immobilized heteropolyacid (HPA) catalysts under microwave-assisted heating. Brønsted acidic ILs catalysts, $[DMA]^+[CH_3SO_3]^-$ and $[NMP]^+[CH_3SO_3]^-$, were found to be effective enabling maximum 58 and 52 wt% HMF yields, respectively, from foxtail weed [164].

4.1.2. Levulinic Acid (LA)

Conversion of carbohydrates into HMF with ionic liquids as solvents or catalysts is effective as discussed above. In acidic aqueous solutions, HMF can undergo further transformation into LA and formic acid (FA) via rehydration. LA has been identified as a promising green, biomass derived platform chemical and it can be produced from hexose sugars by acid-catalyzed hydrolysis. LA can be used for preparation of various high value organic chemicals, polymers, resin, flavor substances, and fuel additives with wide potential industrial applications. [165] The acid-catalyzed route from sugars to LA has recently been proposed as one of the key steps in future biorefineries.

The hydrolysis of cellulose using $MnCl_2$-containing ionic liquids as catalysts was studied and microcrystalline cellulose (MCC) conversion rates of 91.2% and selectivities for HMF, furfural and LA of 45.7%, 26.2% and 10.5%, respectively were achieved. It was found that SO_3H-functionalized ionic liquids showed higher activities than non-functionalized ILs. [138] One-pot synthesis of LA directly from cellulose was realized using heteropolyacid (HPA) ionic liquids $[C_4H_6N_2(CH_2)_3SO_3H]_{3-n}H_nPW_{12}O_{40}$ as catalysts in a water–MIBK biphasic system. The LA from cellulose and starch was realized by this catalyst with yields of 63.1% and 48.7% for reactions of 12 h and 5 h, respectively. [166] A highly selective approach to produce LA from cellulose in SO_3H-functionalized ILs assisted by microwave was developed. [167] Results showed that investigated the effects of reaction conditions and ILs structures on the yield of LA and the highest yield was 55.0%. The catalytic activities of ILs depended on the anions and decreased in the order $HSO_4^- > CH_3SO_3^- > H_2PO_4^-$, which was in good agreement with their acidity order. [167] Studies on the hydrolysis of bamboo shoot shell to LA in the presence of acidic ionic liquid were carried out. A LA yield of 38.2%, were obtained under the optimum conditions [168].

4.1.3. 5-Ethoxy-Methyl-2-Furfural

5-Ethoxy-methyl-2-furfural (EMF), one of the promising next-generation biofuels, was synthesized from biomass substrates. It has a high energy density of 8.7 kwh L^{-1}, which is comparable with those of standard gasoline (8.8 kwh L^{-1}) and diesel fuel (9.7 kwh L^{-1}), and significantly higher than the widely familiar bio-ethanol (6.1 kwh L^{-1}). [169] Surprisingly, the synthesis of EMF from carbohydrates has remained largely unexplored, despite its good prospect as a biofuel.

A previous literature report has shown the production of EMF from corn stover by treating with LiCl in the presence of HCl in chlorinated solvent, followed by further treatment with ethanol. [170] In this process, 72% 5-chloromethylfurfural (CMF) was produced as intermediate from corn stover using concentrated HCl catalyst under high pressure. CMF product was then converted to EMF in nearly quantitative yield by treatment with ethanol. [171] However, there are fewer studies on the conversion of carbohydrates into EMF utilizing ionic liquid as solvent or catalyst. Kim et al., reported the conversion of macro-algae derived agar to a mixture of EMF and EL (5:2) with a total of 30% yield using Dowex 50WX8 solid acid catalyst in the presence of $CrCl_2$ in [EMIM]Cl solvent. [172] Riisager et al., demonstrated that EMF formation depends on the acid strength of ILs catalysts when using a series of task-specific hydrogen sulfate anion, $-SO_3H$, containing ILs. The authors reported that EMF was formed as a major product within 2 h from fructose, glucose and sucrose substrates, but continuing the reaction for 25 h resulted in the formation of EL as the major product. [173] Brønsted acidic ILs, $[DMA]^+[CH_3SO_3]^-$ and $[NMP]^+[CH_3SO_3]^-$, were utilized as catalysts for the direct conversion of waste plant-materials (weeds) to HMF and EMF. It was found that both ILs catalysts were effective for high purity EMF production from HMF and weeds. HMF was quantitatively converted to EMF in 2 h. In the case of weed substrates, EMF was formed as the major product. The ratio of EMF and EL in the isolated product was 7:1. [174].

4.2. Chemicals Derived from Hemicelluloses

As one of the three major components of lignocelluloses, hemicelluloses have been used as plant gum for thickeners, adhesives, protective colloids, emulsifiers, and stabilizers. [175] Recently, a very promising application is the degradation of hemicelluloses into high valued added chemicals such as oligosaccharides, monomeric sugars, xylose, and furfural. [164, 176-179] Monomeric sugar, xylose, can be fermented to a sweetener, xylitol. [180-182] Furfural is a solvent used for the production of lubricants, coatings, adhesives, furan resin, etc. [175, 183] Effective fermentation of xylose to ethanol or organic acids can solve the problems of limited market size for xylose and xylose derivatives. [184-186]

Ionic liquids have been shown to be an excellent reaction media for many chemical reactions in terms of their transcendent advantages. Wang et al., [187] reported that pre-hydrolysis liquor (PHL) was hydrolyzed with 1-methylimidazolium hydrogen sulfate ([HMIM]HSO_4) to prepare furfural and 7.5% yield could be obtained. [HMIM]HSO_4 was conducted as a catalyst instead of traditional reagent concentrated sulfuric acid in the preparation of the furfural process. In order to define a new green catalytic pathway for the production of furfural, the catalyzed conversion of xylan into furfural in [BMIM]Cl was studied by using $AlCl_3$ as catalyst under microwave irradiation. The furfural yield was 84.8% at 170 ^{o}C for 10 s. [188] The effect of $AlCl_3$ on the conversion efficiency of D-xylose and untreated lignocellulosic biomass was also investigated, the yields of furfural from corncob, grass and pine wood catalyzed by $AlCl_3$ in [BMIM]Cl were in the range of 16-33%. [BMIM]Cl and $AlCl_3$ could be recycled for four runs with stable catalytic activity. $AlCl_3$ is less corrosive than mineral acids, and the use of ionic liquid as reaction medium will no longer generate toxic wastewater, thus this reaction system is more ecologically viable. Zhang reported [189] a furfural production in [BMIM]Cl under the presence of MI.$H_3PW_{12}O_{40}$ and a furfural yields were up to 93.7% from xylan at 160 ^{o}C for 10 min. The IL could be easily recycled and reused with stable solvent capacity for multiple runs (5×) after the product furfural was extracted with ethyl acetate. Zhang and Zhao demonstrated $CrCl_3$ could promote biomass (corn stalk, rice straw and pine wood) effectively to produce furfural in yields of 23.0-31.0% under microwave irradiation in ionic liquids. [155].

4.3. Chemicals Derived from Lignin

Lignin is a potential renewable source for many low molecular weight chemicals like benzene, phenol, guaiacol, vanillic acid, methanol, acetic acid, and DMSO. [190] Even though lignin can be applied to various fields, significant hurdles remain before they can be fully realized. One is that lignin is susceptible to self-condensation reactions which counteract depolymerization, and the other is the realization of an efficient, cost-effective, and scalable means of fractionating lignocelluloses into polysaccharide- and lignin-rich streams that enable downstream conversion into fuels and chemicals. Recently, some ionic liquids (ILs) have been shown to be very effective in the delignification of lignocellulosic biomass and depolymerization of lignin [142, 143, 191, 192].

Cox and Ekerdt showed that during ILs dissolution, lignin depolymerization occurred through breakdown of alkyl-aryl ether linkages. [144] Through electro-catalytic oxidative cleavage of lignin. Reichert et al., produced various aromatic compounds like guaiacol, vanilic acid, vanillin, acetovanillone, syringol, syringaldehyde, and syringic acid from alkali lignin. [193] Appropriate ionic liquid triethylammonium methanesulfonate was explored for dissolution of alkali-lignin and used for electrolysis of 5wt.% lignin solution. As appropriate anode material, oxidation-stable ruthenium–vanadium–titanium mixed oxide electrodes were prepared and explored for their electro-catalytic activity. A wide range of aromatic fragments was produced. Syringaldehyde and 2,6-dimethoxy-1,4-benzoquinone (DMBQ) from oxidative depolymerization of Beech lignin were produced in 1-ethyl-3-methylimidazolium trifluoromethanesulfonate ([EMIM][CF_3SO_3]). [192] $Mn(NO_3)_2$ in [EMIM][CF_3SO_3] proved to be the most effective reaction system. A larger scale batch reaction with this system in a 300 mL autoclave (11 g lignin starting material) resulted in a maximum conversion of 66.3 % (24h at 100 °C, 84×10^5 Pa air). By adjusting the reaction conditions and catalyst loading, the selectivity of the process could be shifted from syringaldehyde as the predominant product to DMBQ. Surprisingly, the latter could be isolated as a pure substance in 11.5 wt % overall yield by a simple extraction/crystallization process. Moreover, various lignin derivatives such as guaiacol, syringyl, vanillin, polar lignin were prepared by using ionic liquid [EMIM]Ac as a reaction medium. [194] The different lignin derivatives with low M_w could obtained in ionic liquids.

5. CHALLENGES

Ionic liquids have proved to be efficient solvents for the conversion of lignocellulosic biomass. However, many drawbacks still exist. High cost still limits their industrial application. Purity of ionic liquids also has a significant impact the conversion efficiency of biomass. High viscosity affects the dissolution of biomass and the addition of some organic solvents such as DMSO and DMF can decrease viscosity of ionic liquids. Hydrophobic reagents in hydrophilic ionic liquids easily form heterogeneous reaction. In the process of dissolution and derivatization reaction, biomass could be degraded in ionic liquids at severe condition. In some cases, ionic liquid can participate in the derivazation reaction of biomass, alter the reactivity, or induce the formation of unexpected products. Efficient recycling of ionic liquids is a key issue in industrial scale processes. After modification, some small molecules are difficult to separate. High energy consumption occurs in the recycle procedure. Furthermore, the need to replace volatile organic solvents by preferably less harmful solvents with low vapor pressure is one of the driving forces for the rapid developments in the field of IL research. The evaluation for the toxicity of ionic liquids becomes very important for man and environment. However, ionic liquids have more great features and deserve to being further explored because their structure could be designed and adjusted through molecular design for various applications.

CONCLUSION

In summary, ionic liquids are green and efficient solvents for the derivatives of lignocellulosic biomass and the manufacture of bio-materials and high-value added chemicals, which highly meets the desire of economical and environmental issues. Products with promising properties prepared in ILs systems can be used in different industries such as films, coating, pharmaceutical, food, papermaking, petrochemical engineering and other industries. However, there are still many potential pathways that can be developed. Cost-efficient as well as environmental-friendly reaction approaches will hold considerable promise in the future.

REFERENCES

[1] Dodds D. R.; Gross R. A. *Science* 2007, 318, 1250-1251.
[2] Romaní A.; Garrote G.; Alonso J. L.; Parajó J. C. *Bioresour. technol.* 2010, 101, 8706-8712.
[3] Zhu J.; Pan X.; Zalesny Jr R. S. *Appl. Microbiol. Biotechnol.* 2010, 87, 847-857.
[4] Deng H.; Lin L.; Liu S. *Energy Fuels* 2010, 24, 4797-4802.
[5] Sun R.; Lawther J.; Banks W. *Ind. Crops Prod.* 1997, 6, 1-8.
[6] Kim S. H.; Lee C. M.; Kafle K. Korean *J. Chem. Eng.* 2013, 30, 2127-2141.
[7] Feng P.; Junli R.; Feng X.; RunCang S.; Zhu J.; Zhang X.; Pan X. Sustainable production of fuels, chemicals, and fibers from forest biomass; Chemicals from hemicelluloses: a revies 9780841226432; ACS: Washington, US, 2011, Vol. 1067, pp 219-259.
[8] Hespell R.B. *J. Agric. Food Chem.* 1998, 46, 2615-2619.
[9] Schindler T. M. *Soil Sci. Plant Natr.* 1998, 161, 499-508.
[10] Sun R.; Lawther J. M.; Banks W. *Carbohydr. Poly.* 1996, 29, 325-331.
[11] Sun S. N.;Li M. F.;Yuan T. Q.;Xu F.; Sun R. C. *J. Agric. Food Chem.* 2012, 60, 11120-11127.
[12] Ray D.; Mittal N.; Chung W. *J. Carbohydr. Res.* 2011, 346, 2145-2148.
[13] Ståhlberg T.; Sørensen M. G.; *Riisager A. Green Chem.* 2010, 12, 321-325.
[14] Swatloski R. P.; Spear S. K.; Holbrey J. D.; Rogers R. D. *J. Am. Chem. Soc.* 2002, 124, 4974-4975.
[15] Hassan E. S. R.; Mutelet F.; Pontvianne S.; Moïse *J. C. Environ. Sci. Technol.* 2013, 47, 2809-2816.
[16] Zakrzewska M. E.; Bogel-Łukasik E.; Bogel-Łukasik R. *J. Microencapsul.* 2010, 111, 397-417.
[17] Bonini C.; D'Auria M.; Emanuele L.; Ferri R.; Pucciariello R.; Sabia A. R. *J. Appl. Polym. Sci.* 2005, 98, 1451-1456.
[18] Kim T. H.; Lee Y. *Bioresour. Technol.* 2006, 97, 224-232.

[19] Liu C.; Wyman C. E. *Bioresour. Technol.* 2005, 96, 1978-1985.
[20] Song J.; Zhang B.; Shi J.; Ma J.; Yang G.; Han B. *Chinese J. Chem.* 2012, 30, 2079-2084.
[21] Li Y.; Liu H.; Song C.; Gu X.; Li H.; Zhu W.; Yin S.; Hang C. *Bioresour. Technol.* 2013, 133, 347-353.
[22] Ning H.; Song J.; Hou M.; Yang D.; Fan H.; *Han B. Sci. China Chem.* 2013, 56, 1578-1585.
[23] Kim S.; Holtzapple M. T. *Bioresour. Technol.* 2006, 97, 778-785.
[24] Kim S.; Holtzapple M. T. *Bioresour. Technol.* 2005, 96, 1994-2006.
[25] Chundawat S. P.; Venkatesh B.; Dale B. E. *Biotechnol. Bioengin.* 2007, 96, 219-231.
[26] Teymouri F.; Laureano-Perez L.; Alizadeh H.; Dale B. E. *Bioresour. Technol.* 2005, 96, 2014-2018.
[27] Kim T. H.; Lee Y. Y. *Bioresour. Technol.* 2005, 96, 2007-2013.
[28] Cox B. J.; Ekerdt J. G. *Bioresour. Technol.* 2013, 134, 59-65.
[29] da Costa Lopes A. M.; João K. G.; Bogel-Łukasik E.; Roseiro L. B.; Bogel-Łukasik R. *J. Agric. Food Chem.* 2013, 61, 7874-7882.
[30] FitzPatrick M.; Champagne P.; Cunningham M. F.; Whitney R. A. *Bioresour. Technol.* 2010, 101, 8915-8922.
[31] Sáez F.; Ballesteros M.; Ballesteros I.; Manzanares P.; Oliva J.; Negro M. J. *Chem. Technol. Biotechnol.* 2012, 88, 937-941.
[32] Lee S. H.; Doherty T. V.; Linhardt R. J.; Dordick J. S. *Biotechnol. Bioengin.* 2009, 102, 1368-1376.
[33] Sant'Ana da Silva A.; Lee S.; Endo T.; PS Bon E. *Bioresour. Technol.* 2011, 102, 10505-10509.
[34] Torr K. M.; Love K. T.; Çetinkol Ö. P.; Donaldson L. A.; George A.; Holmes B. M.; Simmons B. A. *Green Chem.* 2012, 14, 778-787.
[35] Kilpeläinen I.; Xie H.; King A.; Granstrom M.; Heikkinen S.; Argyropoulos D. S. *J. Agric. Food Chem.* 2007, 55, 9142-9148.
[36] Wu H.; Morapale M.; Miao J.; Doherty T. V.; Linhardt R. J.; Dordick J. S. *Biotechnol. Bioengin.* 2011, 108, 2865-2875.
[37] Ang T. N.; Ngoh G. C.; Chua A. S. M.; Lee M. G. Biotechnol Biofuels 2012, 5, 67-77.
[38] Barthel S.; Heinze T. *Green Chem.* 2006, 8, 301-306.
[39] Hu L.; Zhao G.; Tang X.; Wu Z.; Xu J.; Lin L.; Liu S. *Bioresour. Technol.* 2013, 148, 501-507.
[40] Jadhav A. H.; Kim H.; Hwang I. T. *Catal. Commun.* 2012, 21, 96-103.
[41] Zhu S.; Wu Y.; Chen Q.; Yu Z.; Wang C.; Jin S.; Ding Y.; Wu G. *Green Chem.* 2006, 8, 325-327.
[42] Zhang H.; Wu J.; Zhang J.; He *J. Macromolecules* 2005, 38, 8272-8277.
[43] Doherty T. V.; Mora-Pale M.; Foley S. E.; Linhardt R. J.; Dordick J. S. *Green Chem.* 2010, 12, 1967-1975.
[44] Tan S. S.; MacFarlane D. R.; Upfal J.; Edye L. A.; Doherty W. O.; Patti A. F.; Pringle J. M.; Scott J. L. *Green Chem.* 2009, 11, 339-345.
[45] Abe M.; Fukaya Y.; Ohno H. *Green Chem.* 2010, 12, 1274-1280.
[46] Ren J.; Sun R.; Liu C.; Chao Z.; Luo W. *Polym. Degrad. Stab.* 2006, 91, 2579-2587.
[47] Darvill J. E.; McNeil M.; Darvill A. G.; Albersheim P. *Plant Physiol.* 1980, 66, 1135-1139.

[48] Tao F.; Song H.; Chou L. *ChemSusChem* 2010, 3, 1298-1303.
[49] Tao F.; Song H.; Chou L. *Carbohydr. Res.* 2011, 346, 58-63.
[50] da Costa Lopes A. M.; João K. G.; Rubik D.; Bogel-Łukasik E.; Duarte L. C.; Andreaus J.; Bogel-Łukasik R. *Bioresour. Technol.* 2013, 61, 7874-7882.
[51] Zhang J. F.; Gong J. X.; Li Q. J.; Yao S. K.; Zhang W. *Adv. Mater. Res.* 2012, 535, 2349-2352.
[52] Kunchornsup W.; Sirivat A. *J. Sol-gel Sci. Technol.* 2010, 56, 19-26.
[53] Xiao S.; Xin T.; He J. *Forestry Studies in China* 2011, 13, 245-251.
[54] Xiao S.; Yuan T. Q.; Cao H. B.; Lin D.; Shen Y.; He J.; Wang B. *Bioresources* 2012, 7, 1748-1759.
[55] Lin C. X.; Zhan H. Y.; Liu M. H.; Fu S. Y.; Huang L. H. *J. Appl. Polym. Sci.* 2010, 118, 399-404.
[56] Hao Y.; Peng J.; Li J. Q.; Zhai M. L.; Wei G. S. *Carbohydr. Polym.* 2009, 77, 779-784.
[57] Sui X. F.; Yuan J. Y.; Zhou M.; Zhang J.; Yang H. J.; Yuan W. Z.; Wei Y.; Pan C. Y. *Biomacromolecules* 2009, 9, 2615-2620.
[58] Meng T.; Gao X.; Zhang J.; Yuan J. Y.; Zhang Y. Z.; He J. S. *Polym.* 2009, 50, 447-454.
[59] Xin T. T.; Yuan T. Q.; Xiao S.; He J. *Bioresources* 2011, 6, 2941-2953.
[60] Lin C. X.; Zhan H. Y.; Liu M. H.; Fu S. Y.; Zhang J. *J. Carbohydr. Polym.* 2009, 78, 432-438.
[61] Zheng Y.; Pan Z.; Zhang R. *Int. J. Agric. Biolog. Engin.* 2009, 2, 51-68.
[62] Peng X. W.; Ren J. L.; Zhong L. X.; Peng F.; Sun R. C. *J. Agric. Food Chem.* 2011, 59, 8208-8215.
[63] Peng X. W.; Zhong L. X.; Ren J. L.; Sun R. C. *J. Agric. Food Chem.* 2012, 60, 3909-3916.
[64] Sun X. F.; Wang H. H.; Jing Z. X.; Mohanathas R. *Carbohydr. Polym.* 2013, 92, 1357-1366.
[65] Voepel J.; Sjöberg J.; Reif M.; Albertsson A. C.; Hultin U. K.; Gasslander U. *J. Appl. Polym. Sci.* 2009, 112, 2401–2412.
[66] Roos A. A.; Edlund U.; Sjoberg J.; Albertsson A. C.; Stalbrand H. *Biomacromolecules* 2008, 9, 2104–2110.
[67] da Costa Lopes A. M.; Joao K. G.; Morais A. R.; Bogel-Lukasik E.; Bogel-Lukasik R. *Sustainable Chemical Processes* 2013, 1, 3-34.
[68] Yang D.; Zhong L. X.; Yuan T. Q.; Peng X. W.; Sun R. C. *Ind. Crops Prod.* 2013, 43, 141-149.
[69] Buxton D.; Russell *J. Crop Sci.* 1988, 28, 553-558.
[70] Sun N.; Rahman M.; Qin Y.; Maxim M. L.; Rodríguez H.; Rogers R. D. *Green Chem.* 2009, 11, 646-655.
[71] Liu C. F.;Zhang A. P.;Li W. Y.;Yue F. X.; Sun R. C. *J. Agric. Food Chem.* 2009, 57, 1814-1820.
[72] Forsyth S. A.; MacFarlane D. R.; Thomson R. J.; Von Itzstein M. *Chem. Commun.* 2002, 714-715.
[73] Turner M. B.; Spear S. K.; Holbrey J. D.; Rogers R. D. *Biomacromolecules* 2004, 5, 1379-1384.
[74] Duchemin B. J.; Mathew A. P.; Oksman K. Compos. Part A *Appl. Sci. Manuf.* 2009, 40, 2031-2037.

[75] Ebner G.; Schiehser S.; Potthast A.; Rosenau T. *Tetrahedron Lett.* 2008, 49, 7322-7324.

[76] Plechkova N. V.; Seddon K. R. *Chem. Soc. Rev.* 2008, 37, 123-150.

[77] Froschauer C.; Hummel M.; Iakovlev M.; Roselli A.; Schottenberger H.; Sixta H. *Biomacromolecules* 2013, 14, 1741-1750.

[78] Takeda T.; Miller J. G.; Fry S. C. *Planta* 2008, 227, 893-905.

[79] Feng X.; Simpson A. J.; Wilson K. P.; Williams D. D.; Simpson M. *J. Nat. Geosci.* 2008, 1, 836-839.

[80] Ambrose K.; Hurisso B. B.; Singer R. D. *Can. J. Chem.* 2013, 91, 1258-1261.

[81] Zakeski J.; Jongerius A. L.; Weckhuysen B. M. *Green Chem.*, 2010, 2010, 1225-1236.

[82] Zakeski J.; Bruijninex P. C. A.; Weckhuysen B. M. *Green Chem.* 2011, 13, 671-680.

[83] Kumar A.; Jain N.; Chauhan S. M. S. *Synlett.* 2007, 3, 411-414.

[84] Gericke M.; Fardim P.; Heinze T. *Molecules* 2012, 17, 7458-7502.

[85] Barthel S.; Heinze T. *Green Chem. 2006*, 8, 301–306.

[86] Cao Y.; Wu, J.; Meng T.; Zhang J.; He J. S.; Li H. Q.; Zhang Y. *Carbohydr. Polym.* 2007, 69, 665–672.

[87] Fidale L. C.; Possidonio S.; El Seoud O. A. *Macromol. Biosci.* 2009, 9, 813–821.

[88] Schlufter K.; Schmauder H. P.; Dorn S.; Heinze T. *Macromol. Rapid Commun.* 2006, 27, 1670–1676.

[89] Granstrom M.; Kavakka J.; King A.; Majoinen J.; Makela V.; Helaja J.; Hietala S.; Viranen T.; Maunu S. L.; Argyropoulos D. S.; Kilpelainen I. *Cellulose* 2008, 15, 481-488.

[90] Xu D.; Li B.; Tate C.; Edgar K. J. *Cellulose* 2011, 18, 405–419.

[91] Barthel S.; Heinze T. *Green Chem. 2006*, 8, 301–306.

[92] Zhang J.; Wu J.; Cao Y.; Sang S.; Zhang J.; He J. *Cellulose* 2009, 16, 299–308.

[93] Liu C. F.; Zhang A. P.; Li W. Y.; Yue F. X.; Sun R. C. *Ind. Crop. Prod.* 2010, 31, 363–369.

[94] Li W. Y.; Jin A. X.; Liu C. F.; Sun R. C.; Zhang A. P.; Kennedy J. F. *Carbohydr. Polym.* 2009, 78, 389–395.

[95] Xu X.; Duan W.; Huang M.; Li G. *Carbohyd. Res.* 2011, 346, 2024–2027.

[96] Possidonio S.; Fidale L. C.; El Seoud O. A. *J. Polym. Sci. Part A* 2010, 48, 134–143.

[97] El Seoud O. A.; da Silva V. C.; Possidonio S.; Casarano R.; Arêas E. P. G.; Gimenes P. *Macromol. Chem. Physic.* 2011, 212, 2541–2550.

[98] Heinze T.; Schwikal K.; Barthel S. *Macromol. Biosci.* 2005, 5, 520-525.

[99] Peng F.; Ren J; Peng B.; Xu F.; Sun R; Sun *J. Carbohydr. Res.* 2008, 343, 2956-2962.

[100] Ren J.; Xu F.; Sun R.; Peng B.; Sun J.. *J. Agric. Food Chem.* 2008, 56, 1251-1258.

[101] Ren J.; Sun R.; Liu C.; Cao Z.; Luo W. *Carbohydr. Polym.* 2007, 70, 406-414.

[102] Sun R.; Sun X.; Bing X. *J. Appl. Polym. Sci.* 2002, 83, 757-766.

[103] Sun R.; Sun X. F.; Zhang F. Y. *Polym. Int.* 2001, 50, 803-811.

[104] Peng X.; Ren J.; Zhong L.; *Sun R. Carbohydr. Polym.* 2011, 86, 1768-1774.

[105] Cheng H. L.; Lin C. X.; Zhu P. A.; Zhan H. Y.; Fu S. Y. *Transations of China Pulp and Paper* 2010, 25, 44-46.

[106] Peng X. W.; Ren J. L.; Sun, R. C. *Biomacromolecules* 2010, 11, 3519-3524.

[107] Wang H. T.; Yuan T. Q.; Meng L. J.; She D.; Geng Z. C.; Sun R. C. *Polymer Degradation and Stability* 2012, 97, 2323-2330.

[108] Cerrutti B. M.; de Souza C. S.; Castellan A.; Ruggiero R.; Frollini E. *Ind. Crops Prod.* 2012, 36, 108-115.

[109] Zhu Y.; Yang S. H., Tham Y. 3rd Eu-CheMS Chemistry Congress, 29 August-2 September, 2012, Numberg, Germany.
[110] Heinze T, Schwikai K, Barhel S. *Macromol. Biosci.* 2005, 5, 520-625.
[111] Mikkola J. P.; Kirilin A.; Tuuf J. C.; Pranovich A.; Holmbom B.; Kustov L.; Murzina D.; Salmia T. *Green Chem.* 2007, 9, 1229- 1237.
[112] Köhler S.; Liebert T.; Heinze T.; Vollmer A.; Mischnick P.; Möllmann E.; Becker W. *Cellulose* 2010, 17, 437–448.
[113] Erdmenger T.; Haensch C.; Hoogenboom R.; Schubert U. S. *Macromol. Biosci.* 2007, 7, 440–445.
[114] Granstrom M.; Olszewska A.; Makela V.; Heikkinen S.; Kilpelainen I. *Tetrahedron Letters* 2009, 50, 1744-1747.
[115] Mormann W.; Wezstein M. *Macromol. Biosci.* 2009, 9, 369–375.
[116] Kostag M.; Kohler S.; Liebert T.; Heinze T. *Macromolecular Symposia*. 2010, 294, 96-106.
[117] Chang C.; Zhang L. *Carbohydr. Polym.* 2011, 84, 40-53.
[118] Rouilly A.; Rigal L. *J. Macromol. Sci. Part C: Polym. Rev.* 2002, 42, 441-479.
[119] Wang J.; Wei L.; Ma Y.; Li K.; Li M.; Yu Y.; Wang L.; Qiu H. *Carbohydr. Polym.* 2013, 98, 736-743.
[120] Izawa H.; Kaneko Y.; Kadokawa J.. *J. Mater. Chem.* 2009, 19, 6969-6972.
[121] Román-Leshkov Y.; Barrett C. J.; Liu Z. Y.; Dumesic J. A. Nat. 2007, 447, 982-985.
[122] Yang Y.; Hu C.; Abu-Omar M. M. *Bioresour. Technol. 2012*, 116, 190-194.
[123] Chang C.; Xu G.; Jiang X. *Bioresour. Technol.* 2012, 121, 93-99.
[124] Dedsuksophon W.; Faungnawakij K.; Champreda V.; Laosiripojana N. *Bioresour. Technol.* 2011, 102, 2040-2046.
[125] Nie J.; Xie J.; Liu H. *Chinese J. Catal.* 2013, 34, 871-875.
[126] Villa A.; Schiavoni M.; Campisi S.; Veith G. M.; *Prati L. Angew. Chem.* (Int. Ed. in Engl.) 2013, 6, 609-612.
[127] Ya'aini N.; Amin N. A. S.; Endud S. *Microporous Mesoporous Mater.* 2013, 171, 14-23.
[128] Wu S.; Fan H.; Xie Y.; Cheng Y.; Wang Q.; Zhang Z.; Han B. *Green Chem.* 2010, 12, 1215-1219.
[129] Caes B. R.; Raines R. T. *ChemSusChem* 2011, 4, 353-356.
[130] Wang P.; Yu H.; Zhan S.; Wang S. *Bioresour. Technol.* 2011, 102, 4179-4183.
[131] Guo X.; Cao Q.; Jiang Y.; Guan J.; Wang X.; Mu X. *Carbohydr. Res.* 2012, 351, 35-41.
[132] Román-Leshkov Y.; Chheda J. N.; Dumesic J. A. *Science* 2006, 312, 1933-1937.
[133] Tong X.; Li Y. *ChemSusChem* 2010, 3, 350-355.
[134] Xie H.; Zhao Z. K.; Wang Q. *ChemSusChem* 2012, 5, 901-905.
[135] Carrera G. V.; Ponte M. N. d.; Branco L. C. *Tetrahedron* 2012, 68, 7408-7413.
[136] Rouch A.; Castellan T.; Fabing I.; Saffon N.; Rodriguez J.; Constantieux T.; Plaquevent J. C.; Génisson Y. *RSC Adv.* 2013, 3, 413-426.
[137] Moreau C.; Finiels A.; Vanoye L. *J. Mol. Cat. A Chem.* 2006, 253, 165-169.
[138] Tao F.; Song H.; Chou L. *Bioresour. Technol.* 2011, 102, 9000-9006.
[139] Lansalot-Matras C.; Moreau C. *Catal. Commun.* 2003, 4, 517-520.
[140] Shi J.; Gao H.; Xia Y.; Li W.; Wang H.; Zheng C. *RSC Adv.* 2013, 3, 7782-7790.
[141] Guo F.; Fang Z.; Zhou T. J. *Bioresour. Technol.* 2012, 112, 313-318.

[142] George A.; Tran K.; Morgan T. J.; Benke P. I.; Berrueco C.; Lorente E.; Wu B. C.; Keasling J. D.; Simmons B. A.; Holmes B. M. Green Chem. 2011, 13, 3375-3385.
[143] Arora R.; Manisseri C.; Li C.; Ong M. D.; Scheller H. V.; Vogel K.; Simmons B. A.; Singh S. *Bioenergy Res.* 2010, 3, 134-145.
[144] Cheng G.; Kent M. S.; He L.; Varanasi P.; Dibble D.; Arora R.; Deng K.; Hong K.; Melnichenko Y. B.; Simmons B. A. *Langmuir* 2012, 28, 11850-11857.
[145] Yan H.; Yang Y.; Tong D.; Xiang X.; Hu C. Catal. Commun. 2009, 10, 1558-1563.
[146] Zhao H.; Holladay J. E.; Brown H.; Zhang Z. C. *Science* 2007, 316, 1597-1600.
[147] Hu S.; Zhang Z.; Song J.; Zhou Y.; Han B. *Green Chem.* 2009, 11, 1746-1749.
[148] Chidambaram M.; Bell A. T. *Green Chem.* 2010, 12, 1253-1262.
[149] Zhang Z.; Zhao Z. K. *Bioresour. Technol.* 2011, 102, 3970-3972.
[150] Hu Z.; Liu B.; Zhang Z.; Chen L. *Ind. Crops Prod.* 2013, 50, 264-269.
[151] Qi X.; Watanabe M.; Aida T. M.; Smith R. L. *Bioresour. Technol.* 2012, 109, 224-228.
[152] Binder J. B.; Raines R. T. J. *Ame. Chemi. Soc.* 2009, 131, 1979-1985.
[153] Wang J.; Xu W.; Ren J.; Liu X.; Lu G.; Wang Y. Green Chem. 2011, 13, 2678-2681.
[154] Su Y.; Brown H. M.; Huang X.; Zhou X. D.; Amonette J. E.; Zhang Z. C. *Appl. Catal. A Gen.* 2009, 361, 117-122.
[155] Zhang Z.; Zhao Z. K. *Bioresour. Technol.* 2010, 101, 1111-1114.
[156] Kim B.; Jeong J.; Lee D.; Kim S.;Yoon H. J.;Lee Y. S.; Cho J. K. *Green Chem.* 2011, 13, 1503-1506.
[157] Liu B.; Zhang Z.; Zhao Z. K. *Chem. Engin. J.* 2012, 215-216, 517-521.
[158] Xiao S.; Liu B.; Wang Y.; Fang Z.; Zhang Z. *Bioresour. Technol.* 2014, 151, 361-366.
[159] Yi Y. B.; Ha M. G.; Lee J. W.; Park S. M.; Choi Y. H.; Chung C. H. *J. Ind. Engin. Chem.* 2012, 19, 523-528.
[160] Bao Q.; Qiao K.; Tomida D.; Yokoyama C. *Catal. Commun.* 2008, 9, 1383-1388.
[161] Qu Y.; Huang C.; Song Y.;Zhang J.; Chen B. *Bioresour. Technol.* 2012, 106, 170-172.
[162] Shi J.; Yang Y.; Wang N.; Song Z.; Gao H.; Xia Y.; Li W.; Wang H. *Catal. Commun.* 2013, 42, 89-92.
[163] Song J.; Fan H.; Ma J.; Han B. *Green Chem.* 2013, 15, 2619-2635.
[164] Keegan T. P.; Dritz S. S.; Nelssen J. L.; DeRouchey J. M.; Tokach M. D.; Goodband R. D. *J Swine Health Prod.* 2005, 13, 12-18.
[165] Cox B. J.; Ekerdt J. G. *Bioresour. Technol.* 2012, 118, 584-588.
[166] Sun Z.; Cheng M.; Li H.; Shi T.; Yuan M.; Wang X.; Jiang Z. *RSC Adv.* 2012, 2, 9058-9065.
[167] Ren H.; Zhou Y.; Liu L. *Bioresour. Technol.* 2012, 129, 616-619.
[168] Zhou C.; Yu X.; Yang H.; Zhang Y.; Wang Y.; Lin L.; Vittayapadung S. *Energy Sources, Part A* 2013, 35, 1852-1862.
[169] Lew C. M.; Rajabbeigi N.; Tsapatsis M. *Ind. Engin. Chem. Res.* 2012, 51, 5364-5366.
[170] Mascal M.; Nikitin E. B. *ChemSusChem* 2009, 2, 859-861.
[171] Mascal M.; Nikitin E. B. *Angew. Chem.* 2008, 120, 8042-8044.
[172] Kim B.; Jeong J.; Shin S.; Lee D.; Kim S.; Yoon H. J.; Cho J. K. *ChemSusChem* 2010, 3, 1273-1275.
[173] Saravanamurugan S.; Nguyen Van Buu O.; Riisager A. *ChemSusChem* 2011, 4, 723-726.
[174] Alam M. I.; De S.; Dutta S.; Saha B. *RSC Adv.* 2012, 2, 6890-6896.
[175] Kamm B.; Kamm M. *Appl. Micro. Biotechnol.* 2004, 64, 137-145.

[176] Wu J.; Zhang J.; Zhang H.; He J.; Ren Q.; Guo M. *Biomacromolecules* 2004, 5, 266-268.
[177] Gröndahl M.; Eriksson L.; Gatenholm P. *Biomacromolecules* 2004, 5, 1528-1535.
[178] Hartman J.; Albertsson A. C.; Lindblad M. S.; Sjöberg J. *J. Appl. Polym. Sci.* 2006, 100, 2985-2991.
[179] Fernandez F.; Hinton M.; Gils B. V. *Avian Pathol.* 2002, 31, 49-58.
[180] Mussatto S. I.; Dragone G.; Roberto I. C. *Biotechnol. Progr.* 2005, 21, 1352-1356.
[181] Walther T.; Hensirisak P.; Agblevor F. A. *Bioresour. Technol.* 2001, 76, 213-220.
[182] Buhner J.; Agblevor F. *Appl. Biochem. Biotechnol.* 2004, 119, 13-30.
[183] Kamm B.; Kamm M.; Schmidt M.; Hirth T.; Schulze M. Biorefineries-industrial processes and products: status quo and future directions; Lignocellulose-based chemical products and product family trees 10.1002/9783527619849; Wiley-VCH Verlag GmbH, Weinheim, Germany, 2006; pp 97-149.
[184] Zhang M.; Eddy C.; Deanda K.; Finkelstein M.; Picataggio S. *Science* 1995, 267, 240-243.
[185] Hahn-Hägerdal B.; Galbe M.; Gorwa-Grauslund M.; Liden G.; Zacchi G. *Trends Biotechnol.* 2006, 24, 549-556.
[186] Wyman C. E. *Trends Biotechnol.* 2007, 25, 153-157.
[187] Wang Y.; Song H.; Hou J. P.; Jia C. M.; Yao S. *Separ Sci. Technol.* 2013, 48, 2217-2224.
[188] Zhang L.; Yu H.; Wang P.; Dong H.; Peng X. *Bioresour. Technol.* 2013, 130, 110-116.
[189] Zhang L.;Yu H.; Wang P. *Bioresour. Technol.* 2013, 136, 515-521.
[190] Gandini A. *Green Chem.* 2011, 13, 1061-1083.
[191] Singh S.; Simmons B. A.; Vogel K. P. *Biotechnol. Bioengin.* 2009, 104, 68-75.
[192] Stärk K.; Taccardi N.; Bösmann A.; Wasserscheid P. *ChemSusChem* 2010, 3, 719-723.
[193] Reichert E.; Wintringer R.; Volmer D. A.; Hempelmann R. *Phys. Chem. Chem. Phy.* 2012, 14, 5214-5221.
[194] Varanasi P.; Singh P.; Auer M.; Adams P. D.; Simmons B. A.; Singh S. *Biotechnol. Biofuels* 2013, 6, 14-21.

In: Ionic Liquids: Synthesis, Characterization and Applications ISBN: 978-1-63117-716-3
Editor: Anson Brooks

Chapter 7

APPLICATION OF IONIC LIQUIDS IN LIGNOCELLULOSE BIOMASS DISSOLUTION, FRACTION AND CHEMICAL MODIFICATION

Wenjiao Ge and Xiaohui Wang*
State Key Laboratory of Pulp and Paper Engineering, School of Light Industry and Food Science, South China University of Technology, Guangzhou, China

ABSTRACT

In respond to the growing need for sustainable development, the efficient utilization of biomass resources is becoming increasingly important. Lignocellulose biomass mainly composed of cellulose, hemicellulose and lignin is an abundant but highly underutilized renewable resource. The limited types of solvents that readily dissolve ligocellulose, the lack of an environmental-friendly method to extract lignocellulose components, and the difficulty in efficiently modifying lignocellulose are the major reasons that limit the further application of lignocellulose biomass. Since ionic liquid had emerged as the functional solvent and reaction media of biomass, the dissolution, fraction and modification of lignocellulose biomass in ionic liquid have been intensively studied in the past decade. To date, hundreds of papers have been published on these topics. ILs have provided a new platform for the "green" and efficient conversion and utilization of lignocellulose resources. In this chapter, the applications of ILs in the dissolution, fraction and chemically modification of lignocellulose biomass will be systematically introduced. The scientific mechanisms behind the dissolution and fraction of lignocellulose will be explained. The latest progress in applications of ILs in modification of lignocellulose will be introduced to the readers.

Keywords: Lignocellulose; Ionic liquids; Dissolution; Fraction; Modification

* Corresponding author: State Key Laboratory of Pulp and Paper Engineering, South China University of Technology, Guangzhou 510640, Guangdong, China. Email: fewangxh@scut.edu.cn.

1. INTRODUCTION

Since the "industrial revolution", fossil fuels have been largely used and the derived energy and chemicals make invaluable contributions to the quality of people's lives. However, serious environmental problems have been emerged, such as global warming, caused by "man-made CO_2" from overuse of fossil fuels. Increased concern about these problems has put pressure on society to find alternative energy sources and feedstocks[1, 2]. It is well known that lignocellulose biomass is the most abundant renewable resource in the world. It consists of three types of polymers (cellulose, hemicellulose, and lignin), which are strongly intermeshed and chemically bonded by non-covalent forces and by covalent cross-linkages [3]. Recently, efficient and comprehensive utilization of lignocellulose has received much attention from researchers and governments because it provides a practical route to maintain social sustainable development by reducing the dependence of traditional chemical and energy industries on non-renewable fossil resources [4]. However, there are a limited number of common solvents in which the lignocellulose is readily soluble, and the efficient modification of lignocellulose to obtain desirable properties is hardly achievable [5]. For example, Heinze and Liebert [6] discovered several aqueous and nonaqueous cellulose solvents, but all of these solvents suffer either from high environmental toxicity or from insufficient salvation power.

In the past few decades, ILshave a large variety of applications in various areas of chemical industries including biomass engineering. In brief, ILs are a group of salts that exists as liquids at relatively low temperatures (<100°C). They were first discovered in 1914 by Walden, however, their huge potential in industry and research was only realized within the last few decades [6]. Compared to the conventional organic solvents, ILs are attractive as potential solvents for lignocellulose due to their numerous advantages. One of the main advantages of ILs is their very low vapor pressure compared to volatile and hence hazardous organic solvents. For example, the vapor pressure of 1-butyl-3-methylimidazolium hexafluorophosphate ([BMIM][PF_6]) at 298.15 K was only 10^{-11}Pa [7]. As a result, ILs may be recycled and repeatedly reused. These properties make ILs environmentally benign "green" solvent. In the meantime, ILs are good solvents for abroad spectrum of inorganic, organic and polymeric materials and are miscible with numerous organic solvents. Therefore, many reactions that were used to be very hard to perform with lignocellulose can now be efficiently carried out homogeneously in ILs. Another prominent advantage of ILs is that they are composed of two parts, which endows them a large synthetic flexibility that is not available for single component molecularsolvents. ILs are also readily prepared from commercially available reagents. Up to now, ILs have been commercially provide from a number of suppliers in a range of different qualities [8].

This chapter aims to provide a summary of our current state of knowledge on the applications of ILs in the dissolution, fraction and chemically modification of lignocellulose biomass, the scientific mechanisms behind the dissolution and fraction of lignocellulose, and the latest progress in applications modification of lignocellulose using ILs as reaction media will be discussed.

2. LIGNOCELLULOSE BIOMASS

Lignocellulose, the major component of biomass, derives from wood, grass, agricultural residues, forestry wastes and municipal solid wastes [3]. It consists of three different types of polymers, the semi-crystalline polysaccharidecellulose, the amorphous multicomponent polysaccharidehemicellulose, and the amorphous phenylpropanoid polymer lignin [9, 10]. Their chemical structures are shown in Figure 1.

Cellulose is the most abundant organic polymer in the world and it has been widely studied during the past decades because it is a biodegradable material and a renewable resource. This polymer makes up about 45% of the dry weight of wood, which is the main structural component of the primary cell walls of plants [3, 11]. Although wood pulp remains the most important raw material for cellulose, it can also be extracted from algae, bacteria, and annual crops. Structurally, cellulose is composed of D-glucose subunits linked by β-1, 4-glycosidic bonds. These form long chains linked together by hydrogen bonds and van der Waals forces. In addition, cellulose can appear in crystalline form, called crystalline cellulose, whereas there area small percentage of non-organized cellulose chains, which form amorphous cellulose. In addition, cellulose is insoluble in water and most common organic so as to make a challenge to find suitable solvents for dissolution and further application.

Hemicellulose is a complex carbohydrate composite that consists of different polymers like pentoses (like xylose and arabinose), hexoses (like mannose, glucose and galactose), and sugar acids. The principal component of hemicellulose from hardwood and agricultural plants is glucuronoxylan, whereas glucomannan is predominant in softwood [9]. In contrast to cellulose, hemicellulose has a small molecular weight and branches with short lateral chains that consist of different sugars, which are easily hydrolyzable polymers [12]. They do not form aggregates, even when they are co-crystallized with cellulose chains [3]. The solubility of the different hemicellulose compounds is in descending order: mannose, xylose, glucose, arabinose, and galactose. And the solubility of higher molecular weight hemicellulose could not be predicted because of unknown melting points [13]. For example, Bobleter et al., [14] concluded that the solubilization of hemicellulose components into water started around 180°C under neutral conditions, while that already from 150°C parts of hemicellulose dissolution according to Garrote et al., [15].

Lignin (after cellulose and hemicellulose) is one of the most abundant polymers in nature. It is present in the cellular wall, conferring structural support, impermeability, and resistance against microbial attack and oxidative stress. In molecular structure, lignin is an amorphous heteropolymer consisting of three different phenylpropane units (*p*-coumaryl, coniferyl and sinapyl alcohol) (Figure 1) that are held together by different kind of linkages [3]. The amorphous heteropolymer is also non-water soluble and optically inactive [12]. According to Bobleter, lignin normally starts to dissolve into water around 180°C under neutral conditions [14]. However, the solubility of lignin in acid, neutral or alkaline environments depends on the precursor (*p*-coumaryl, coniferyl, sinapyl alcohol or combinations of them) of the lignin [16].

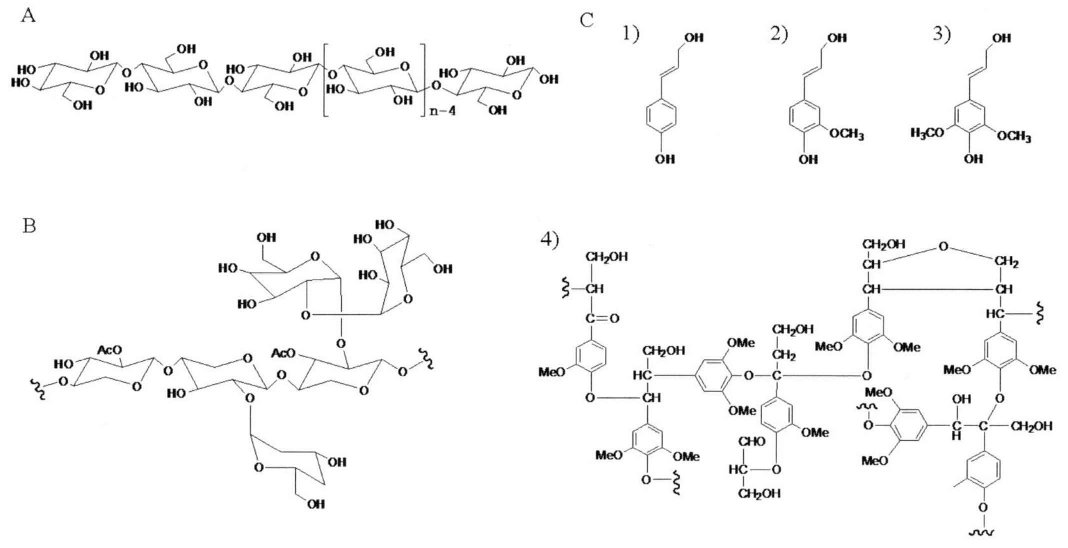

Figure 1. Chemical structures of (A) cellulose; (B) hemicellulose; and (C) lignin. Lignin monomers are: (1) p-coumaryl alcohol; (2) coniferyl alcohol; and (3)sinapylalcohol.

3. ILs

ILs aregenerally defined as salts that melt at or below 100°Cto afford liquids. At least one ion has a delocalized charge and one component is organic, which prevent the formation of a stable crystal lattice. The number of cations and anions composing ILs is very large. The most commonly used ILs [3] are composed by organic cations, such as alkylimidazolium ($[R_1R_2IM]^+$), alkylpyridinium ($[RPy]^+$), tetraalkylammonium ($[NR4]^+$), or tetralkylphosphonium ($[PR_4]^+$), and anions, such as chloride ($[Cl]^-$), bromide ($[Br]^-$), and iodide ($[I]^-$), as well as hexafluorophosphate ($[PF_6]^-$), tetrafluoroborate($[BF_4]^-$), nitrate ($[NO_3]^-$), methanesulfonate($[CH_3SO_3]^-$), trifluoromethane sulfonate(triflate) ($[CF_3SO_3]^-$), and bis-(trifluoromethane sulfonyl)amide ($[Tf_2N]^-$) (Figure 2). ILs have been known for a long time, but their extensive useas solvents hasrecently become significant [17]. The first room temperature IL ethylammonium nitrate ($[EtNH_3][NO_3]$) was discovered in1914 by Walden. However, interest did not develop until the discovery of binary ILs made from mixtures of aluminum(III) chloride and N-alkyl pyridinium or 1,3-dialkylimidazolium chloride. In the early 1970s, Wilkes [18] tried to develop better batteries fornuclear warheads and space probes which required molten saltsto operate. But the molten salts were too hot to keep the nearby materials unaffected. Lately, the chemists found one salt that is liquid at room temperature, which is IL. They continued to improve their IL for use as batteryelectrolytes and then a small community of researchers began to make ILs and test their properties. In the late 1990s, ILsbecame one of the most promising chemicals as solvents. Recently, researchers havediscovered that ILs are more than just green solvents and theyhave found several applications such as replacing them withvolatile organic solvents, making new materials, conducting heateffectively, homogenous and heterogeneous catalysis, biological reactions media and removal ofmetal ions.

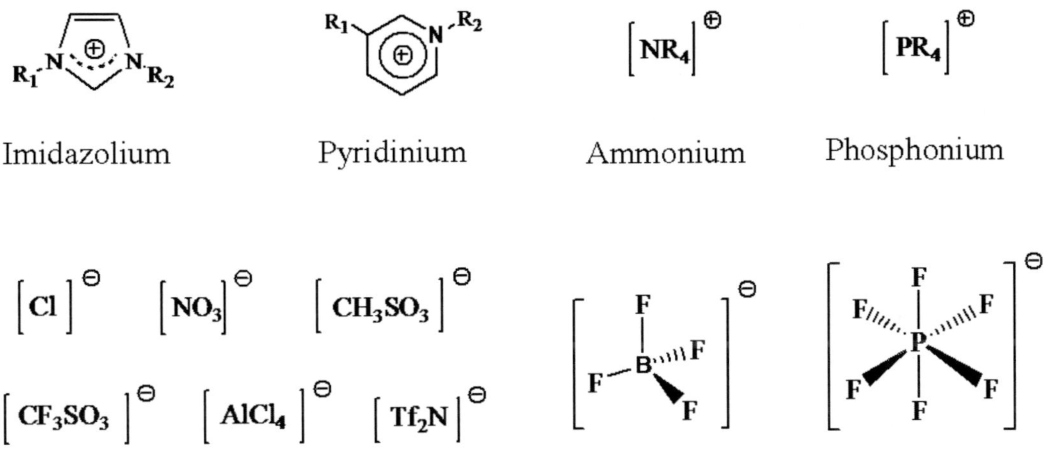

Figure 2. Commonly used cations and anions.

ILs have a range of unique physicochemical properties, such as negligible vapor pressure, good thermal stability, non-flammability and high ionic conductivity, which discriminate them from the traditional volatile organic solvents. The melting point, density, viscosity, thermal stability, polarity, miscibility with water and common organic solvents and the toxicity of ILs are usually concerned when they are referred as the solvent or reaction media of lignocellulose. Some of these properties, such as melting point, polarity, density, hydrophilicity and thermal stability, can be adjusted though modifications of the cation and/or anion of ILs. For example, the meltingpoint (Tm) of ILs decreases when the size and asymmetry of the cation increase [19]. The presence of alkyl ether groups or the increase of alky chain in imidazolium based ILs tend to provoke higher density. The melting point of ILs may also be influenced by the presence or absence of strong hydrogen bonding. The anions tend to form strong hydrogen bonding with the proton of the cations, such as the halide in imidazolium-based ILs, usually contribute to a higher melting point [20].

The viscosity of ILs is higher than that of water and keeps constant with increasing shear rate, which is a major barrier to the commercial application of ILs [19]. However, the viscosity of ILs can be reduced by increasing temperature or decreasing the van der Waals interactions and hydrogen bonding in the ILs molecules [21]. The reported decomposition temperatures of ILs are generally >400°C, with minimal vapor pressure below their decomposition temperatures. The thermal stability of ILs andthe miscibility of ILs with water or organic solvents varies with side chain lengths on the cation and the choice of anion [22, 23].

In a summary, many of the important physico-chemical properties of ILs are structural correlated. For the applications in biomass engineering, the ILs with low melting point, low viscosity, low toxicity and high stability are usually anticipated. Based on the improved understanding to the structure-property relationship, many effective ILs ideal for the biomass dissolution and functionalization have been identified in the past couple of decades.

4. Dissolution and Regeneration

4.1. Dissolution of Lignocellulose Biomass

ILs have been shown to be highly effective at solvating lignocellulose and a wide variety of materialsincluding salts, fats, proteins, amino acids, surfactants and sugars due to their ionic character. The first reports on the solubility of cellulose in ILs were published in the 1930s. Graenacher discovered that molten *N*-ethylpyridinium chloride, in the presence of nitrogen-containing bases, could be used to dissolve cellulose. Thismight be the first example of cellulose dissolution using ILs. However, this case was thought of little practical valuebecause of the relatively high melting point of the salt (120°C) and the volatility of the co-solvents [24]. In 2002, Rogers and his coworkers [25] carried outcomprehensive studies on cellulose dissolution and regeneration in ILs. Cellulose, whether it is refined or natural, can be dissolvedin some hydrophilic ILs such as1-butyl-3-methylimidazoliumchloride ([BMIM][Cl]) and 1-allyl-3-methylimidazolium chloride ([AMIM][Cl]) [25, 26]. Li and Zhao [27] reported cellulose hydrolysis in [BMIM][Cl] with mineral acids as catalyst and up to 44% glucose yield and 77%total reducing sugars were achieved. Amarasekara and Owereh [28] reported the use of Brønsted acidic ILs to dissolveand hydrolyze cellulose under mild conditions, where theILs act as both the solvent and catalyst.

Cellulose solubility and the solution properties can be controlled by selection of the ILs constitutes and ionic parameters, such as cation and anion size as well as hydrogen bond basicity [24, 29, 30]. Relatively small cations are often efficient in dissolving cellulose. It has been reported that $[AMIM]^+$, is more powerful in dissolution of cellulose than $[BMIM]^+$due to its smaller size [31]. Furthermore, 1-ethyl-3-methylimidazolium chloride ([EMIM][Cl]) was better than[BMIM][Cl] since $[EMIM]^+$is smaller than $[BMIM]^+$[32]. However, when $[EMIM]^+$was combined with $[Br]^-$, poor dissolution effect of cellulose wasachieved [33]. Functional groups in the cation can also be crucial for dissolution of cellulose. The presence of a hydroxyl end-group in the cation decreases the solubility of cellulose in the corresponding IL [34]. Usually, a good solvent for cellulose is an IL containing a small polarizing cation combined with $[Cl]^-$, whichis beneficial upon interaction with cellulose via attacking celluloseat its oxygen atoms in the hydroxyl groups [31]. Thethermochemical radii of different anions increase in the following order: $[Cl]^- < [Br]^- < [I]^- < [PF_6]^-$, which is inaccordance with the decreasing cellulose dissolution properties ofthe following ILs: [BMIM][Cl]> [BMIM][I] and [BMIM][Br] (partial solubility). Moreover, [BMIM] $[PF_6]$ and [BMIM]$[BF_4]$ arenot capable of dissolving cellulose. On the other hand, formate and acetate were also foundto be quite powerful in cellulose dissolution when combined with a proper cation [35]. When searching for good ILs for cellulose, it is natural to compare the properties of efficient, non-aqueoussolvents of cellulose with the properties of ILs. Typically, [BMIM][Cl] was applied as the solvent for cellulose dissolution, i.e., up to 25 wt% of cellulose was dissolved in[BMIM][Cl] [25]. Moreover, when applying high power acoustic cavitation ultrasound to the IL and cellulose system, up to 28% of cellulose could be dissolved in [BMIM] [Cl] [36]. The reason for this can be related to the high hydrogenbond basicity of [BMIM] [Cl] as well as its high dipolarity in comparison to other ILssuch as [BMIM] $[BF_4]$ or [BMIM] $[PF_6]$ [37].

In addition, when ILs are considered as cellulose solvents, it is also important to investigate whether the dissolution conditions impact the solubility of cellulose. It has been reported that microwave heating can significantlyaccelerate its dissolution. For example, the solubility of cellulose with a DP of 1000 could be increased by 150% [25]. It has also been reported that solutions containing up to 25 wt% cellulose can be prepared in [BMIM][Cl] under microwave heating. The high chloride concentration in [BMIM][Cl], plays an important role in its dissolution. The active chloride is considered highly effective in breaking theextensive hydrogenbonding network present in cellulose [38]. However, sonication seems only to have little benign influence on the dissolution [25]. Moreover, the presence of water in [BMIM][Cl] greatly decreases the solubility of cellulose through competitively hydrogen bonding its microfibrils. Cellulose is disordered in its [BMIM][Cl] solution and this cellulose solvation process was confirmed at atomic levelby high-resolution ^{13}C NMR studies [39].

Hemicellulose is the second most common polysaccharides in nature and represent about 20–35% of lignocellulosic biomass [40]. Recently, the use ofILs as solvent of hemicellulose andenvironmentally benign reactionmedia starts attracting interest. Zhao et al., [41] examines the effectof IL-pretreatment on the enzymatic hydrolysis of purified xylan (as a model system of hemicellulose). In his study, the xylan was first dissolved in ILs containing [BMIM][Cl],[EMIM][OAc]and [BMIM][OAc]. However, the IL-pretreatment resulted in little improvement in the hydrolysis of xylan. Liu et al., [42] have successfully prepared a series ofnew ILs from renewable biomaterials by simple neutralizationreactions. Most of these ILs dissolved lignin efficiently and selectively (with solubilities of 140–220 mg of lignin perg of IL). The solubility of xylan in these ILs (which rangedfrom<1to85mgg^{-1}) depended on the nature of the anion, while cellulose was scarcely soluble (<5mgg^{-1}).

The dissolution of lignin was investigated in several imidazolium salts containing methyl-, ethyl-, allyl-, butyl-, hexyl- andbenzyl groups in the imidazolium ring coupled to a number of common anions, such as [Cl]$^-$[Br]$^-$, [BF_4]$^-$, [Ac]$^-$,[Tf_2N]$^-$[43, 44]. A recent interesting trial is represented byTan et al., in which [EMIM]$^+$ was coupled to a mixture of alkylbenzene sulfonates with xylene sulfonateas the main anion for the extraction of lignin from sugarcane plantwaste (bagasse). An extraction efficiency of over 93% was reported. Furthermore, polycyclic amidine bases containing different cations have found application for the efficient dissolve of lignin. The two most powerful solventsfor lignin were [BMIM][Tf_2N] and 1,3-dimethylimidazolium methyl sulphate ([EDMIM][CH_3SO_3]) [44], followed by several nearly equallygood solvents, such as [BMIM][CH_3SO_3] [45], [EMIM][OAc] and [AMIM][Cl] [44]. The direct comparison ofthe dissolution power is difficult in some cases, e.g., for [BMIM][Tf_2N] and [OMIM][Tf_2N][45] because both the dissolution temperatures and the lignin raw material weredifferent. In spite of this, however, it has been reported that the latter onewas efficient in lignin dissolution already at 50°C [45]. 1-benzyl-3-methylimidazolium chloride [BZMIM][Cl] and [BMIM][Cl] are reported to be relatively poor solvents for lignin. The former one has a benzyl group containingunsaturated bonds, which are similar to those in lignin. This fact should help in the interaction of ILs with lignin, but actually the effect wasminor. The latter one is a good solvent for cellulose, but not very good for lignin [44]. Other poor ILs solventsfor lignin are [BMIM][Br], [BMIM][BF_4], [BDMIM][BF_4]and [BMIM][PF_6]. When attempting to generalize the lignin dissolution power of ILs, It can be claimed that the large non-coordinating anions, such as [BF_4]$^-$and [PF_6]$^-$as well as [Br]$^-$ were not favorable, while the chloride anion was much more powerful. Since an allyl chain together with a chloride anion in [AMIM][Cl] rendered the IL a goodsolvent for lignin, the unsaturated group in the allyl chain was probably able to interact with lignin. Additionally, the solubilities of lignin were rationalized by comparing its Hildebrandsolubility parameterto those of the ILs. The Hildebrand solubility parameter for lignin was determined to be 24.6. When thevalue of this parameter is close to that of the IL, convenientdissolution of lignin takes place. The Hildebrand solubility parameter for [BMIM][OTf] was measured as 24.9, which is close to thatof lignin [44, 46].

4.2. Dissolution Mechanisms

The mechanism of cellulose dissolution has not been thoroughly understood. There are a few publications in which cellulosedissolution has been studied by NMR techniques [47, 48]. For the dissolution of cellulose, it is generally considered that the basicity of the anion and its ability to accept hydrogen bonds and thus disrupt the interpolymer attractions in cellulose were the main factors to be considered in the design of ILs to dissolve cellulose [5]. The use of anions resulted in primarily hydrophilic salts, and therefore the regeneration of cellulose from the IL solution was simply carried out by the addition of water to the solution. As a further proof of complete dissolution, the dissolved cellulose was reconstituted in amorphous or cellulose II form, which was distinct from the native cellulose I form. Figure 3 shows the proposed dissolution mechanism of cellulose in ILs [49, 50]. The oxygen and hydrogen atoms of the cellulose form electron donor-electron acceptor complexes with the charged species of the IL. It has been suggested that the donor-acceptor interaction occurs primarily between the C-6 and C-3 hydroxyl groups of neighbored cellulose chains [32]. This interaction results in the separation of hydroxyl groups from the different cellulose chains leading to dissolution of the cellulose in the IL [31, 49]. The dissolution of celluloseinvestigated by^{13}C NMR and$^{35/37}$Cl NMR spectrometric techniquesrevealed that the hydrogen bonds in cellulose were disrupted when it was dissolved in [BMIM][Cl][48]. According to these NMR results, it was concluded that chloride anion in [BMIM][Cl] can interact with the hydroxyl groups of thesugar units at 1:1 stoichiometry. Furthermore, cellobiose solvation was studied in [EMIM][OAc] by in situ and variable-temperature NMR [47]. Cellobiose is the repeating unit of cellulose. Theresults revealed that the main reason for cellobiose solvation was the formation of hydrogen bonds between the hydroxyls and the aromatic protons in [EMIM]$^+$cation [51]. Compare to that of cellulose, the study on the mechanism ofhemicellulose and lignin dissolution in ILs has been scarcely reported.

4.3. Regenearation

Cellulose could be precipitated from the ILs by the addition of excess water, or other precipitating antisolvents like ethanol, acetone, chloroform, and tetrahydrofuran (THF) [52-54]. The antisolvents are usually polar solvents. When an antisolvent, such as water is added, the ions of the IL are extracted into the aqueous phase. IL molecules become shielded by water molecules, which form hydrodynamic shells around the IL molecules and thus the direct interactions between cellulose and IL molecules are inhibited [35]. The regeneration process of cellulose from ILs is similar to those of cellulose from other solvents.

Figure 3. Proposed dissolution mechanism of cellulose in[BMIM]Cl.

The regenerated cellulose from ILs may have almost the same degree of polymerization and polydispersity as the initial one [55]. However, its morphology is significantly changed and its microfibrils are fused into a relatively homogeneous macrostructure. Figure 4 shows scanning electron micrographs (SEM) of dissolving pulp and regenerated cellulose [25]. The dissolving pulp shows fibers at 300× magnification in the SEM. After regeneration, the morphology of the material was significantly changed. There appeared a rough, but conglomerate texture, in which the fibers are fused into a relatively homogeneous macrostructure. Recent studies indicate that the regenerated cellulose from their solution of ILs, such as [AMIM][Cl] or [BMIM][Cl] [44] exhibited typically lower degrees of crystallinity than native cellulose. The crystallinity index determined from infraredspectra was investigated for the regenerated celluloses obtainedvia dissolving cellulose in different ILs [56]. The highest value was obtained when [BMIM][Cl] was the IL ofchoice [57]. On the other hand, almost identical values for the crystallinity indexes were achieved in the case of [EMIM][OAc] as well as several alkylalkoxy containing acetates [34]. The regenerated cellulose generally displayed about 58~75% lower values for crystallinity when compared with native cellulose. The XRD measurements revealedthat the regenerated cellulose obtained from its dissolution in [AMIM][Cl] contained cellulose type II, corresponding to a 2θvalueof 20.3° [31]. In contrast, the values obtained for native cellulose I in dissolved pulp were at 2θof14.8°, 16.3°, and 22.6° [58].

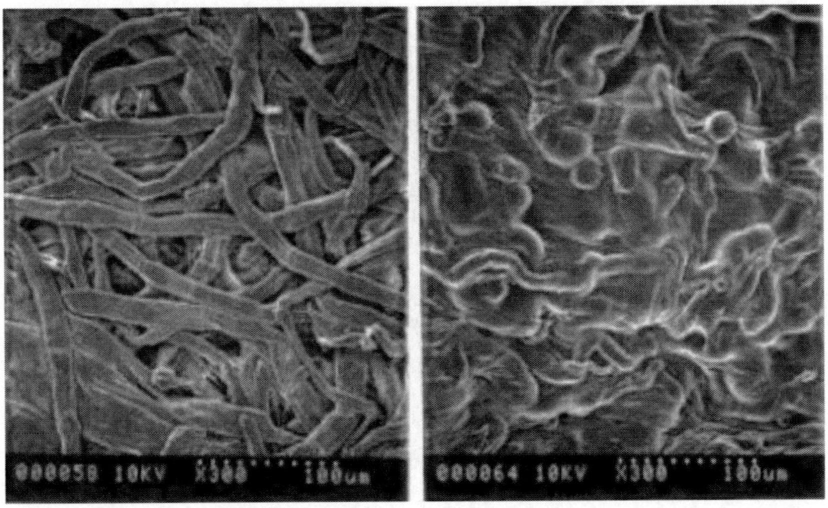

Figure 4. SEM micrographs of the initial dissolving pulp (left) and after dissolution in [BMIM][Cl] and regeneration into water (right).

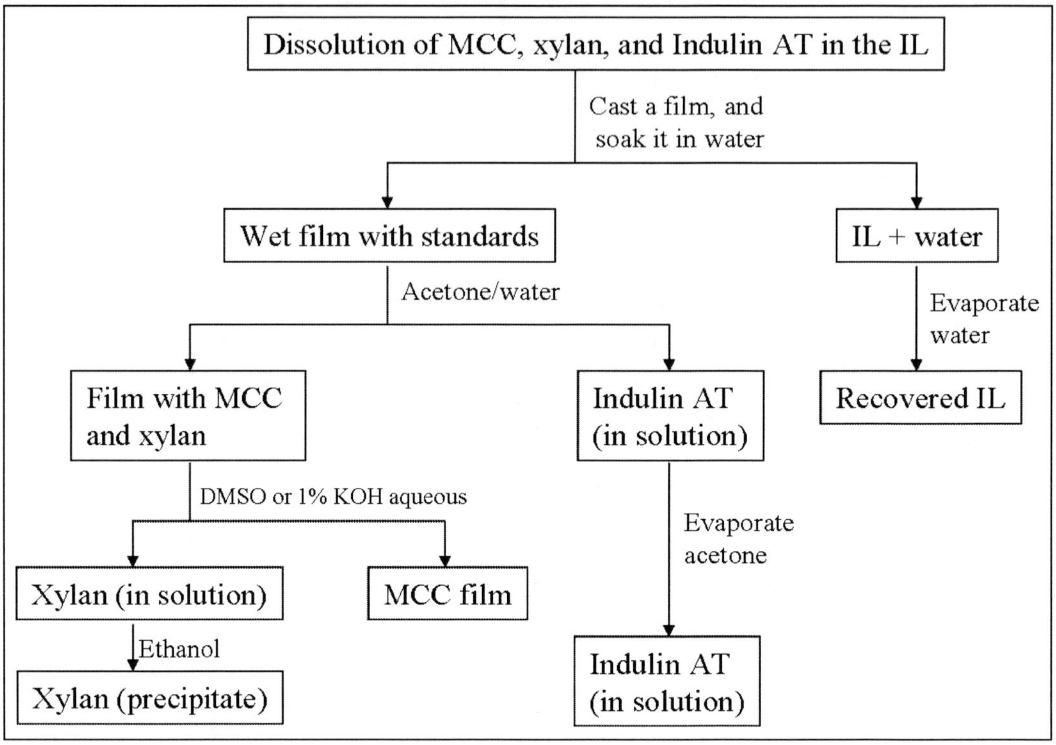

Figure 5. Processes for separation of the mixed biomass standards froman IL-based solution.

The regenerated cellulose can be in a variety of structural forms, such as flocs, beads, film, etc., simply by changing the regeneration methods and conditions. Thin films of (0.1to 0.2) mm are obtained by casting the cellulose solution onglass plates prior to leaching with water [59]. Duchemin et al., [60] sprayed water on the surface of the gel in order to fix theshape of the final sample before immersing it in water forseveral days. They characterized the all-cellulosecomposites and investigated the influence of process parameters, such as dissolution time, precipitation rate, or the initialcellulose concentration of the solvent, on the resultingmaterial properties. All these parameters were foundcorrelated with the mechanical properties of all-cellulose composites.

In addition, Aaltonen et al., [61] reported that nanofibrillar cellulose aerogels can be preparedif the regenerated gel is washed with liquid carbon dioxideprior to supercritical drying. The releasedcarbon dioxide leaves a porous polymer structure resultingin an aerogel. Stiff gels can also be obtained by a slow and gentle precipitation process [50]. However, if theorder of adding the solutions is reversed that is the coagulationsolvent is added to the cellulose solution, the cellulose can be recovered in the form of flocs that can be easily separated by filtration [56].

5. Lignocellulose Biomass Fraction

It is well that known lignocellulose biomass has been proposed as large renewable resources for chemicals and sugars to reducesociety's dependence on nonrenewable petroleum-based feedstocks [62, 63]. However, the components of raw biomass resource with very different properties are firmly connected with each other to form a stable network. And fraction of lignocellulosic materials has become one of most headache problems in the ‘‘green’’ comprehensive utilization of lignocellulosic materials [55, 64]. Since ILs have good solubility to the lignocellulose biomass, they can be applied as a new method of fraction.

The major challenge in processing lignocellulose biomass isassociated with cleaving the primary bonds between thebiopolymer components, prior to their separation [5]. But beforeproceeding to this more difficult step, it is necessary to investigate whether it is availableto carry out theseparation of the free biopolymers from a common ILsolution if those bonds were already cleaved. Thus, Rahmante al. [65]designedand carried out experiments to dissolve and separate microcrystalline cellulose (MCC), xylan (a model hemicellulose), and Indulin AT (a model lignin) using appropriate, simple, recyclable regenerated solvents under mild conditions(room temperature and atmospheric pressure). The biopolymer separation steps described above areillustrated in Figure 5. All the extraction steps were conductedat room temperature, contributing to a lower energy process. Furthermore, all the recovered reagents (acetone, ethanol, the1 wt% KOH solution, the IL and, of course, water) can berecycled [65, 66].

Based on the fact that cellulose can bedissolved in ILs and can be easily regenerated fromthe ILs, cellulose is precipitated from ILs solution by the addition of water, while otherorganic compounds, such as lignin and extractives, still remain in the solution. Myllymaki and Aksela [67] have found a facilemethod for fraction of lignocellulosic materials although their object is to remove lignin from lignocellulosic materials for pulping process. Dadi et al., reported that the dissolve of maple wood powder by [EMIM][OAc]. The crystalline structure of cellulose in wood powder was disruptedwhen the experiment was performed at high temperature of 130°C [52]. Sun et al., [68] obtained cellulose rich material after theregeneration of the IL solution by wood dissolution with [EMIM][OAc]. Yellow southern pine chips containcellulose I, which is the natural form of crystalline cellulose. Theregenerated cellulose rich material exhibited a low crystallinity with the peaks corresponding to that of cellulose II, which is more stable than the naturally occurringcellulose I [68]. The process has also been used for the pretreatment and conversion of lignocellulosic materials [52, 69]. Materials treated by this process had a much higher rate of enzymatic hydrolysis and yield of reducing sugars than the materials treated by traditional pretreatment methods such as steam explosion and chemical pretreatment. Further studies [55] on the simultaneous saccharification and fermentation of this pretreated lignocelluloseindicated that the yield of ethanol or lactic acidwas also much higher. It should be emphasized that pretreatment of lignocellulosic materials using ILs is more environment-friendly thanconventional pretreatment methods because the ILs can be recovered and reused. Moreover, byfractionating lignocellulose with ILs, it ispossible to extract cellulose cleanly, which establishes aplatform for the development of cellulose compositesand derivatives [70].

In pulping process, it is of great importance to reduce the content of lignin [9]. Lignin, which acts as the “glue”binding allcomponents together, is a key sugar responsible for theintricacy of the whole system. Most methods describedin the literature lead to either the degradation of the polymers or a substantial loss in the content of fermentablesugars of the remaining carbohydrates [24]. Lee et al., [44] discovered ILs that dissolve lignin relatively well butexhibit a limited affinity toward cellulose and other constituents of wood flour. They distinguished that [MMIM][CH_3OSO_3], [BMIM][CF_3SO_3], [EMIM][OAc], [AMIM][Cl],

[BMIM][Cl], or [BZMIM][Cl] are particularly good solvents for lignin. Thehighest solubility of lignin is obtained in [MMIM][CH_3OSO_3]and [BMIM][CF_3SO_3][24, 44]. ILs containing $[Cl]^-$ anions are, on the other hand, the most appropriate candidates for cellulose dissolution. For example, [BMIM][Cl] is capable of dissolving cellulose completely within 20 min. At 80°C, [BMIM][Cl] dissolves even high-molecular-mass bacterial cellulose with a degree ofpolymerization of about 6500 without degradation [71].

A goodcompromise between the solubility of lignin and cellulose is achieved for [EMIM][OAc][44]. The use of this IL leads to aretention of the separated lignin in its natural structure [72]. In the same time, cellulose shows a greateraccessibility to the modifications and its crystallinity is disrupted compared to lignin [24]. The separated fraction of the regenerated cellulose and theprocessed lignin from the wood being dissolved in [EMIM][OAc]for 16 h at 110°C were investigated by^{13}C NMR spectroscopy [68]. The results revealed that the regenerated cellulose contained the peaks originating fromcellulose, lignin and hemicelluloses, i.e., peaks at 55.7 [73] and 63.3 ppm. Prior to the^{13}C NMR experiments, the reconstructed material was dissolved in DMSO-d6. However, the lignin fractionwasfree fromcarbohydrates. The amount of lignin present inthe regenerated cellulose rich material was lower than that in wood. However, the main conclusion after these studies was that it is difficult to dissolve native lignin present in wood. Thus, it came up that thekey question in wood dissolution by ILs is to reveal howto cleave the covalent bonds between lignin and biopolymer components [51]. In addition to the above mentioned characterization of regeneratedcellulose obtained from dissolution of wood, attempts were also madeto prepare cellulose membranes from the regenerated celluloseobtained from pine wood dissolution with [BMIM][Cl] at 100°C. Nevertheless, it turned out that no proper films could be preparedfrom this regenerated cellulose [51]. However, whenan analogous experiment was performed starting from MCC after treatment with [BMIM][Cl], membranes could be successfully prepared. These results indicate that the presence oflignin and/or hemicelluloses together with cellulose inhibited theproper preparation of cellulose membranes.

6. Biomass Modification in ILs

Chemical modification of lignocellulose aims at alerting its physicochemical properties to desired need. Up to now, homogeneous reaction has found great attention for its specific advantages, for example, creating more options to induce novel functional groups; opening new avenues for the design of products; and opening up to anopportunity to control the total degree of substitution [74]. However, traditional homogeneous reaction of lignocellulose, such as cellulose, is performed in relatively harsh conditions and requires the use of unusual solvents typically with high ionic strength [6, 31]. Moreover, these processes sometimes cause serious environmental problems because the solvent is usually volatile and cannot be recovered and reused [31]. Fortunately, ILs, as the "green" solvents, can be applied to the homogeneous and efficient chemical modification of lignocellulose biomass [74-76]. And great progress has been made in homogeneous modification of lignocellulose using ILs as reaction media. Figure6 provides a simplified overview about somecellulose functionalization routes. The hydroxyl group offers a broad spectrum of possible reactions with other functional groups [50].

It has been reported that acylatate, benzoylate, and carbamate derivatives of cellulose were successfully prepared in [BMPy][Cl], [BMIM][Cl], [BDMIM][Cl], [AMIM][Cl], [ADMIM][Br], [EMIM][Cl], and [EMIM][OAc] [50]. Zhang and his coworkers [77] reported that the homogeneous acetylation of cellulose couldbe carried out in [AMIM][Cl] without any catalysts, and celluloseacetates with a wide range of degree of substitution could beobtained. Handaet al. [76] found the efficientO-acetylation of cellulose could be accomplished using a zincbased IL. Heinze and Barthel [75] successfully carried out the acylation and carbanylation reactions of cellulosein BMIMCl without any catalysts under mild conditions, lowexcess of reagent and a short reaction time. According to all thesereports, the reaction media ILs could be easilyrecycled and reused [75, 76]. In addition, thedegree of substitution (DS) can be controlled by the choice of acylating agent, the reaction time, and the mole ratios ofthe reagents, while it is also dependent on the dissolution state ofthe cellulose [49, 75, 78]. In the case of acylations, acyl chloridestend to be more effective than the corresponding anhydridesfor the modification [79]. The presence of bases such as pyridineor 1-methylimidazolium can increase the cellulose modification efficiency [80-82]. Furthermore, the different reactivityof the three hydroxyl groups on the glucose monomer unit(C-6> C-3> C-2) allows the regioselective introduction of protecting groups such as ethers [80]. For example, it has been reported that cellulose selectively protected with 4-methoxytrityl in [AMIM][Cl] resulted in an increased reactivity of themodified cellulose [83]. Such regioselective functionalizationsof cellulose are very important for the controlled formation ofcolloidal suprastructures [84].

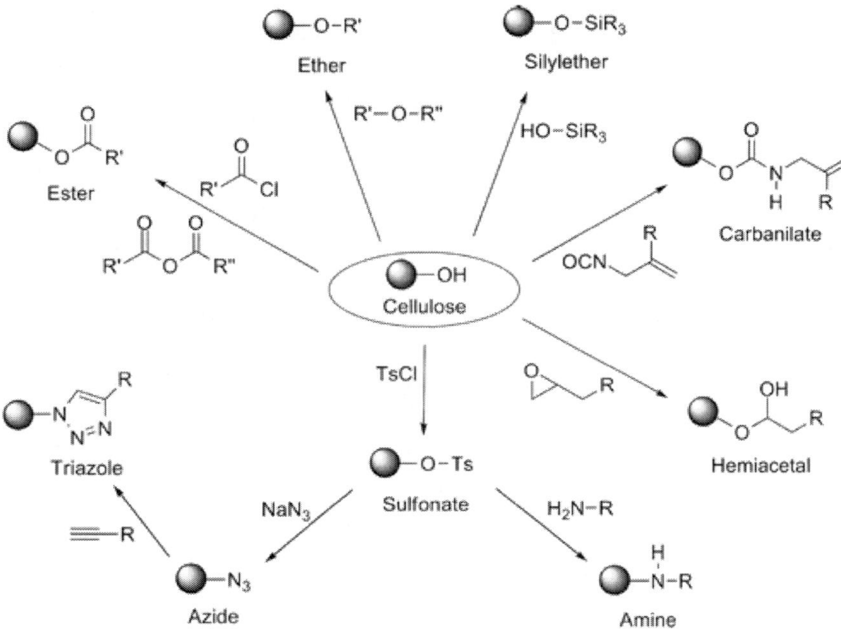

Figure 6. Functionalization routes for cellulose.

Usually, cellulose graft copolymers can be prepared in ILs via typical polymerization methods including radical polymerization, ringopeningpolymerization (ROP), and atomtransferradicalpolymerization (ATRP). Hattori et al. [85] immobilized carbamates via silane coupling on celluloseand copolymerized them with methacrylic acid or ethyleneglycol dimethacrylate upon UV radiation. The concept ofgrafting molecules with tertiary halides suitable for formingradicals to react with polymerizable double bonds tocomposites has been exploited by others. Guo et al., [86, 87] successfully synthesized a series of amphiphilic cellulose graft copolymers with various molecular factors via ROP in [BMIM][Cl]. These cellulose derivatives include cellulose-graft-poly(ε-caprolactone), microcrystalline cellulose-graft- poly(L-lacticacid), etc., which can self-assemble into spherical nanomicellesin aqueous solution. Meng et al. [88]synthesized a cellulose-based macroinitiator for ATRP by direct homogeneous acylation of cellulose in [AMIM][Cl], without using any catalysts and protecting group chemistry. Another group [89] reacted the cellulose hydroxyl groups with2-bromoisobutyryl bromide and copolymerized it with methacrylate or styrene via ATRP, obtaining living polymers that were still active even after one year. Meng and coworkers [90] activated the cellulose surface with allylbromide or divinylsulfone prior to ATRP grafting with polymethyl methacrylateor polystyrene.

Besides ethers and esters, modifications to lignocellulose also include sulfates, sulfonates, deoxy derivatives, silylates, amines, carbanilates, tritylates, hemiacetals, or azides [50]. Water-soluble cellulose sulfates have been prepared with different sulfating agents in [BMIM][Cl], [AMIM][Cl], and[AMIM][OAc]. This is the first report of a non-derivatizing cellulose solvent that allows direct sulfation of the biopolymer under completely homogeneous reaction conditions [91]. Koehler et al., [82] synthesized trimethylsilyl cellulose in [BMIM][Cl],[EMIM][Cl], and [EMIM][OAc]. Phthalated cellulose has been prepared in [AMIM][Cl], and furorates have been obtainedthrough homogeneous modification of cellulose in [BMIM][Cl] [92]. In addition, some more interesting and noteworthy biopolymer modifications include the use of dendrons, boronic acid derivatives, or acetobacter xylinum bacteria [50]. Within the past fewyears, several groups successfully introduced dendriticmacromolecules, up to their fourth generation, onto the cellulose surfaces [93-95]. Dendritic immobilized materials are currently used for enhanced separations in high resolution capillary gas chromatography and capillary microextractions [96]. Liebert exploited the affinity of boric and boronic acid to react with polyols in ILs and showed that interactions may occur at C-2 and C-3 of glucose derivatives. The formation of a seven-membered ringwas confirmed leading to the assumption that a comparablestructure might be formed in the glucose based macromolecules cellulose and starch [97]. Finally, Pommet et al. [98, 99] modifiednatural cellulose by utilizing bacteria to deposit nanosized bacterial cellulose around the fibers, which enhanced theiradhesion to renewable polymers.

In addition, it is analogous to what was described above for cellulose, lignocellulose is susceptible to chemical modification in the IL medium under mild andhomogeneous conditions. Ren and coworkers [100] firstly demonstrated that ILs could be used as an environmentally friendly solvent for the chemical modification of hemicellulose. In their study, wheat straw hemicellulose was acetylated with acetic anhydride using iodine as a novel catalyst in [BMIM][Cl]. Results show that the thermal stability of the acetylated hemicellulose increased upon chemical modification. Moreover, Xie et al., [79, 101] dissolved wood (spruce thermomechanical pulp) in chloride ILs, followed by theapplication of a variety of homogeneous reactions. Bothacylation and carbanilation reactions were carried out, and highly substituted lignocellulosic esters were obtained undermild conditions in the presence of an added base such aspyridine in some cases. The resulting wood derivativescompletely lacked any fiber characteristics, showing ahomogeneous porous powdery appearance instead. At the same time, they found that benzoylation or lauroylation reactions can improve the compatibility between the wood-based lignocellulosic materials and conventional syntheticpolymers such as polystyrene and polypropylene [5, 101].

CONCLUSION

In summary, ILs show good physicochemical properties and numerous advantages compared to conventional solvents. And many laboratory applications have already established ILs, as powerful non-derivatizing solvents for the dissolution of lignocellulose biomass, provide a new and "green" platform for lignocellulose utilization. They alsofacilitate biorefinery by fractionation of the lignocellulosic resources via "green" means. Moreover, ILs can be the ''green'' homogeneous reactionmediato prepare various advanced materials. In fact, the modified lignocellulose has great potential to be applied in the field of biomedicine, materials, food industry, etc. The rapid increase in the number of publications on the use of ILs for the dissolution andtreatment of lignocellulose biomass can betaken as a good indicator of the promising future of ILs inthis area [5]. However, the current studies focuson cellulose as the only active participant, while neglecting the potential of hemicellulose and lignin. In the future, the efficient comprehensive use of lignocellulose biomass will play an important role in reforming traditional chemical and energy industries based on biorenewable resources and environmentally friendly processing [102].

In addition, ILs processing of lignocellulosic biomass could have a number of advantages over current technologies, by enablingthe recovery of all three of the major natural lignocellulose biomass, thus allowing their direct use incomposites and other advanced materials. Nevertheless, progress in several key aspects is needed for any of thesepotential ILs processes to be realized at industrial scale [55]. Forexample, low energy recycle of the ILs is critical to theeconomics of such a process; particularly in a world whereenergy usage is increasingly coming under major scrutiny andenergy costs are increasing rapidly. At the same time, the ILs mustbecome available at lower cost and should ultimately be of lowtoxicity and be biodegradable. Although ILs will not evaporate and thus will not cause air pollution, it does not mean that they will not harm the environment. Most of ILs are water soluble and they may enter the aquatic environment by accidental spills or effluents [17]. The most commonly used ILs [BMIM][PF_6] and [BMIM][BF_4] are known to decompose in the presence of water and as a result hydrofluoric and phosphoric acids are formed [103]. Therefore, both toxicity and ecotoxicity information which provide metabolism and degradability of ILs are also required to label them as green solvents or investigate their environmental impact. Furthermore, the search for even more environmentallyfriendly ILsshould be one of the drivers for research in the area, and fullybiodegradable and non-toxic ILs should be a major goal.

Finally, athorough understanding of the influence of the ionstructure and ion typeon the physical behavior of the ILs isthe key to unlocking the potential of this solvent class [50]. Themaintenance of a well-managed and freely accessible database is themost promising tool to achieve this understanding. Such adatabase allows the development of new structural propertyrelationships and comparison with results from molecularmechanics based calculations. Then a deeper scientific knowledgeof the mode of action of the IL and its interactions with thedifferent lignocellulose biomass will be required [5]. Significant advancement is needed especiallyin regard to the role played by the cation in the dissolution process, which remains largely unknown at present.

REFERENCES

[1] Mosier, N., Wyman, C., Dale, B., Elander, R., Lee, Y., Holtzapple, M., et al., (2005). Features of promising technologies for pretreatment of lignocellulosic biomass. *Bioresource technology,96*(6),673-686.

[2] Guo, M., Stuckey, D.& Murphy, R. (2013). Is it possible to develop biopolymer production systems independent of fossil fuels? Case study in energy profiling of polyhydroxybutyrate-valerate (PHBV). *Green Chem,15*(3),706-717.

[3] Pérez, J., Munoz-Dorado, J., de la Rubia, T.& Martinez, J. (2002). Biodegradation and biological treatments of cellulose, hemicellulose and lignin: an overview. *International Microbiology,5*(2),53-63.

[4] Kilpeläinen, I., Xie, H., King, A., Granstrom, M., Heikkinen, S.& Argyropoulos, D.S. (2007). Dissolution of wood in ionic liquids. *J Agric Food Chem,55*(22),9142-9148.

[5] Sun, N., Rodríguez, H., Rahman, M.& Rogers, R.D. (2011). Where are ionic liquid strategies most suited in the pursuit of chemicals and energy from lignocellulosic biomass? *Chemical Communications,47*(5),1405-1421.

[6] Heinze, T.& Liebert, T. (2001). Unconventional methods in cellulose functionalization. *Progress in Polymer Science,26*(9),1689-1762.

[7] Kabo, G.J., Blokhin, A.V., Paulechka, Y.U., Kabo, A.G., Shymanovich, M.P.& Magee, J.W. (2004). Thermodynamic properties of 1-butyl-3-methylimidazolium hexafluorophos-phate in the condensed state. *J Chem Eng Data,49*(3),453-461.

[8] Welton, T. (2004). Ionic liquids in catalysis. *Coordin Chem Rev,248*(21),2459-2477.

[9] Hendriks, A.& Zeeman, G. (2009). Pretreatments to enhance the digestibility of lignocellulosic biomass. *Bioresource technology,100*(1),10-18.

[10] Mora-Pale, M., Meli, L., Doherty, T.V., Linhardt, R.J.& Dordick, J.S. (2011). Room temperature ionic liquids as emerging solvents for the pretreatment of lignocellulosic biomass. *Biotechnol Bioeng,108*(6),1229-1245.

[11] Guo, Y.Z., Wang, X.H., Li, D., Du, H., Wang, X. Y. & Sun, R.C. (2012). Synthesis and characterization of hydrophobic long-chain fatty acylated cellulose and its self-assembled nanoparticles. *Polymer Bulletin,69*(4),389-403.

[12] Fengel, D.& Wegener, G.(1983). Wood: chemistry, ultrastructure, reactions: Walter de Gruyter.

[13] Gray, M.C., Converse, A.O.& Wyman, C.E. (2003). Sugar monomer and oligomer solubility - Data and predictions for application to biomass hydrolysis. *Appl Biochem Biotech,105*(179-193).

[14] Bobleter, O. (1994). Hydrothermal degradation of polymers derived from plants. *Progress in Polymer Science,19*(5),797-841.

[15] Garrote, G., Dominguez, H.& Parajo, J. (1999). Hydrothermal processing of lignocellulosic materials. *European Journal of Wood and Wood Products,57*(3), 191-202.

[16] Grabber, J.H. (2005). How do lignin composition, structure, and cross-linking affect degradability? A review of cell wall model studies. *Crop Sci.*,*45*(3),820-831.
[17] Keskin, S., Kayrak-Talay, D., Akman, U.& Hortaçsu, Ö. (2007). A review of ionic liquids towards supercritical fluid applications. *The Journal of Supercritical Fluids*,*43*(1),150-180.
[18] Wilkes, J.S.& Zaworotko, M.J. (1992). Air and water stable 1-ethyl-3-methylimidazolium based ionic liquids. *J Chem Soc, Chem Commun*,*13*,965-967.
[19] Chiappe, C.& Pieraccini, D. (2005). Ionic liquids: solvent properties and organic reactivity. *J Phys Org Chem*,*18*(4),275-297.
[20] Pringle, J.M., Forsyth, C.M., Forsyth, M.& MacFarlane, D.R. (2003). The zwitterion 1-butylimidazolium-3-(n-butanesulfonate). *Acta Crystallographica Section E-Structure Reports Online*,*59*(0),O1759-O1761.
[21] Pringle, J.M., Golding, J., Baranyai, K., Forsyth, C.M., Deacon, G.B., Scott, J.L., et al., (2003). The effect of anion fluorination in ionic liquids—physical properties of a range of bis (methanesulfonyl) amide salts. *New J Chem*,*27*(10),1504-1510.
[22] Baranyai, K.J., Deacon, G.B., MacFarlane, D.R., Pringle, J.M.& Scott, J.L. (2004). Thermal degradation of ionic liquids at elevated temperatures. *Aust J Chem*,*57*(2), 145-147.
[23] Sudhir, N. (2001). How polar are room-temperature ionic liquids? *Chemical Communications*,*0*(5),413-414.
[24] Zakrzewska, M.E., Bogel-Łukasik, E.& Bogel-Łukasik, R. (2010). Solubility of carbohydrates in ionic liquids. *Energ Fuel*,*24*(2),737-745.
[25] Swatloski, R.P., Spear, S.K., Holbrey, J.D.& Rogers, R.D. (2002). Dissolution of cellose with ionic liquids. *Journal of the American Chemical Society*,*124*(18),4974-4975.
[26] Zhang, H., Wu, J., Zhang, J.& He, J.S. (2005). 1-Allyl-3-methylimidazolium chloride room temperature ionic liquid: A new and powerful nonderivatizing solvent for cellulose. *Macromolecules*,*38*(20),8272-8277.
[27] Li, C.& Zhao, Z.K. (2007). Efficient acid-catalyzed hydrolysis of cellulose in ionic liquid. *Adv Synth Catal*,*349*(11-12),1847-1850.
[28] Amarasekara, A.S.& Owereh, O.S. (2009). Hydrolysis and Decomposition of Cellulose in Bronsted Acidic Ionic Liquids Under Mild Conditions. *Ind Eng Chem Res*,*48*(22),10152-10155.
[29] Cruz, H., Fanselow, M., Holbrey, J. D. & Seddon, K.R. (2012). Determining relative rates of cellulose dissolution in ionic liquids through in situ viscosity measurement. *Chemical Communications*,*48*(45),5620-5622.
[30] Ohira, K., Abe, Y., Kawatsura, M., Suzuki, K., Mizuno, M., Amano, Y., et al., (2012). Design of cellulose dissolving ionic liquids inspired by nature. *ChemSusChem*,*5*(2),388-391.
[31] Zhang, H., Wu, J., Zhang, J.& He, J. (2005). 1-Allyl-3-methylimidazolium chloride room temperature ionic liquid: a new and powerful nonderivatizing solvent for cellulose. *Macromolecules*,*38*(20),8272-8277.
[32] Kosan, B., Michels, C.& Meister, F. (2008). Dissolution and forming of cellulose with ionic liquids. *Cellulose*,*15*(1),59-66.
[33] Vitz, J., Erdmenger, T., Haensch, C.& Schubert, U.S. (2009). Extended dissolution studies of cellulose in imidazolium based ionic liquids. *Green Chem*,*11*(3),417-424.

[34] Zhao, H., Baker, G.A., Song, Z., Olubajo, O., Crittle, T.& Peters, D. (2008). Designing enzyme-compatible ionic liquids that can dissolve carbohydrates. *Green Chem,10*(6),696-705.

[35] Zavrel, M., Bross, D., Funke, M., Büchs, J.& Spiess, A.C. (2009). High-throughput screening for ionic liquids dissolving (ligno-) cellulose. *Bioresource technology,100*(9),2580-2587.

[36] Mikkola, J.P., Kirilin, A., Tuuf, J.C., Pranovich, A., Holmbom, B., Kustov, L.M., et al., (2007). Ultrasound enhancement of cellulose processing in ionic liquids: from dissolution towards functionalization. *Green Chem,9*(11),1229-1237.

[37] Anderson, J.L., Ding, J., Welton, T.& Armstrong, D.W. (2002). Characterizing ionic liquids on the basis of multiple solvation interactions. *Journal of the American Chemical Society,124*(47),14247-14254.

[38] Feng, L.& Chen, Z.-I. (2008). Research progress on dissolution and functional modification of cellulose in ionic liquids. *J Mol Liq,142*(1-3),1-5.

[39] Moulthrop, J.S., Swatloski, R.P., Moyna, G.& Rogers, R.D. (2005). High-resolution C-13 NMR studies of cellulose and cellulose oligomers in ionic liquid solutions. *Chemical Communications,0*(12),1557-1559.

[40] Saha, B.C. (2003). Hemicellulose bioconversion. *Journal of Industrial Microbiology and Biotechnology,30*(5),279-291.

[41] Zhao, H., Baker, G.A.& Cowins, J.V. (2010). Fast Enzymatic Saccharification of Switchgrass After Pretreatment with Ionic Liquids. *Biotechnol Prog,26*(1),127-133.

[42] Liu, Q.P., Hou, X.D., Li, N.& Zong, M.H. (2012). Ionic liquids from renewable biomaterials: synthesis, characterization and application in the pretreatment of biomass. *Green Chem,14*(2),304-307.

[43] Remsing, R.C., Swatloski, R.P., Rogers, R.D.& Moyna, G. (2006). Mechanism of cellulose dissolution in the ionic liquid 1-n-butyl-3-methylimidazolium chloride: a 13C and 35/37Cl NMR relaxation study on model systems. *Chemical Communications,0*(12),1271-1273.

[44] Lee, S.H., Doherty, T.V., Linhardt, R.J.& Dordick, J.S. (2009). Ionic liquid-mediated selective extraction of lignin from wood leading to enhanced enzymatic cellulose hydrolysis. *Biotechnol Bioeng,102*(5),1368-1376.

[45] Pu, Y., Jiang, N.& Ragauskas, A.J. (2007). Ionic liquid as a green solvent for lignin. *J Wood Chem Technol,27*(1),23-33.

[46] Swiderski, K., McLean, A., Gordon, C.M.& Vaughan, D.H. (2004). Estimates of internal energies of vaporisation of some room temperature ionic liquids. *Chemical Communications,19*),2178-2179.

[47] Zhang, J., Zhang, H., Wu, J., Zhang, J., He, J.& Xiang, J. (2010). NMR spectroscopic studies of cellobiose solvation in EmimAc aimed to understand the dissolution mechanism of cellulose in ionic liquids. *Phys Chem Chem Phys,12*(8),1941-1947.

[48] Moulthrop, J.S., Swatloski, R.P., Moyna, G.& Rogers, R.D. (2005). High-resolution 13C NMR studies of cellulose and cellulose oligomers in ionic liquid solutions. *Chemical Communications,0*(12),1557-1559.

[49] Feng, L.& Chen, Z.-l. (2008). Research progress on dissolution and functional modification of cellulose in ionic liquids. *J Mol Liq,142*(1),1-5.

[50] Pinkert, A., Marsh, K.N., Pang, S.& Staiger, M.P. (2009). Ionic liquids and their interaction with cellulose. *Chem Rev,109*(12),6712-6728.

[51] Mäki-Arvela, P., Anugwom, I., Virtanen, P., Sjöholm, R.& Mikkola, J.-P. (2010). Dissolution of lignocellulosic materials and its constituents using ionic liquids—a review. *Ind Crop Prod,32*(3),175-201.

[52] Dadi, A.P., Varanasi, S.& Schall, C.A. (2006). Enhancement of cellulose saccharification kinetics using an ionic liquid pretreatment step. *Biotechnol Bioeng,95*(5),904-910.

[53] Fort, D.A., Remsing, R.C., Swatloski, R.P., Moyna, P., Moyna, G.& Rogers, R.D. (2007). Can ionic liquids dissolve wood? Processing and analysis of lignocellulosic materials with 1-n-butyl-3-methylimidazolium chloride. *Green Chem,9*(1),63-69.

[54] Kilpeläinen, I., Xie, H., King, A., Granstrom, M., Heikkinen, S.& Argyropoulos, D.S. (2007). Dissolution of wood in ionic liquids. *J Agr Food Chem,55*(22),9142-9148.

[55] Zhu, S., Wu, Y., Chen, Q., Yu, Z., Wang, C., Jin, S., et al., (2006). Dissolution of cellulose with ionic liquids and its application: a mini-review. *Green Chem,8*(4), 325-327.

[56] Zhao, H., Jones, C.L., Baker, G.A., Xia, S., Olubajo, O.& Person, V.N. (2009). Regenerating cellulose from ionic liquids for an accelerated enzymatic hydrolysis. *J Biotechnol,139*(1),47-54.

[57] Swatloski, R.P., Holbrey, J.D., Memon, S.B., Caldwell, G.A., Caldwell, K.A.& Rogers, R.D. (2004). Using Caenorhabditis elegans to probe toxicity of 1-alkyl-3-methylimidazolium chloride based ionic liquids. *Chemical Communications,0*(6), 668-669.

[58] Takahashi, Y.& Matsunaga, H. (1991). Crystal structure of native cellulose. *Macromolecules,24*(13),3968-3969.

[59] Bagheri, M., Rodríguez, H., Swatloski, R.P., Spear, S.K., Daly, D.T.& Rogers, R.D. (2007). Ionic Liquid-Based Preparation of Cellulose– Dendrimer Films as Solid Supports for Enzyme Immobilization. *Biomacromolecules,9*(1),381-387.

[60] Duchemin, B.J., Newman, R.H.& Staiger, M.P. (2009). Structure–property relationship of all-cellulose composites. *Compos Sci Technol,69*(7),1225-1230.

[61] Aaltonen, O.& Jauhiainen, O. (2009). The preparation of lignocellulosic aerogels from ionic liquid solutions. *Carbohydr Polym,75*(1),125-129.

[62] Zhu, S.D., Wu, Y.X., Yu, Z.N., Wang, C.W., Yu, F.Q., Jin, S.W., et al., (2006). Comparison of three microwave/chemical pretreatment processes for enzymatic hydrolysis of rice straw. *Biosystems Engineering,93*(3),279-283.

[63] Zhu, S.D., Wu, Y.X., Yu, Z.N., Zhang, X., Wang, C.W., Yu, F.Q., et al., (2006). Production of ethanol from microwave-assisted alkali pretreated wheat straw. *Process Biochem,41*(4),869-873.

[64] Sun, Y.& Cheng, J.Y. (2002). Hydrolysis of lignocellulosic materials for ethanol production: a review. *Bioresource technology,83*(1),1-11.

[65] Daly, D.T., Rahman, M., Rodriguez, H., Rogers, R.D., Sun, N.& Swatloski, R.P. Fractioning biomass, comprises providing fractionation composition comprising biomass e.g., wood pulp, ionic liquid, and fractionation polymer e.g., polyethyleneimine and adjusting the temperature to provide biphasic composition. Univ Alabama.

[66] Sun, N., Rodriguez, H., Rahman, M.& Rogers, R.D. (2011). Where are ionic liquid strategies most suited in the pursuit of chemicals and energy from lignocellulosic biomass? *Chemical Communications,47*(5),1405-1421.

[67] Myllymaeki, V., Aksela, R.& Myllymaki, V. Method for dissolving lignocellulosic material, such as wood, involves mixing lignocellulosic material with ionic liquid solvent under microwave irradiation and/or pressure in absence of water for complete dissolution. Kemira Oyj.

[68] Sun, N., Rahman, M., Qin, Y., Maxim, M.L., Rodríguez, H.& Rogers, R.D. (2009). Complete dissolution and partial delignification of wood in the ionic liquid 1-ethyl-3-methylimidazolium acetate. *Green Chem,11*(5),646-655.

[69] Liu, L.& Chen, H. (2006). Enzymatic hydrolysis of cellulose materials treated with ionic liquid [BMIM] Cl. *Chin Sci Bull,51*(20),2432-2436.

[70] Zhu, S. (2008). Use of ionic liquids for the efficient utilization of lignocellulosic materials. *J Chem Technol Biotechnol,83*(6),777-779.

[71] Schlufter, K., Schmauder, H.P., Dorn, S.& Heinze, T. (2006). Efficient Homogeneous Chemical Modification of Bacterial Cellulose in the Ionic Liquid 1-N-Butyl-3-methylimidazolium Chloride. *Macromol Rapid Commun,27*(19),1670-1676.

[72] Chakar, F.S.& Ragauskas, A.J. (2004). Review of current and future softwood kraft lignin process chemistry. *Ind Crop Prod,20*(2),131-141.

[73] Lu, F.& Ralph, J. (2003). Non-degradative dissolution and acetylation of ball-milled plant cell walls: High-resolution solution-state NMR. *The Plant Journal,35*(4), 535-544.

[74] Wu, J., Zhang, J., Zhang, H., He, J., Ren, Q.& Guo, M. (2004). Homogeneous acetylation of cellulose in a new ionic liquid. *Biomacromolecules,5*(2),266-268.

[75] Barthel, S.& Heinze, T. (2006). Acylation and carbanilation of cellulose in ionic liquids. *Green Chem,8*(3),301-306.

[76] Thomas, J. (2005). O-Acetylation of cellulose and monosaccharides using a zinc based ionic liquid. *Green Chem,7*(10),705-707.

[77] Turner, M.B., Spear, S.K., Holbrey, J.D.& Rogers, R.D. (2004). Production of bioactive cellulose films reconstituted from ionic liquids. *Biomacromolecules,5*(4),1379-1384.

[78] Köhler, S.& Heinze, T. (2007). Efficient synthesis of cellulose furoates in 1-N-butyl-3-methylimidazolium chloride. *Cellulose,14*(5),489-495.

[79] Xie, H., King, A., Kilpelainen, I., Granstrom, M.& Argyropoulos, D.S. (2007). Thorough chemical modification of wood-based lignocellulosic materials in ionic liquids. *Biomacromolecules,8*(12),3740-3748.

[80] Heinze, T., Schwikal, K.& Barthel, S. (2005). Ionic liquids as reaction medium in cellulose functionalization. *Macromol Biosci,5*(6),520-525.

[81] Li, C.& Zhao, Z.K. (2007). Efficient Acid-Catalyzed Hydrolysis of Cellulose in Ionic Liquid. *Adv Synth Catal,349*(11-12),1847-1850.

[82] Köhler, S., Liebert, T.& Heinze, T. (2008). Interactions of ionic liquids with polysaccharides. VI. Pure cellulose nanoparticles from trimethylsilyl cellulose synthesized in ionic liquids. *Journal of Polymer Science Part A: Polymer Chemistry,46*(12),4070-4080.

[83] Granström, M., Olszewska, A., Mäkelä, V., Heikkinen, S.& Kilpeläinen, I. (2009). A new protection group strategy for cellulose in an ionic liquid: simultaneous protection of two sites to yield 2, 6-di-O-substituted mono-< i> p</i>-methoxytrityl cellulose. *Tetrahedron Lett,50*(15),1744-1747.

[84] Klemm, D., Heublein, B., Fink, H.P.& Bohn, A. (2005). Cellulose: fascinating biopolymer and sustainable raw material. *Angewandte Chemie International Edition*,*44*(22),3358-3393.

[85] Hattori, K., Hiwatari, M., Iiyama, C., Yoshimi, Y., Kohori, F., Sakai, K., et al., (2004). Gate effect of theophylline-imprinted polymers grafted to the cellulose by living radical polymerization. *J Membrane Sci*,*233*(1),169-173.

[86] Guo, Y., Wang, X., Shu, X., Shen, Z.& Sun, R.-C. (2012). Self-assembly and paclitaxel loading capacity of cellulose-graft-poly (lactide) nanomicelles. *J Agr Food Chem*,*60*(15),3900-3908.

[87] Guo, Y., Wang, X., Shen, Z., Shu, X.& Sun, R. (2012). Preparation of cellulose-< i> graft</i>-poly (< i>ε</i>-caprolactone) nanomicelles by homogeneous ROP in ionic liquid. *Carbohydr Polym*.

[88] Meng, T., Gao, X., Zhang, J., Yuan, J., Zhang, Y.& He, J. (2009). Graft copolymers prepared by atom transfer radical polymerization (ATRP) from cellulose. *Polymer*,*50*(2),447-454.

[89] Lindqvist, J.& Malmström, E. (2006). Surface modification of natural substrates by atom transfer radical polymerization. *Journal of Applied Polymer Science*,*100*(5),4155-4162.

[90] Xia, H.-F., Lin, D.-Q., Wang, L.-P., Chen, Z.-J.& Yao, S.-J. (2008). Preparation and evaluation of cellulose adsorbents for hydrophobic charge induction chromatography. *Ind Eng Chem Res*,*47*(23),9566-9572.

[91] Gericke, M., Liebert, T.& Heinze, T. (2009). Interaction of ionic liquids with polysaccharides, 8–synthesis of cellulose sulfates suitable for polyelectrolyte complex formation. *Macromol Biosci*,*9*(4),343-353.

[92] Liu, C.F., Sun, R.C., Zhang, A.P., Qin, M.H., Ren, J.L.& Wang, X.A. (2007). Preparation and characterization of phthalated cellulose derivatives in room-temperature ionic liquid without catalysts. *J Agr Food Chem*,*55*(6),2399-2406.

[93] Cuissinat, C., Navard, P.& Heinze, T. (2008). Swelling and dissolution of cellulose. Part IV: Free floating cotton and wood fibres in ionic liquids. *Carbohydr Polym*,*72*(4),590-596.

[94] Pohl, M., Schaller, J., Meister, F.& Heinze, T. (2008) *Novel bulky esters of biopolymers: dendritic cellulose.* Macromol Symp:Wiley Online Library 119-128.

[95] Heinze, T., Pohl, M., Schaller, J.& Meister, F. (2007). Novel bulky esters of cellulose. *Macromol Biosci*,*7*(11),1225-1231.

[96] Hassan, M.L., Moorefield, C.N., Kotta, K.& Newkome, G.R. (2005). Regioselective combinatorial-type synthesis, characterization, and physical properties of dendronized cellulose. *Polymer*,*46*(21),8947-8955.

[97] Liebert, T. (2008). Innovative concepts for the shaping and modification of cellulose. *Macromol Symp*,*262*(0),28-38.

[98] Pommet, M., Juntaro, J., Heng, J.Y., Mantalaris, A., Lee, A.F., Wilson, K., et al., (2008). Surface modification of natural fibers using bacteria: depositing bacterial cellulose onto natural fibers to create hierarchical fiber reinforced nanocomposites. *Biomacromolecules*,*9*(6),1643-1651.

[99] Juntaro, J., Pommet, M., Kalinka, G., Mantalaris, A., Shaffer, M.S.& Bismarck, A. (2008). Creating hierarchical structures in renewable composites by attaching bacterial cellulose onto sisal fibers. *Adv Mater*,*20*(16),3122-3126.

[100] Ren, J., Sun, R., Liu, C., Cao, Z.& Luo, W. (2007). Acetylation of wheat straw hemicelluloses in ionic liquid using iodine as a catalyst. *Carbohydr Polym*,*70*(4), 406-414.

[101] Xie, H., Jarvi, P., Karesoja, M., King, A., Kilpelainen, I.& Argyropoulos, D.S. (2009). Highly compatible wood thermoplastic composites from lignocellulosic material modified in ionic liquids: Preparation and thermal properties. *Journal of Applied Polymer Science*,*111*(5),2468-2476.

[102] Galbe, M.& Zacchi, G. (2012). Pretreatment: The key to efficient utilization of lignocellulosic materials. *Biomass and Bioenergy*,*46*(0),70-78.

[103] Cammarata, L., Kazarian, S., Salter, P.& Welton, T. (2001). Molecular states of water in room temperature ionic liquids. *Phys Chem Chem Phys*,*3*(23),5192-5200.

In: Ionic Liquids: Synthesis, Characterization and Applications ISBN: 978-1-63117-716-3
Editor: Anson Brooks

Chapter 8

RECENT DEVELOPMENTS IN IONIC LIQUID TOXICITY ASSESSMENT

E. Mulkiewicz[*] and P. Stepnowski
Institute for Environmental and Human Health Protection,
Faculty of Chemistry, University of Gdansk, Gdansk, Poland

ABSTRACT

The on-going development and use of ionic liquids (ILs) require better understanding of their effects on the environment. A considerable amount of data on the toxicity of ILs towards the aquatic and terrestrial environments and also to different trophic levels, including primary producers (e.g., bacteria, algae and plants), primary consumers (e.g., invertebrates) and secondary consumers (e.g., fish) has been published over the past decade. Toxicity has been evaluated on the basis of test systems at different levels of biological complexity, from molecules and cells to organisms. A wide variety of IL species with different head groups, alkyl and functionalized side chains and anions has been investigated to establish not only their overall toxic effect but also the influence of structural elements on their biological activity. This chapter reviews the achievements and current status of environmental risk assessment of ILs. The problem how ecotoxicological experiments that consider diverse mixtures of organisms, abiotic conditions and additional stressors verify environmentally "safe" concentrations of ILs is discussed.

Research into ionic liquids (ILs) has been a very rapidly growing discipline in recent years. ILs have attracted very considerable attention because of their unique properties, which may be useful in new processes and technologies. They have already been studied for a variety of applications such as alternative solvents in organic synthesis and catalysis, reaction media for biocatalysis and biotransformation, in the separation sciences, electrolytes in batteries and solar cells, alternative lubricants, and as media for tissue preservation. ILs have excellent solvation properties, are thermally, chemically and electrochemically stable, and

[*] Corresponding author's email: ewa.mulkiewicz@ug.edu.pl.

their vapour pressure is negligible (Petkovic et al., 2011). Nevertheless, the same properties that make them attractive replacements for other volatile industrial compounds may render them hazardous to ecosystems. Although at present there are no known instances of environmental pollution by ILs, the possibility of their accidental discharge, causing environmental contamination, should be taken into consideration. It is unquestionable that the vast majority of ILs do not pose any hazard to the atmosphere, but they could well endanger waters and soils. The current limited use of ILs makes it difficult to predict what levels they may reach in the environment. The long-term consequences of their presence in the environment are also unknown.

With the new European legislation on chemicals (REACH – registration, evaluation, authorization and restriction of chemicals), knowledge of the potential toxicological hazards of chemicals towards animals, plants and the environment in general is taking on a greater importance. Before large-scale technical applications can be implemented, these hazards should be identified. The on-going development and use of ILs require a better understanding of their effects on the environment. A considerable amount of data has been published on the toxicity of ILs towards the aquatic and terrestrial environments and also to different trophic levels, including primary producers (e.g., bacteria, algae and plants), primary consumers (e.g., invertebrates) and secondary consumers (e.g., fish). Toxicity has been evaluated on the basis of test systems at different levels of biological complexity, from molecules and cells to organisms. Most of the toxicological data presented here were obtained with model organisms in standard toxicological and ecotoxicological tests, using the standard methods recommended by regulatory agencies like OECD (Organization for Economic Co-operation and Development). In acute tests the LC_{50} (lethal concentration – the concentration causing 50% mortality), or if an effect other than mortality was investigated, the EC_{50} (effect concentration – the concentration causing a 50% increase or decrease in the effect) were determined. The standard metrics are comparable across organisms and chemicals and provide baseline information about IL toxicity that will be useful for future studies.

Cytotoxicity, Toxicity at Molecular Level, Models of Toxicity

It is widely assumed that the first detectable responses to environmental perturbation are changes at the molecular, biochemical or cellular level (Hugget et al., 1992). The basic modes of toxic action of substance, including unspecific interaction with membrane structures (*i.e.,* baseline toxicity/necrosis), interactions with proteins (e.g., enzyme inhibition or binding to receptor proteins) and chemical reactivity towards the molecules of life, can be screened with cellular assays. Thus, the importance of *in vitro* systems such as cell lines has increased in ecotoxicology. Although *in vitro* systems cannot replicate the complex interactions in organisms *in vivo*, they do provide important predictive information that allows for reductions in animal testing when appropriate. The applicability of *in vitro* cell systems for the assessment of three toxicological modes of action – cytotoxicity, dioxin-like toxicity and oestrogen receptor-mediated activity – was discussed and exemplified by Fent (2001).

With all their advantages and limitations, cellular test systems have been employed in the studies of biological effects of ILs. The most frequently used cell lines have been the rat leukaemia cell line IPC-81 (Ranke et al., 2004, 2006; Stolte et al., 2006, 2007; Stasiewicz et al., 2008; Pernak et al., 2011, Steudte et al., 2012; Stolte et al., 2012; Peric et al., 2013), the rat glioma cell line C6 (Ranke et al., 2004), the human carcinoma epithelial cell line HeLa (Stepnowski et al., 2004; Wang et al., 2007), the murine macrophage cell line J774 (Weaver et al., 2010), the human colon carcinoma cell lines HT-29 and CaCo-2 (Frade et al., 2007), the Chinese hamster lung fibroblast cell line V79 (Schaffran et al., 2009), the human breast cancer cell line MCF7 (Kumar et al., 2009) and the murine fibroblast NIH 3T3 cells (McLaughlin et al., 2011).

As the term "ionic liquid" encompasses a very heterogeneous class of chemicals with different physical and chemical properties, their toxicity can vary over a broad range: EC_{50}s obtained in studies with mammalian cells differ by several orders of magnitude, from the upper mM range, which is not of biological significance, down to nM concentrations, which indicate high acute toxicity. The cytotoxicity of a wide variety of IL species with different head groups, alkyl and functionalized side chains and anions has been investigated to establish not only their overall toxic effect but also the influence of structural elements on their biological activity. Professor Jastorff's team postulated an approach that allows the (eco)toxicity of individual structural elements (toxicophore/ecotoxicophore) of ILs to be investigated. To find out how these individual structural variables may elicit effects on the molecules of life, cells and organisms, they systematically selected and analysed them according to the T-SAR concept (thinking in terms of structure–activity relationships). The different structural elements were combined based on the test kit concept, in a set of ILs in which only one parameter (head group, side chain length or anion) was changed. All the cytotoxicity and acetylcholinesterase inhibition data obtained by this group was summarized by Ranke et al., (2007), who presented a comprehensive set of IPC-81 cytotoxicities for 253 ILs and AChE inhibition values for 292. A substantial amount of IL toxicity and ecotoxicity data is available online on the UFT/Merck Ionic Liquid Biological Database http://www.il-eco.uft.uni-bremen.de/.

Stolte et al., (2007) followed this approach and investigated the effects of seven different head groups, including three aromatic (4-(dimethylamino)pyridinium, pyridinium and imidazolium), three non-aromatic heterocycles (4-methylmorpholinium, 1-methylpiperidinium, 1-methylpyrrolidinium) and one non-cyclic quaternary ammonium head group (*N,N*-dimethylethylammonium) on the toxicity of ILs to the rat leukaemia cell line IPC-81. The seven head groups were combined with eight different aliphatic side chains containing ether (in different positions), terminal hydroxyl and nitrile groups. The toxicity was very low ($EC_{50} > 10$ mM) in 45 of the 100 compounds tested, moderate ($EC_{50} \sim 1000$ μM) in 52 and significant ($EC_{50} < 100$ μM) in only 4. The results obtained for ILs with the same fixed side chain and halides as counter anions indicated the minimal influence of the head group on IL cytotoxicity; a clear head group effect was found only for butyl-substituted head groups, especially 4-(dimethylamino)pyridinium. The EC_{50} of this latter compound was distinctly lower (by 1–2 orders of magnitude) than that of the corresponding imidazolium head group. This was explained by very high lipophilicity of the 4-(dimethylamino) pyridinium moiety due to the broadly delocalized positive charge and lipophilicity of dimethylamino group. In contrast, the cytotoxicities of morpholium and ammonium head groups were very low. The EC_{50}s between 0.15 and 14 mM, obtained in another study for a series of 4-benzyl-4-methylmorpholinium salts with different inorganic and organic anions with the same cell line, also enabled to classify the compounds as moderate or low toxic

(Pernak et al., 2011). In the study performed by Stolte et al., (2012) ammonium- and pyrrolidinium-based cations combined with methylsulphate ($1OSO_3^-$), methylsulphonate ($1SO_3^-$) and bis(trifluoromethylsulfonyl)imide ($(CF_3SO_2)_2N^-$) anion were no cytotoxic up to the highest tested concentration of 1000 mg/l. Kumar et al., (2009) used the human breast cancer cell line MCF7 to measure the cytotoxicity of piperidinium and pyrrolidinium cations, new alternatives to imidazolium and pyridinium, with different anions. The EC_{50}s of the ILs towards the MCF7 cells ranged from 8 μM to 44 mM. The authors showed that piperidinium and pyrrolidinium were about 10 times less toxic than their pyridinium homologues. Moreover, the effect of increasing the number of carbon atoms in the aromatic ring by replacing the pyrrolidinium cation with piperidinium resulted in a 2 – 6 fold increase in the toxicity of cation rings containing propyl, butyl or octyl substituents. The influence of the aromatic cation ring on the toxicity of the IL depended upon the length of the alkyl chain substituents: when the octyl-substituted aromatic ring of the pyridinium cation was replaced by the saturated piperidinium cation, the EC_{50} increased 25-fold. These observations are consistent with those of Stolte et al., (2006), who showed that cation rings with long, but not short, alkyl chains affected IL toxicity. Finally, the presence of a methyl group attached to the pyridinium cation had only a minor influence on IL toxicity, resulting in a maximum three-fold change in toxicity between 4-methyl-1-octylpyridinium bromide (Py8-4Me Br) (0.024 mM) and 1-octylpyridinium bromide (Py8 Br) (0.008 mM). Py8-4Me Br was the least toxic of the four ILs containing a pyridinium cation and an octyl chain. However, the change in toxicity depended on the substitution position; the cytotoxic effects of 2-methyl-1-octylpyridinium bromide and 3-methyl-1-octylpyridinium bromide were the same as that of Py8 Br. Wang et al., (2007) demonstrated that phosphonium bis(trifluoromethylsulfonyl)imide salts exhibited the highest toxicity towards HeLa cells, followed by, in decreasing order, 1-alkyl-3-methylimidazolium, alkylpyridinium, alkyltriethylammonium and *N*-alkyl-*N,N*-dimethyl-*N*-(2-hydroxyethyl)ammonium salts. The toxicity of methylimidazolium with different substituent chains, choline, Aliquat® (tri-n-octylmethylammonium), P666-14 (trihexyltetradecylphosphonium) and DMG (tetra-n-hexyldimethylguanidinium) cations combined with different types of anions, was evaluated in two colon carcinoma HT-29 and CaCo-2 cell lines. Aliquat® dicyanoamide, Aliquat® acesulfamate and Aliquat® saccharinate drastically reduced cell viability even at low doses: at concentrations of 100 μM these ILs reduced viability to 20%. The toxicity of the cations seemed to follow the choline/methylimidazolium < DMG < phosphonium < Aliquat® trend (Frade et al., 2007). The cytotoxicity of a group of redesigned ILs containing 1-alkoxymethyl-3-hydroxypyridinium cations combined with acesulfamate, saccharinate and chloride anions was also studied: the EC_{50}s from IPC-81 leukaemia cell line assays ranged from 0.03 to 5.2 mM, indicating the low toxicity of these compounds (Stasiewicz et al., 2008). According to the authors the reason for the relatively low biological activity must be at the cationic rather than the anionic site: additional oxygen-containing functional groups decreased the overall hydrophobicity of the molecule and concomitantly the toxicity. Jodynis et al., (2010) investigated the cytotoxicity of didecyldimethylammonium saccharinate ([DDA][Sac]) towards six types of cell lines originated from the human gastrointestinal tract, liver, lung and skin to provide information from different tissues. Differences in cell sensitivity to [DDA][Sac] were found, the compound was more toxic to skin and lung cells than to the cell lines derived from the gastrointestinal tract. Although the EC_{50}s varied from 1.44 μM to 5.47 μM, [DDA][Sac] killed all cells at a concentration of 10 μM, regardless of cell line type.

The general trend that the more hydrophobic aromatic head groups like pyridinium and imidazolium exert stronger effects on cell viability than non-aromatic and hence the more hydrophilic pyrrolidinium and piperidinium head groups was observed for every cell line tested. The cytotoxicities of morpholium and choline cations were very low.

All the above studies reported decreased viability (increase in cytotoxicity) as a result of the increasing alkyl chain length of the substituents for each cation class. Kumar et al., (2009) concluded that the short-chain alkyl groups exhibit a critical concentration window, beyond which ILs jump from nearly non-toxic to completely toxic. Depending on the mammalian cell line used, the observed relationship differs slightly in its intensity. McLaughlin et al., (2011) studied the cytotoxicity of 1-alkylquinolinium bromide (Quin*n* Br (n = 8, 10, 12, 14, or 18)) in murine fibroblast NIH 3T3 cells and found a significant, tenfold, decrease in EC_{50} as a result of the alkyl chain being lengthened from 8 (137 μM) to 10 (14 μM) carbon atoms. Further increases in alkyl chain length resulted in diminishing reductions in EC_{50}s with toxicity reaching a plateau at Quin14–Quin18 Br. This phenomenon is well known from the literature as "cut off" effect for highly hydrophobic substances (log Kow > 5). Different explanations for this effect have been presented, based on either insufficient solubility (nominal concentration deviating from real test concentration) or kinetic aspects (uptake is slowed down because of the steric effects for compounds with large molecules) (Ranke et al., 2007).

As far as the structural design of IL cations is concerned, the side chain is the main effector altering cytotoxicity. Stolte et al., (2006) reported the strongest cytotoxic influence of the butyl side chain in all head groups. Lipophilicity and concomitantly the cytotoxicity of investigated cations was affected mainly by the polarity of the side chain. In accordance with their relatively low lipophilicity, short functionalized side chains can diminish cytotoxicity to a greater extent than non-polar alkyl chains. The introduction of an oxygen atom into the side chain reduced cytotoxicity. Additionally, the terminal hydroxyl groups were found to be more polar than the ether groups. The presence of hydroxyl, nitrile, dimethyldisulfide and methyl nitrile functional groups in the side chains decreased toxicity by nearly one order of magnitude compared to ILs without these groups. The presence of these functional groups presumably hinder cellular uptake by membrane diffusion and reduce lipophilicity-based interactions with the cell membrane (Stolte et al., 2006, 2007; Frade et al., 2007; Kumar et al., 2009).

Although it has been asserted that modifications of the anion lead to changes in the chemical and physical properties of ILs (Sheldon, 2001), no general systematic effect of the anion on IL toxicity has been observed (Ranke et al., 2004). The cytotoxicity results obtained for the 1-butyl-3-methylimidazolium (IM14) cation associated with chloride (Cl^-), tetrafluoroborate (BF_4^-) and hexafluorophosphate (PF_6^-) towards HeLa cells were probably dependent on the type of anion. The lowest effective concentration was found for tetrafluoroborate species (0.63 mM), whereas hexafluorophosphate and chloride inhibited HeLa cell growth with comparatively high concentrations of > 10 mM. Surprisingly, when the anion effects were compared, PF_6^- was found to have the strongest inhibition (Stepnowski et al., 2004). A similar effect was observed by Ranke et al., (2006) in IPC-81 leukaemia cells, where the lower toxicity of IM14 PF_6 compared to the hexafluorophosphate anion alone was explained by reduced anion uptake due to the formation of an ion pair. The anion in this ion pair can, however, also partly decompose, as shown by Swatloski et al., (2003). The influence of the Br^-, BF_4^-, AsF_6^- (hexafluoroarsenate) and $(CF_3SO_2)_2N^-$ anions on the cytotoxicity of

ILs towards MCF7 was studied (Kumar et al., 2009). These authors observed that the $(CF_3SO_2)_2N^-$ anion combined with piperidinium, pyrrolidinium, pyridinium or imidazolium cations was more toxic than the bromide anion. It has been suggested that the increase in toxicity of ILs with the $(CF_3SO_2)_2N^-$ anion could be due to the hydrolytic cleavage that results in the formation of free fluoride ions. However, ILs containing the AsF_6^- or BF_4^- anion were just as toxic as their bromide homologues, indicating that it was the cation that controlled the toxicity in both cases. They also observed that the toxicity of the IL due to the hydrophobic anion was particularly significant when the alkyl substituent in the pyrrolidinium or piperidinium cation was short. However, replacing bromide with the $(CF_3SO_2)_2N^-$ anion in ILs with 1-butyl-1-methylpyrrolidinium or 1-butyl-1-methylpiperidinium cations increased the toxicity by one order of magnitude, whereas in ILs containing 1-methyl-1-octylpyrrolidinium or 1-methyl-1-octylpiperidinium cations no increase in toxicity was observed. Interestingly, the fluorinated anions did not have as much influence in the case of pyridinium ILs, and substituting bromide for the $(CF_3SO_2)_2N^-$ ion in IL based on 4-methyl-1-octylpyridinium cation had no influence on its toxicity. This implies that the aromatic ring of the pyridinium cation influences the toxic mechanism much more significantly than the anion does. The presence of the $(CF_3SO_2)_2N^-$ anion substantially raised the toxicity of ILs towards HT-29 and CaCo-2 cell lines, independently of the cation used. The least toxic anion was acesulfame, while the others (BF_4^-, PF_6^-, saccharinate, dicyanamide $(CN)_2N^-$) induced a slight decrease in viability (Frade et al., 2007). Wang et al., (2007) confirmed the low toxicity of Cl^- and Br^- and the high toxicity of $(CF_3SO_2)_2N^-$ anions paired with different imidazolium, pyridinium, ammonium and phosphonium cations. Salts containing the tetrafluoroborate anion exhibited the highest EC_{50}s, which is in contrast to the results obtained with the same cell line by Stepnowski et al., (2004). However, the effect of changing the anion was smaller than that of changing the alkyl substituent, e.g., while 1-ethyl-3-methylimidazolium tetrafluoroborate (IM12 BF_4) was observed to have an EC_{50} of 9.9 mM, the corresponding $(CF_3SO_2)_2N^-$ salt had an EC_{50} of 1.8 mM, both being considerably less toxic than 3-methyl-1-octylimidazolium bromide (IM18 Br) (EC_{50} = 0.3 mM). A similar effect was observed for 1-alkoxymethyl-3-hydroxypyridinium cations with acesulfamate, saccharinate and chloride anions. For ILs with short alkoxyl chains EC_{50}s for chlorides were 2–3 times lower than the corresponding values for acesulfamates and saccharinates. Nevertheless, all the cations with the longest alkoxyl chains (C11) inhibited cell growth the most effectively, regardless of the anion (Stasiewicz et al., 2008). An anion effect on IL cytotoxicity was demonstrated for 10 out of 27 anions (including borates, methides, amides, imides, phosphates, phosphinates, antimonates, sulfates and sulfonates) tested by Stolte et al., (2006). The report confirmed previous observations that halides did not exhibit an intrinsic anion effect up to a concentration range of 5000 μM and that all observed cytotoxic effects for compounds with these anions could be attributed exclusively to the cation. The $(CF_3SO_2)_2N^-$, octylsulfate ($8OSO_3^-$) and bis-[1,2-benzenediolato(2-)]borate anions exhibited an enhanced cytotoxic effect. The toxicity of the reactive bis(trifluoromethyl)imide ($(CF_3)_2N^-$) anion was surprisingly low in IPC-81 cells. Small, hydrophilic anions like Cl^-, Br^-, PF_6^- and BF_4^- contributed less to the toxicity of ILs, whereas bulky, hydrophobic and fluorinated species like $(CF_3SO_2)_2N^-$ or $8OSO_3^-$ exhibited evident cytotoxic effects and had a strong influence on cell viability. As far as chemical stability is concerned, hydrolysis reactions are the main type of chemical transformations that certain anions undergo. The hydrolysis of some anions has recently been investigated and confirmed. According to Stolte et al., (2006) compounds such as bis-[1,2-benzenediolato (2-)] borate, bis-[oxalato(2-)]-borate, bis(trifluoromethyl)imide, and hexafluoroantimonate (SbF_6^-) are also vulnerable to hydrolytic cleavage and degradation. Some of the chemical species formed by hydrolysis may exhibit a high intrinsic reactivity with cytotoxic effects. In particular, the hydrolysis of fluorine-containing anions is of interest since they may form toxic, free fluoride ions. The fluoride ion is a potent inhibitor of Na^+K^+ATPase, located on the cell surface, and may interfere with essential cellular processes. Also, the mode of action based on compound lipophilicity was taken into account in the case of anion toxicity, where highly fluorinated alkyl side chains were present. The influence of the degree of fluorination on cytotoxicity was supported by the case of IM14 CH_3SO_3, the cytotoxicity of which (EC_{50} = 3200 mM) was significantly lower than that of IM14 CF_3SO_3 (EC_{50} = 1000 mM). Since trifluoromethanesulfonate is a stable anion under physiological conditions, the increase in cytotoxicity in this case cannot be due to the formation of hydrogen fluoride, but must be a consequence of the greater lipophilicity of the anion. Anion lipophilicity facilitates interaction with cell membranes and hydrophobic protein domains, potentially disrupting essential physiological functions. Stolte et al., (2006) used the concentration addition model to estimate IL cytotoxicity. The cytotoxicity of most of the cation and anion combinations tested could be reduced to the respective intrinsic cytotoxicities of the cation and anion species. Nevertheless, the overadditive effect was observed for the $(CF_3SO_2)_2N^-$ anion in combination with different imidazolium cations (the experimental EC_{50}s were three times lower than the calculated ones). This effect may be due to the temporary formation of direct ion pairs in aqueous solution, which could result in a significantly higher bioavailability of the particular IL, amplifying membrane interactions and entailing stronger cytotoxic effects. The hazard potential of *N,N,N*-trialkylammoniododecaborates, which are anions in a new class of ILs, was investigated using the V79 cell line. A group of seven different anions was tested: *N,N,N*-trimethylammonio-undecahydro-*closo*-dodecaborate (MeAB), *N,N,N*-triethylammonio-undecahydro-*closo*-dodecaborate (EtAB), *N,N,N*-tripropylammonio-undecahydro-*closo*-dodecaborate (PrAB), *N,N,N*-tributylammonio-undecahydro-*closo*-dodecaborate (BuAB), *N,N,N*-triisopentylammonio-undecahydro-*closo*-dodecaborate (iPnAB), *N,N,N*-trihexylammonio-undecahydro-*closo*-dodecaborate (HxAB), *N,N*-diethyl-*N*-benzylammonio-undecahydro-*closo*-dodecaborate (Et2BnAB) and the structurally identical but unalkylated ammonio-undecahydro-dodecaborate (B12NH3). The EC_{50}s obtained ranged between 9.1 mM for the trimethylammonio derivative (MeAB) and 0.19 mM for the trihexyl derivative (HxAB). The incorporation of benzene instead of one alkyl chain in the asymmetrically substituted derivative Et2BnAB resulted in an increase of toxicity (EC_{50} = 0.72 mM) relative to EtAB (EC_{50} = 1.66 mM). In this study increasing lipophilicity (measured by the Kow value) and cytotoxicity were observed to be correlated (Schaffran et al., 2009). The cytotoxicity levels of five phosphate-based anion moieties, namely, $H_2PO_4^-$ (DHP), dibutyl phosphate (DBP), bis(2-ethylhexyl) phosphate (BEH), bis(2,4,4-trimethylpentyl) phosphinate (TMP) and *O,O*'-diethyl dithiophosphate (DEP), all paired with the choline [Ch] cation, were established using the murine macrophage cell line J774. The EC_{50}s ranged from 33.7 mM ([Ch][Cl]) through 20 mM ([Ch][DHP]) to 9.1 mM ([Ch][DBP]) and 8.2 mM [Ch][DEP]). The EC_{50}s of [Ch][TMP] and [Ch][BEH] were considerably lower – 0.25 and 0.3 mM respectively (Weaver et al., 2010). Nitrate (NO_3^-), hydrogen sulphate (HSO_4^-), formate, acetate, methoxyacetate, 2-(2-methoxyethoxy)acetate, hexanoate, 2-ethylbutyrate, lactate, crotonate and saccharinate anions did not influence the cytotoxicity of ILs containing the 4-benzyl-4-methylmorpholinium cation. However, the combination with BF_4^-, octanoate, dodecyl sulphate ($C_{12}OSO_3^-$), maleate and salicylate yielded lower EC_{50}s. The acute toxicities indicate that substitution of chloride by an organic anion containing two oxygen atoms decreases toxicity. However, the presence of three or more oxygen atoms, a linear alkyl group containing five or more carbon atoms, or an aromatic ring in the anion is associated with an increase in toxicity (Pernak et al., 2011).

Cytotoxicity studies using mammalian cell lines have clearly documented the correlation between the lipophilicity (hydrophobicity) of ILs (driven mainly by the substituents at the cationic head group) and their toxicity, a well-known phenomenon in (eco)toxicology. This mode of action is referred to as "necrosis" in mammals and as "baseline toxicity" with respect to aquatic systems. The increasing lipophilicity of a chemical facilitates its uptake into an organism, resulting in a higher internal concentration and hence more serious toxic effects. Moreover, there is a stronger impact on the membrane system because cations with enhanced hydrophobicity adsorb onto or intercalate into membranes, resulting in membrane perturbation (expansion or swelling, increase in fluidity, lowering of the phase transition temperature, alteration of the ion permeability of the membrane and interference with membrane proteins) (Ranke et al., 2007). The correlation between the hydrophobicity of ILs and their cellular sorption, distribution and toxicity was demonstrated by Ranke et al., (2006). Imidazolium tetrafluoroborates with longer alkyl chains were sorbed with higher enrichment factors and sorption coefficients. The concentration of IM18 in the cells was 17 times higher than in the culture medium, whereas IM16 and IM14 were enriched in the cell volume to a much lesser extent – 3.5 and 2.3 times respectively. Those authors found that imidazolium cations were accumulated by the cells with the largest fraction of IM18 found in the cytosol (about 80%), 8% in the nuclear fraction and about 10% in the membrane fraction. The correlation found between cytotoxicity and the HPLC-derived lipophilicity parameters ($\log k_0$) confirmed necrosis/baseline toxicity as the mode of action for ILs containing small, lesser toxic anions. Studies of the interaction between ILs and biomembranes using the Langmuir technique with lipid monolayers consisting of three different types of phosphatides demonstrated that IM18 BF_4 produced significant dose-dependent modifications of the surface pressure, which suggests that the integrity of the lipid layer was disrupted with potential consequences to biomembrane function. In contrast, IM14 BF_4 caused only slight changes in the properties of the mechanical lipid layer (Ranke et al., 2006). The mode of toxicity of long alkyl chain ILs *via* destabilization of the outer plasma membrane was confirmed using fluorescence microscopy. The total loss of cellular integrity, indicating rapid, putatively necrotic cellular breakdown due to the disruption of cellular membrane integrity, was observed at 1 h with 1 mM 1-tetradecylquinolinium bromide (McLaughlin et al., 2011). It is not clear whether the disruption of membrane integrity was a direct result of IL interaction, or an indirect by-product of rapid cellular death through another mechanism (Ranke et al., 2007). In another study of the possible mechanisms of the toxic actions of ILs on cells, liposomes were used as a model for cell membranes (Schaffran et al., 2009). Disturbance of the liposome membrane or disordering of the bilayer caused leakage of the liposomal content. All the trialkylammonio derivatives were able to induce leakage of the liposomal contents. The concentrations required to induce around 80% leakage after 8 minutes decreased with increasing alkyl chain length from 10 mM for MeAB and 0.5 mM for BuAB, to 0.05 mM for HxAB. In contrast, B12NH3, with no alkyl chains, was unable to induce leakage, even at a very high concentration (100 mM). Much higher concentrations of compounds with iso-alkyl chains necessary to provoke the release of liposomal content indicates that additional factors, such as steric effects, may have to be invoked to explain the action of the compounds. Wang et al., (2007) attempted to determine the possible sources of the observed toxicity of ILs using different microscopic methods (IPCM, SEM, CFM). Fluorescence microscopy showed early stages of apoptosis, which turn into late stages of apoptosis on exposure to higher concentrations of IM12 BF_4. Cell necrosis was observed, possibly due to direct membrane destruction by the compound. The observed decrease in mitochondrial membrane potential, enhanced production of oxidative oxygen species (ROS) and elevated concentration of intracellular calcium clearly indicated that the IM12 BF_4 treated cells undergo apoptosis.

The molecular toxicity of ILs was investigated using different *in vitro* enzymatic assays, the most frequently used one being the acetylcholinesterase (AChE) inhibition assay. This enzyme is an essential part of the nervous system. Widely used as a biomarker detecting pesticides, it is also inhibited by different classes of chemicals, including metals and surfactants (Herbert et al., 1995; Guilhermino et al., 1998). More recently, it was used to assess the environmental risk of perfluorinated carboxylic acids (Mulkiewicz et al., 2007). The acetylcholinesterase inhibition assay was used to assess different groups of ILs in the systematic studies performed by Jastorff's group. The enzyme was chosen because of the similarity of the chemical structure of many IL cations and the choline part of the enzyme substrate (acetylcholine). Both are cationic and the positive charge is masked by lipophilic residues (Raves et al., 1997).

Imidazolium ILs showed acetylcholinesterase inhibition with IC_{50}s between 13 and 189 μM. A correlation between the alkyl chain length of R2 and inhibitory potency was found: a longer chain length resulted in stronger inhibition of the enzyme. No influence of the elongation of R1 (1-butyl-3-ethylimidazolium tetrafluoroborate compared to IM14 BF_4) was noticed. This indicated that toxicity was influenced not only by the lipophilic properties but also by the steric position of the lipophilic side chains. The inhibitory concentration of 1-decyl-3-methylimidazolium tetrafluoroborate (13 μM) was almost as low as for aldicarb (5 μM), an insecticide designed to kill insects *via* mechanism of acetylcholinesterase inhibition. The ILs with pyridinium as the cationic core structure inhibited the enzyme slightly more strongly than the imidazolium analogues: for comparison, the IC_{50} obtained for 1-butyl-3-methylpyridinium tetrafluoroborate (Py4-3Me BF_4) was three times lower than for the corresponding 1-butyl-3-methylimidazolium salt. Likewise, as in the case of imidazolium ILs, the inhibitory potency depended on the alkyl chain length in position R2. Phosphonium ILs did not inhibit AChE activity (Stock et al., 2004). A study by Arning et al., (2008) with a group of ILs with a large variety of head groups, aromatic pyridinium and imidazolium head groups, as well as heterocyclic but non-aromatic piperidinium and pyrrolidinium moieties, showed their high inhibitory potency towards acetylcholinesterase. The positively charged nitrogen atom, a widely delocalized aromatic system, and the lipophilicity of the side chains in the cationic head groups were identified as the key structural elements in binding to the enzyme's active site. The most striking inhibitory effect was achieved with the dimethylaminopyridinium and the quinolinium head groups with respective IC_{50}s of 4 μM and 6 μM. Methylated pyridinium cations also significantly inhibited AChE activity, but the inhibitory potential was strongly dependent on the position of substitution: whereas the IC_{50} of 1-butyl-4-methylpyridinium was 27 μM, that of 1-butyl-2-methylpyridinium was only 5 μM. The pyridinium and imidazolium head groups, as well as the piperidinium and pyrrolidinium moieties, displayed a moderate AChE inhibition potency (IC_{50}s from 50 to 83 μM). In contrast, the polar and non-aromatic morpholinium head group, as well as the sterically bulky tetrabutylammonium and tetrabutylphosphonium cations, exhibited the lowest inhibitory potential towards the enzyme. It was shown that the introduction of polar hydroxy, ether or nitrile groups into the alkyl side chain of IL lowered its inhibitory potential towards the enzyme. Furthermore, it was demonstrated that for the strongly inhibiting dimethylaminopyridinium and quinolinium head groups, the side chain effect was marginal compared to the less inhibitory imidazolium and pyridinium cations. Stasiewicz et al., (2008) investigated the influence of 1-alkoxymethyl-3-hydroxypyridinium cations with three different anions (acesulfamate, saccharinate and chloride) on AChE activity. No significant enzyme inhibition was observed: IC_{50}s ranged from 0.3 to > 7.8 mM, the lowest ones being attributed to the compounds with the longest (C11) alkyl chains. According to these authors is possible that an additional hydroxyl group in the pyridinium ring prevents interaction between the quaternary nitrogen in the IL cation and the anionic enzyme site. No influence of the anion on the inhibitory potential of the tested ILs could be ascertained. The EC_{50} values obtained for cholinium amino acids ILs to acetylcholinesterase, approximately an order of magnitude higher than that of the traditional IL 1-butyl-3-methylimidazolium tetrafluoroborate, confirmed low inhibitory potential of choline-based ILs (Hou et al., 2013).

The influence of ILs on the activity of other enzymes, namely AMP-deaminase, catalase and the monooxygenases P450 BM-3 and CYP3A4 was also investigated. Składanowski et al., (2005) demonstrated that some imidazolium ILs were able to inhibit AMP-deaminase, the enzyme essential for regulation of adenine nucleotide pools in the cell. The anion was shown to have an influence: the IC_{50}s of ILs containing fluorine PF_6^- and BF_4^- were slightly lower (5μM) than those of chlorine and *p*-tosylate (10μM). In contrast, imidazolium ILs exhibited a weak inhibitory effect on catalase *in vitro*, with the IC_{50}s between 2.77 M (IM12 CH_3SO_3) and 0.15 M (IM12 BF_4) thousands of times higher than those of sodium azide - positive control for catalase inhibition. The effective concentrations were very high considering the possible environmental threshold values. Regarding the cationic head group, it was observed that ILs containing the non-aromatic pyrrolidinium and phosphonium groups exhibited lower IC_{50}s than imidazolium cations with the same anion (Pinto et al., 2011). The effects of ten ILs consisting of four 1-alkyl-3-methylimidazolium derivatives differing in hydrophobic chain length, two altered head groups (pyridinium and pyrrolidinium) and an additional functional group (nitrile) in the side chain on cytochrome P450 BM-3 activity were investigated. The results indicated that varying the chain length of the 1-alkyl-3-methylimidazolium had evident effects on the resistance of P450 BM-3 (based on IC_{50}), with inhibition increasing in the order IM12 Cl (232 mM), IM14 Cl, (148 mM), IM16 Cl (32 mM) and IM18 Cl (9 mM). No influence of the head group on inhibitory properties towards the enzyme under investigation was observed, and changing the cation to pyridinium and pyrrolidinium or the addition of a nitrile functional group did not cause obvious changes in IC_{50} (Tee et al., 2008). The activity of CYP3A4 in the presence of 1,3-dimethylimidazolium tetrafluoroborate (IM11 BF_4), IM14 BF_4 and 1-butylpyridinium tetrafluoroborate (Py4 BF_4) as a potential solvents (in a biphasic system with potassium phosphate buffer) for biocatalysis was tested. The reported decrease in CYP3A4 activity was comparable to that observed in the presence of organic co-solvents. Enzymatic activity in buffer/ IM14 PF_6 emulsions was slightly higher. The addition of 25% of this IL was necessary to decrease the activity to 15%. The volumes tolerated were, however, insufficient to dissolve large amounts of non-polar substrates (Chefson and Auclair, 2007).

The influence of IL anions on the inhibition of acetylcholinesterase was tested with IM14 combined with several anions (BF_4^-, Br^-, Cl^-, PF_6^-, $8OSO_3^-$, $CF_3SO_3^-$, $(CN)_2N^-$, $MDEGSO_4$, α-methyl[poly(oxy-1,2-ethanediyl)]sulfate). The IC_{50}s varied between 80 (IM14 Br) and 140 mM (IM14 PF_6). For the pyridinium cation the influence of the anion was different than for the imidazolium cation. While IM14 PF_6 was less inhibitory than IM14 BF_4, the corresponding results for Py4-3Me were reversed. The IC_{50}s of 28 – 34 μM obtained for alkylpyridinium ILs with fluoroanions were some of the lowest IC_{50}s in the pool of ILs tested (Stock et al., 2004). The IC_{50}s of IM12 with different anions (Cl^-, $CF_3SO_3^-$, CF_3COO^- and BF_4^-) showed the lowest tolerance of P450 BM-3 towards the BF_4^- anion (Tee et al., 2008). ILs with BF_4 also noticeably inhibited catalase, whereas the $CF_3SO_3^-$ and $CH_3SO_3^-$ anions did not affect enzyme activity (Pinto et al., 2011). In the study performed by Matzke et al., (2007) ILs with $(CF_3)_2N^-$, $(CF_3SO_2)_2N^-$ and $(2\text{-}OPhO)_2B^-$ anions inhibited enzyme activity more strongly. A broad set of anion species (inorganic, organic and complex borate anions), commonly used as IL counter ions, was tested by Arning et al., (2008) and the vast majority exhibited no effect on AChE activity up to the highest tested concentration range of 1000 μM. Only for fluoride and fluoride containing hexafluoroantimonate and hexafluorophosphate a significant inhibition was found, with respective IC_{50}s of 575 μM (F^-), 219 μM (SbF_6^-) and 145 μM (PF_6^-). AChE inhibition of *N,N,N*-trialkylammoniododecaborates, which are anions in a new class of ILs, was reported. The longer the alkyl chain length, the greater the enzyme inhibition. The MeAB derivative inhibited enzyme activity to 50% at a concentration of 21.9 mM; in comparison, only 32.5 μM of HxAB was required for a 50% inhibition. Interestingly, Et2BnAB inhibited AChE at a concentration half that of PrAB, which was unexpected given its effect on V79 cells (Schaffran et al., 2009).

To acquire knowledge of the interaction between ILs and DNA, a surface electrochemical micromethod was applied and revealed that the aromatic group of imidazolium-based ILs interacted strongly with isolated DNA molecules (Xie et al., 2008). The mutagenic potency of imidazolium, pyridinium and quaternary ammonium bromides was investigated using the Ames test. This utilizes *Salmonella typimurium* strains to provide information on the ability of a chemical to cause frameshift and missense mutations in bacteria, which in turn can be related to carcinogenicity in higher organisms. It was observed that some ILs (IM14 Br and IM18 Br) displayed trends towards potential mutagenicity at high doses but did not meet the USEPA criteria for classification as a mutagen (Docherty et al., 2006).

TOXICITY TOWARDS MICROORGANISMS

Microorganisms play a fundamental role as major decomposers in many terrestrial and aquatic ecosystems and therefore are an important link in biogeochemical and nutrient cycles and trophic chains. Photosynthetic bacteria are important primary producers in the aquatic environment. Fungi have great relevance from the environmental point of view as they are the sole microorganisms capable of decomposing lignin. Symbiotic bacteria are also important for many higher organisms. Moreover, microorganisms play an important role in biodegradation processes. Microorganisms represent an ideal starting point for IL hazard assessment and structure–activity relationship investigations. Most of the compounds described below were designed, synthesized and tested as antimicrobial agents for their potential medical application or alternatively as disinfectants. One should be aware of the fact that the excellent antimicrobial properties of some ILs towards pathogenic microorganisms may at the same time represent a significant risk to other relevant microorganisms, among others those in wastewater treatment systems.

Docherty and Kulpa (2005) investigated the antimicrobial effects of butyl-, hexyl- and octyl- imidazolium and pyridinium bromide ILs (at a concentration of 1000 ppm) on a group of microorganisms representing a variety of physiological and respiratory capabilities. All ILs were most inhibitory to *Bacillus subtilis*, a Gram positive rod, whereas *Staphylococcus aureus*, a Gram positive cocci, and *Saccharomyces cerevisiae*, a yeast, were least affected by the IL treatments. *Escherichia coli* and *Pseudomonas fluorescens*, both Gram negative rods, were also sensitive to the IL treatments. Pernak et al., synthesized and studied the antimicrobial activity of different ILs, namely, 3-alkoxymethyl-1-methylimidazolium salts (Pernak et al., 2003), pyridinium and benzimidazolium chlorides (Pernak et al., 2001b), 1,3-dialkoxymethylimidazolium chlorides (Pernak et al., 2004b), 1-alkoxymethyl-3-carbamoylpyridinium derivatives (Pernak et al., 2001a) against cocci (*Micrococcus luteus, S. aureus, Enterococcus faecalis*), rods (*E. coli, Pseudomonas aeruginosa, Proteus vulgaris*), bacillus (*B. subtilis*) and fungi (*Candida albicans*). Generally, all these compounds were very active against cocci and fungi, and active against rods and bacillus, and their activities were greatly affected by the length of the alkyl or alkoxy side chains. 3-alkoxymethyl-1-methylimidazolium cations containing twelve carbon atoms in the alkoxy group exhibited very high antimicrobial activity with MIC (minimum inhibitory concentration, the lowest concentration at which there was no visible growth after 24 hours of incubation) values for *M. luteus* from 21 to 37 μM (depending on the anion) (Pernak et al., 2003). Although all the pyridinium and benzimidazolium chlorides studied were active against cocci, rods, fungi and bacillus, a wide range of MICs was obtained, from 6.5 μM to more than 1.4 mM. The biological activities of ILs were greatly affected not only by the alkyl chain length in the alkoxymethyl substituent but also by the kind of quaternary ammonium moieties in the molecule. The chlorides most active against these microorganisms were certain pyridinium derivatives. Their activities were similar to that of commercially available benzalkonium chloride (BAC). They were also effective against *Mycobacterium tuberculosis* (MIC < 6.25 g/ml) and were considered potential compounds for use in the treatment of mycobacterial infections. It was noticed that compounds with two quaternary ammonium moieties in the molecule were strongly active against bacteria and fungi (Pernak et al., 2001b). Considerable antimicrobial activity, comparable to that of BAC, and antitubercular activity was also observed in selected 1-alkoxymethyl-3-carbamoylpyridinium derivatives. The antimicrobial activity of the *O*- and *N*-alkoxymethylated products was significantly affected by the alkoxymethyl chain length in positions 1 and 3 (Pernak et al., 2001a). In general, salts with short substituents (propoxymethyl, butoxymethyl and pentyloxymethyl) were inactive against bacteria and fungi. However, it has been shown that elongating the chain in position R by more than twelve carbon atoms or symmetrical chains in positions R1 and R3 by more than 7 - 9 carbons decreased antimicrobial activity (Pernak et al., 2003, 2004b). Some of the possible explanations for this cut-off effect are the insufficient water solubility of highly lipophilic compounds, the tendency to self-assemble and form micelles, and the occurrence of steric effects that may limit interaction with the cell surface.

Chiral imidazolium and ammonium chlorides (with 9 - 11 carbon atoms in the alkyl group) exhibited an extremely high activity against microbes (with MICs < 0.25 μM) and a wide spectrum of activity, whereas chiral pyridinium chlorides demonstrated more specific activity. They were practically inactive against fungi and most Gram negative rods (*S. marcescens*, *P. vulgaris* and *P. aeruginosa*). On the other hand, these compounds proved to be very active against *M. luteus* and, in some cases (1-[(1R,2S,5R)-(-)-methoxymethyl]pyridinium chlorides with a hydroxyl or $N(CH_3)$ group in position R2 or R3), the activity was definitely more pronounced than that of BAC. The MICs and MBCs of chiral imidazolium ILs were comparable, indicating not only a strong biostatic effect, but also strong biocidal activity (Pernak et al., 2007a, Pernak and Feder-Kubis, 2005, 2006). The antimicrobial activity of long alkyl chain quaternary ammonium-based (didecyldimethylammonium (DDA), benzalkonium (BA), benzethonium (HA1622) ILs with lactate (L-lactate and D,L-lactate), NO_3^-, NO_2^-, BF_4^- and $(CF_3SO_2)_2N^-$ anions was tested. The results showed that didecyldimethylammonium and benzalkonium lactates were very effective: the MICs obtained for cocci, fungi and *E. coli* ranged from < 0.1 to 4 μg/ml. They were less effective against Gram negative rods, however. The data showed that DDA lactate was the most useful, regardless of whether it was in the L-lactate or D,L-lactate form (Cybulski et al., 2008). BA, DDA and HA1622 cations with different (NO_3^-, NO_2^-, BF_4^-) anions were also very effective agents against microorganisms ($2 < MIC < 31$ μg/ml). The $(CF_3SO_2)_2N^-$ salts exerted no antibacterial or antifungal effect (Pernak et al., 2006). According to the authors, the very low MIC and MBC values showed that these products had a high disinfection potential. Antimicrobial activity was also demonstrated in quaternary ammonium compounds (DDA, BA) with artificial sweetener anions (saccharinate and acesulfamate). The lack of antimicrobial activity observed in 3-hydroxy-1-octyloxymethylpyridinium chloride as well as in 3-hydroxy-1-octyloxymethylpyridinium saccharinate and 3-hydroxy-1-octyloxymethylpyridinium acesulfamate (Hough-Troutman et al., 2009) did not confirm the dependence of their antimicrobial activities on the substituent in position 3 of the pyridine ring (Pernak and Branicka, 2003). New choline-like quaternary ammonium chlorides and acesulfamates (Pernak and Chwała, 2003; Pernak et al., 2007c) also demonstrated antimicrobial activity: very low MICs of < 1 μM were obtained for these compounds. They were distinctly more active against Gram positive bacteria and fungi than against Gram negative bacteria, the highest resistance against them being exhibited by *P. aeruginosa*. Alkoxymethyl(2-hydroxyethyl) dimethylammonium chlorides were less active. Quaternary derivatives of esters had a higher antimicrobial activity than quaternary derivatives of alcohols. The activities of some alkoxymethyl(2-decanoyloxyethyl) dimethylammonium and alkoxymethyl(2-benzoyloxyethyl) dimethylammonium chlorides were similar to the activity of BAC. In contrast to the large group of compounds with pronounced antimicrobial activity, studies performed with a series of 4-benzyl-4-methylmorpholinium cations with different anions and 4-carbalkoxymethyl-4-methylmorpholinium chlorides revealed insignificant activity of morpholinium salts against bacteria and fungi (Borycki et al., 2010, Pernak et al., 2011). Trigeminal tricationic imidazolium and pyridinium chlorides demonstrated diverse antimicrobial activity. The latter were inactive against bacteria and fungi, but the former proved effective against microbes. The trigeminal tricationic imidazolium chloride salt containing an octyl substituent was very active against rods, cocci, bacillus and fungi, exhibiting an efficiency superior to that of commercially available BAC (Pernak et al., 2007b). Tetrabutylammonium salts paired with long-chain organic esters (formate, acetate, propionate, butyrate, benzoate, nitrobenzoate, cinnamate, salicylate, sulfanilate, linoleate and andoleate) demonstrated antimicrobial activity, and microorganism specificity was suggested. Tetrabutylammonium benzoate, butyrate, acetate, propionate and nitrobenzoate were active only against Gram positive bacteria, but the other compounds were also active against Gram negative bacteria (Saadeh et al., 2009). The antimicrobial activity of alkyltrihexylphosphonium chlorides and bromides was also selective: effects were strong against cocci and bacillus, but were insignificant against fungi. In the case of rods, activity was high only against *E. coli*. The loss of activity against a wide spectrum of microbes can be explained by the presence of four large substituents around the phosphorus atom (Cieniecka-Roslonkiewicz et al., 2005).

Significant differences in IL activity against various microorganisms have been observed. The results obtained with 1-alkoxymethylcarbamoylpyridinium derivatives, for example, showed them to be more active against Gram positive than against Gram negative bacteria (Pernak et al., 2001a). It has been suggested that the sensitivity of different microorganisms to ILs depends on the composition and properties of the cell envelope. Gram positive bacteria have thicker and more hydrophobic cell walls and a much higher peptidoglycan content (about 90%); the cell walls of Gram negative bacteria are chemically more complex and have an additional outer membrane composed mostly of phospholipids and lipopolysaccharides. Hence the outer membrane could act as a barrier against antimicrobial molecules. Gram positive bacteria do not have this kind of protection, so they could adsorb ILs more efficiently and transport them to the inner cell membrane (Samori, 2011). The results obtained by Docherty and Kulpa (2005) did not confirm that the structure of the bacterial envelope was the factor determining the observed differences in sensitivity to ILs.

The toxicity of different classes of ILs to fungi, however limited only to two species, namely *C. albicans* and *Rhodotorula rubra*, was indicated in most of the studies mentioned above. Petkovic et al., (2009) assessed the toxic effect of sixteen ILs (containing an imidazolium, pyridinium, piperidinium or cholinium cation) with different anions (chloride, ethyl sulphate ($C_2H_5OSO_3^-$), thiocyanate (SCN^-), ethanoate, DL-lactate, $(CF_3SO_2)_2N^-$) towards filamentous fungi (*Penicillium* sp.) as model eukaryotic organisms. It was also shown that fungi display a much greater tolerance towards ILs than any other microorganism studied so far. Imidazolium-based ILs had the highest toxicity rankings of all the ILs studied, with an average of three toxic cases. The choline cation elicited the lowest toxic effect, and choline chloride failed to inhibit growth in any of the fungal species. With the exception of 1-ethylpyridinium DL-lactate, which exhibited no toxicity, the remaining ILs lay between these extremes. It was shown that fungal cultures respond to specific ILs (e.g., choline chloride and 1-butyl-1-methylpyrrolidinium chloride) by changing their cell biochemistry, resulting in an altered pattern of secondary metabolites; the induced expression of a distinct set of fungal extracellular proteins was observed.

Those studies showed that the correspondence between microbial activity and cation structure was evident in every case, and in particular that it was related not only to the length of the alkyl and alkoxymethyl substituents, but even to the head group: trigeminal tricationic imidazolium salts, for example, based on a glycerol backbone and containing three quaternary nitrogen atoms connected to a central carbon atom, were more active than the dimeric and monomeric imidazolium analogues. The available data confirmed the low toxicity of the morpholium head group.

The anion effect was not observed with 3-alkoxymethyl-1-methylimidazolium cations combined with Cl^-, BF_4^- and PF_6^- anions (Pernak et al., 2003). Of the Br^-, BF_4^-, PF_6^-, $(CN)_2N^-$, $(CF_3SO_2)_2N^-$ and $8OSO_3^-$ anions tested in another study, the last two contributed more to the overall toxicity of imidazolium ILs, but in most cases the inhibitory end points were not reached (MICs > 70 mM) (Morrissey et al., 2009). The impact of anion size on the antimicrobial activity of the pyridinium salts was demonstrated. The most favourable were metal-containing anions – $MgCl_4^{2-}$, $CoCl_4^{2-}$, $ZnCl_4^{2-}$ and $FeCl_4^-$. For anions not containing a metal, mean MICs increased in the order $CH_3COO^- > Br^- = NO_3^- > Cl^- > ClO_4^- > BF_4^- > I^-$ (Pernak et al., 2001a). In the case of long alkyl chain quaternary ammonium-based ILs, NO_3 and NO_2 salts were very effective against bacteria and fungi, their activity being comparable to that of the original chlorides. The BF_4 salts proved to be slightly less effective, while $(CF_3SO_2)_2N$ salts exerted no antibacterial or antifungal effect at all (Pernak et al., 2006). Unlike the 1-alkyl-3-methylimidazolium analogues, both cation structure and the type of anion in phosphonium salts affected their biological activity. Exchange of halide for other anions (e.g., $CF_3SO_3^-$ $(CF_3SO_2)_2N^-$, $(CN)_2N^-$, NO_3^-, $(^iC_8H_{17})_2PO_2^-$, $C_9H_{19}CO_2^-$, dicyclohexylphosphinate (*cyc*-C_6H_{11}) PO_2^-, $(^iC_4H_9)_2PS_2^-$, BF_4^-, PF_6^-) resulted in the loss of the antimicrobial activity of the salts (Cieniecka-Roslonkiewicz et al., 2005). Lactates belong to a group of organic ions listed as non-toxic, pharmaceutically acceptable anions. As regards the impact of lactate as a counter anion in ILs, its combination with long-chain ammonium cations yields compounds with high biological activity in comparison with the same cation paired with different anions. DDA L-lactate and BA L-lactate displayed not only considerable antimicrobial activity but also acute oral and dermal toxicity (Pernak et al., 2006; Cybulski et al., 2008). Studies of 1-alkylimidazolium and 1-alkoxymethylimidazolium combined with lactate (protic ionic liquids) showed that the bacteriostatic activity of lactates was more pronounced, *i.e.,* the MICs were lower, than their microbiocidal activity, as measured by MBC values. The biological activity of protic ILs was related not only to the length of the substituent but also to the DL- or L-type anion. L-lactate salts were more toxic than D,L-lactate salts. Of all the salts examined, the microorganism growth inhibiting effect was the most pronounced in 1-undecyloxymethylimidazolium L-lactate (MIC *(M. luteus)* = 5.8 μM) against cocci, whereas 1-nonyloxymethylimidazolium DL-lactate proved to be the most effective against rods (MIC *(E. coli)* = 49 μM) and 1-dodecylimidazolium L-lactate against fungi (MIC (*C. albicans*) = 48 μM) (Pernak et al., 2004a). Łuczak et al., (2010) studied the impact of a wide range of anions on the antifungal effects of 1-alkyl-3-methylimidazolium derivatives. Those authors demonstrated a more distinct anion effect for short-chain imidazolium, whereas for ILs with six or more carbons in the side chain the cation effect was the main factor determining their antimicrobial activity. The influence of the anions decreased in the order bis(trifluoromethylsulfonyl)imide > octylsulfate > tetrafluoroborate > *p*-toluenesulfonate > methylsulfate > chloride > trifluoromethanesulfonate. A new group of environmentally friendly ILs was presented by Petkovic et al., (2010) in a paper focusing on the toxic effect of the anion. The ILs to be investigated were designed by combining the benign cholinium cation, with a range of linear alkanoate anions of systematically elongated chains ($C_nH_{2n+1}CO_2^-$, $n = 1 - 9$), as well as two branched isomers, 2-methylpropanoate and 2,2-dimethylpropanoate. Surprisingly, most of the species tested exhibited active growth in media containing extremely high IL concentrations, although they displayed very distinct behaviour; the MICs obtained for filamentous fungi varied from 2.2 mM to 1.5 M. The results showed that the toxicity of the anion was defined by its lipophilicity, as a consequence of the length of the linear chain. The inhibitory and fungicidal effects of the branched isomers were less than those of the linear isomers. Not only are cholinium alkanoates environmentally benign and biodegradable, they are also extraordinarily good solvents for some very recalcitrant plant biocomposites, in some cases reaching dissolution yields of 60 wt % (Garcia et al., 2010). Hou et al., (2013) evaluated toxicity of cholinium amino acids ([Ch][AA]) ionic liquids. Among the ILs tested, [Ch][Phe] and [Ch][Trp] were overall the most toxic to bacterial cells (*E. coli*, *S. aureus*, *Salmonella enteritidis*, and *Listeria monocytogenes*), what suggested that the incorporation of aromatic systems into the anions substantially increased the toxicity of the ILs.

Most published studies of the antimicrobial activity of ILs have focused on the planktonic mode of microbial growth, whereas the majority of microorganisms, both pathogenic and environmental, in fact grow as biofilms (highly-ordered surface-adhered communities enclosed within a self-produced protective extracellular polymeric matrix (glycocalyx)). Carson et al., (2009) evaluated the antimicrobial and antibiofilm activity of 1-alkyl-3-methylimidazolium against different pathogenic biofilm-forming strains, including MRSA. The MBECs (minimum biofilm eradication concentrations) for 3-methyl-1-tetradecylimidazolium chloride were 124 μM for *S. aureus* and 1984 μM for *Proteus mirabilis*. The MBECs were at least one order of magnitude higher than the MICs obtained for the same species. It was observed that the compounds IM1*n* Cl where $n \geq 10$ exhibited good antibiofilm activity. In general Gram positive cocci were the most sensitive (the lowest MICs (7.75 μM) were obtained for *S. epidermidis*) to these compounds whilst Gram negative rods and representative fungi were less susceptible. New ILs based on the alkylquinolinium cation also possessed antimicrobial and antibiofilm activities. For comparison the MBECs of 1-tetradecylquinolinium bromide were 414 μM for *S. aureus* and 1656 μM for *P. mirabilis*. The alkylquinolinium ILs are a good example that modifications to the head group can make a differences to both the potency and spectrum of antimicrobial activity (Busetti et al., 2010).

Many studies have used *Vibrio fischeri* as a model prokaryotic organism to study IL toxicity. *V. fischeri* is a Gram negative marine bacterium with bioluminescent properties related to the metabolic state of the cells. The bioluminescence is reduced when the bacterial cells come into contact with toxic compounds. The *Vibrio fischeri* acute bioluminescence inhibition assay is a standard ecotoxicological bioassay, widely used in environmental toxicity studies to screen aquatic ecosystems and to investigate the toxic potential of chemical compounds. The biological activity of ILs towards *V. fischeri* is well documented in the literature as regards the toxic effects of imidazolium, pyridinium, morpholinium, piperidinium, pyrrolidinium, ammonium and phosphonium head groups. The impact of increasing alkyl chain length on IL toxicity was also confirmed in this bioassay: 1-dodecyl-3-methylimidazolium was 10 thousand times more toxic than 1-ethyl-3-methylimidazolium. However, there is also evidence of a cut-off effect for very lipophilic compounds with long alkyl side chains (C10) (Stolte et al., 2007). The introduction of one oxygen atom in a hydroxyl or ether moiety increases the polarity of the side chain and reduces toxicity by one order of magnitude (Stolte et al., 2007, Samori et al., 2007). The same EC_{50} obtained for cations with ethoxymethyl, methoxyethyl, ethoxyethyl or methoxypropyl substituents shows that the reduction in toxicity to *V. fischeri* is not related to the position of the oxygen in the side chain. A further increase in the number of oxygen atoms and increasing polarity increases toxicity: IM12O2O2O2O1 $(CN)_2N$ was even more toxic than IM14 $(CN)_2N$ (Docherty and Kulpa, 2005). The toxicity of non-aromatic head groups such as morpholium, pyrrolidinium and piperidinium was low (EC_{50} > 10000 μM) to *V. fischeri* whereas that of aromatic ones like imidazolium and pyridinium was higher, though still in the moderate range (EC_{50} > 1000 μM) (Docherty and Kulpa, 2005; Stolte et al., 2007). The toxic effect of ILs with 6 - 8 carbon atoms in the alkyl chain was comparable; the differences became more distinct in ILs with short alkyl chains. The EC_{50} (1 mM) obtained for 1-butyl-3-methylimdazolium was three times higher than that reported for 1-butyl-3-methylpyridinium (Matzke et al., 2007, Stolte et al., 2007). The short-chain asymmetric ammonium salts were the least toxic compounds. The increase in substituent length of the ammonium moiety significantly raised the toxicity, from an effective concentration of 100 mM for tetraethylammonium bromide through 1.86 mM for tetrabutylammonium bromide to 0.3 mM for triethyl(butyl)ammonium bromide. Phosphonium salts were more toxic than ammonium salts (Couling et al., 2006).

In the case of imidazolium based ILs $(2\text{-OPhO})_2B^-$, $(CF_3SO_2)_2N^-$ and $8OSO_3^-$ were the most toxic anions: the EC_{50} values for ILs with $(CF_3SO_2)_2N^-$ anion were one order of magnitude lower than for the same cations with chloride (Couling et al., 2006; Matzke et al., 2007; Stolte et al., 2007). All the other anions – bromide, chloride, tetrafluoroborate, hexafluorophosphate, dicyanamide and *p*-toluenesulfonate – had EC_{50}s of the same order of magnitude (Matzke et al., 2007, Samori et al., 2007, Stolte et al., 2007). Pyridinium, choline and piperidinium ILs with the $(CF_3SO_2)_2N^-$ anion were more toxic than the same ILs with halides, and the difference was even more apparent in the case of morpholium, pyrrolidinium and ammonium head groups with the same anion (Couling et al., 2006; Stolte et al., 2007).

Toxicity to Algae

A considerable number of reports concerning the toxicity of ILs to aquatic organisms have focused on phytoplankton. The most frequently used test organisms were the freshwater green algae *Pseudokirchneriella subcapitata* (formerly known as *Selenastrum capricornutum*) (Wells and Coombe, 2006; Cho et al., 2007, 2008a,b,c; Pharm et al., 2008a,b; Pretti et al., 2009), *Scenedesmus subspicatus, Scenedesmus quadricauda, Chlamydomonas reinhardtii* (Kulacki and Lamberti, 2008), *Scenedesmus vacuolatus* (Matzke et al., 2007, 2008, Stolte et al., 2007; Schaffran et al., 2009) *Scenedesmus obliquus* and *Chlorella ellipsoidea* (Ma et al., 2009) and the seawater green algae *Chlorella vulgaris* and *Oocystis submarina* (Latała et al., 2005, 2009a). Marine diatoms *Cyclotella meneghiniana, Skeletonema marinoi* and *Bacillaria paxillifer* and cyanobacterium *Geitlerinema amphibium* were also used as model organisms for IL toxicity tests (Latała et al., 2009 a,b). The widespread distribution of algae (freshwater, marine and terrestrial environments) makes them ideal test organisms for toxicological studies and, because they have a short life cycle, they respond quickly to environmental change. They are considered to be primary producers in the aquatic environment, so their ecological role is crucial in supplying the energy to sustain other higher trophic levels. The algal tests provide information about the availability of chemical compounds to an organism and their inhibitory effects. Algal photosynthesis is one of the major physiological phenomena contributing to algal growth, and which may also affect the structure and functioning of whole aquatic ecosystems. Microalgal photosynthesis has been shown to be an effective means of estimating the toxicity of toxicants: the results for short-term photosynthetic activity test coincided well with conventional algal growth assays, the EC_{50}s for IM14 BF_4 being 0.115 mM and 0.126 mM respectively (Cho et al., 2008c). Thus two approaches were followed to quantify toxicity of these compounds: cell proliferation and analyses of photosynthetic activity. The influence of structural elements like different head groups, functionalized side chains and the anionic moiety on the toxic potential of ILs towards algae has been explored in many studies.

As far as head groups are concerned, the toxicities of morpholinium and ammonium compounds were very low in the algae assay. In contrast, the dimethylaminopyridinium core had a dramatic effect on the growth of *S. vacuolatus* (Stolte et al., 2007). Pretti et al., (2009) tested the impact of different head groups (imidazolium, pyridinium, pyrrolidinium, ammonium, morpholinium, thiophenium and sulfonium) on the toxicity of ILs to the green algae *P. subcapitata.* They observed that the toxicities of pyrrolidinium- and morpholium-based ILs (EC_{50} > 100 mg/l) were very low in comparison with those of long-chain ammonium salts. The EC_{50}s of the other 15 ILs ranged from 0.12 to 82.13 mg/l. The quaternary ammonium based salts AMMOENG 100 and AMMOENG 130 were more toxic, with EC_{50}s of 0.12 and 0.83 mg/l respectively. The results of that study showed that just a difference in the cationic moiety enables the compounds to be classified from practically harmless (Pyr14 $(CF_3SO_2)_2N$, EC_{50} > 100 mg/ml) through slightly toxic (Py4 $(CF_3SO_2)_2N$), EC_{50} = 7.1 mg/ml) and moderately toxic (IM14 $(CF_3SO_2)_2N$, EC_{50} = 26.5 mg/ml) to highly toxic (AMMOENG, EC_{50} = 0.12 mg/ml) and confirmed the significant contribution of cationic head groups to IL toxicity. Pyridinium salts turned out to be more toxic than imidazolium compounds in both the acute toxicity test, conducted on a biosensor specifically designed to measure microalgal photosynthetic activity (the EC_{50}s of 3-methyl-1-octylpyridinium bromide (Py8-3Me Br) and IM18 Br were 0.055 mM and 3.47 mM respectively) (Pham et al., 2008b), and the chronic growth inhibition assay (the EC_{50}s of Py4 $(CF_3SO_2)_2N$ and IM14 $(CF_3SO_2)_2N$ were 7.1 and 26.5 mg/l respectively) (Pretti et al., 2009). In contrast, other studies showed that pyridinium and imidazolium ILs inhibited the growth of green algae in the same concentration range, no significant differences between alkylimidazolium salts and an alkylpyridinium compound of similar lipophilicity were observed (Wells and Coombe, 2006; Stolte et al., 2007; Cho et al., 2008a; Latała et al., 2009a).

The importance of side chain length as a factor contributing to the overall observed biological effect has also been examined in algae. It has been shown for all algal test systems that the longer the *n*-alkyl chain of imidazolium ILs, the greater their toxicity (Wells and Coombe, 2006, Cho et al., 2007, Kulacki and Lamberti, 2008, Stolte et al., 2007, Pham et al., 2008, Latała et al., 2009a,b, Ma et al., 2010). In the studies of Kulacki and Lamberti (2008) EC_{50}s (groundwater) for *C. reinhardtii* were 4883 μM and 15 μM for IM14 Br and IM18 Br respectively. Stolte et al., (2007) demonstrated possibility to alter IL cation toxicity by 6 - 7 orders of magnitude, ranging from extreme toxicity for IM1-10 Cl (EC_{50} = 0.3 nM) to low/no toxicity for e.g., IM13OH Cl (EC_{50} > 1 mM), simply by varying the lipophilicity of the side chain. A strong correlation of toxicity with alkyl side chain length was also confirmed in pyridinium, phosphonium, and ammonium ILs (Wells and Coombe, 2006). The substitution of one or two carbon atoms in the longer alkyl chain by a more electronegative atom (chlorine or oxygen) substantially reduced acute toxicity to *P. subcapitata* (Pretti et al., 2009).

Investigations into the influence of the anionic component on the toxicity of ILs towards algae have been reported. Cho et al., (2007) demonstrated the effects of several typical anions (Br^-, Cl^-, BF_4^-, PF_6^-, $CF_3SO_3^-$, $C_8H_{17}SO_4^-$, SbF_6^-) on the toxicity of imidazolium-based ILs; SbF_6^- and PF_6^- were identified as the most toxic. Evident intrinsic toxicity has also been demonstrated for $(CF_3SO_2)_2N^-$, $(2OPhO)_2B^-$ and $8OSO_3^-$ anions, which were more toxic to *S.*

vacuolatus than the halides, which did not exhibit an intrinsic anion effect in the tested concentration range of up to 10 mM (Matzke et al., 2007; Stolte et al., 2007). More recently Steudte et al., (2012) investigated (eco)toxicity of different fluoro-organic and cyano-based ionic liquid anions. Especially the Tris(pentafluoroethyl)trifluorophosphate ($(C_2F_5)_3PF_3^-$) moiety demonstrated significant hazard towards aquatic organisms including algae *S. vacuolatus* with EC_{50} of 0.67 μM obtained for IM12 $(C_2F_5)_3PF_3$. Latała et al., (2009 a,b) found that apart from BF_4^-, the presence of $CF_3SO_3^-$ and $(CN)_2N^-$ as counter anions in IL structures gave rise to the most pronounced toxic effects in comparison with the other anions tested. It was demonstrated that anions displayed different modes of action as a consequence of their heterogeneous structure. The elevated toxicity of SbF_6^-, PF_6^- and BF_4^- anions was probably due to the hydrolysis of these entities, which led to the formation of fluoride and consequently a rise in toxicity (Cho et al., 2008b, Matzke et al., 2007, Stolte et al., 2007). Whereas in the case of trifluoromethanesulfonate it is likely caused by relatively high lipophilicity of this anion, additionally known to be strongly associated with alkylimidazolium cations, that in turn may enhance cell wall penetration. Relatively stronger effects with ILs containing α-methyl[poly(oxy-1,2-ethanediyl)]sulfate were also recorded. The main difference between this anion and the others lies in its polymeric nature. Therefore, the generally observed baseline toxicity of lipophilic cations may well be potentiated by a polymeric anion (Latała et al., 2009a,b). *N,N,N*-trialkylammoniododecaborates, which are anions in a new class of ILs, were tested for their hazard potential to aquatic ecosystems. MeAB and EtAB had no inhibitory effect on *S. vacuolatus* reproduction up to a concentration of 3 mM. When one ethyl group was replaced by a benzyl group, Et2BnAB, a 50% inhibition was observed at 1.33 mM. BuAB and HxAB were more effective, and 16 μM of HxAB led to a 50% inhibition of reproduction. iPnAB, with branched chains, displayed a lower inhibitory effect than BuAB, with *n*-alkane chains (Schaffran et al., 2009).

The relationship between the toxicities of ILs and the incubation time was examined. The toxicities of tetrabutylphosphonium and tetrabutylammonium ILs containing bromide were observed to decrease when the incubation time was raised from 48 to 96 h. Conversely, the toxicities of 1-butyl-3-methylimidazolium and 1-butyl-3-methylpyridinium containing bromide were found to increase with increasing incubation time. This indicates that such ILs may become more toxic on being released into a freshwater ecosystem (Cho et al., 2008a).

As evidenced by the investigations of Latała et al., the toxic effects of ILs on algae can be moderated by environmental conditions like salinity. It was found that in more saline waters IL toxicity was significantly reduced, in the case of *C. vulgaris* exposed to 5 μM of IM18 Cl growth inhibition was reduced from 64% (0 PSU) to 5.4% (32 PSU) (Latała et al., 2005, 2010). One of the possible explanations given by these authors is that the lower toxicity is probably due to the reduced permeability of IL cations through the algal cell walls. The main effect is attributed to the presence of chlorides, which are known to be very mobile and capable of stabilizing complexes with cationic microelements, thereby diminishing their bioavailability in water. Higher chloride concentrations offer a good ion-pairing environment for imidazolium cations, therefore preventing interaction between the cation and the negatively charged biological membranes of algal organisms. Moreover, a higher salinity implies an elevated concentration of inorganic monovalent cations, which compete with hydroxyl or silanol functional groups in cell wall structures.

Kulacki and Lamberti (2008) found that the presence of nutrient media had a mitigating effect on the toxicity of these ILs to *C. reinhardtii* in comparison to low-nutrient groundwater. According to these authors the difference in nutrient levels might enable a taxon to cope better with the higher stress levels associated with a particular IL.

Although ILs effectively inhibit algal growth, there are significant differences – of two orders of magnitude – in the responses of the species used in tests. It was shown that the diatom *C. meneghiniana* was more sensitive than the green algae *O. submarina* to IL contamination: the respective EC_{50}s for IM18 Cl were 0.73 μM and 79 μM. Moreover, *O. submarina* appeared to become acclimated to the imidazolium ILs at concentrations of up to 50 μM: after a few days their ability to grow recovered, eventually reaching the same density as the control cultures. In contrast, the growth of *C. meneghiniana* cultures was effectively inhibited throughout the experiment (the 10-day test period), regardless of the IL concentration applied (Latała et al., 2005, 2009a).

Interestingly, the IL toxicity ranged widely not only between taxonomic groups but also between species from the same group. Among the green algae, *C. vulgaris* was twice as sensitive to ILs as *O. submarina*; likewise, among the diatoms, *S. marinoi* was twice as sensitive as *C. meneghiniana*. These differences in sensitivity within taxonomic groups correlate well with cell size (a tenfold difference in cell size results in a 100% difference in sensitivity): the smaller the cell, the greater the sensitivity to ILs. The authors concluded that the relation between cell size and sensitivity again underscores the significance of the cell wall component in the toxic action of ILs. Algal cultures consisting of smaller cells have a relatively larger cell wall surface area, which in turn enables toxins to be transported more efficiently into the cells.

A conventional algal test based on growth rate and a short-term photosynthetic activity test were performed simultaneously to assess the phytotoxic effect of IM14 BF_4 and IM14 Br. It was observed that, as for IM14 BF_4, the results of both assays coincided well, but in the case of algae exposed to IM14 Br, the growth rate was affected more than the photosynthetic response. In addition, with respect to IL incorporating the perfluorinated anion (IM14 BF_4), the EC_{50}s of the previously prepared stock solution were significantly lower than those of the freshly made up solution for both test systems. This was probably due to the hydrolysis of BF_4, which leads with time to the enhanced formation of fluoride and consequently higher toxicity (Pham et al., 2008a,b). This indicates that, after ILs are discharged into an aqueous system, they can become more toxic than laboratory data with fresh ILs would lead us to expect.

The results of the studies carried out by Ma et al., (2010) showed that both primary productivity and chlorophyll content decreased significantly after exposure to 1-alkyl-3-methylimidazolium bromide. According to the authors this could be attributed to the incorporation of IL into the polar head groups of the phospholipid bilayer, which resulted in the disruption of the membrane-bound proteins and the integrity of the chloroplasts. The structural integrity of chloroplasts, especially in the inner membrane, is of vital importance for photosynthesis. As chlorophyll plays a significant part in photosynthesis, any reduction in its content may directly lower primary productivity.

Most of the investigations on the influence of ILs on the aquatic environment have been done with freshwater algae, so knowledge of the fate of ILs in the marine environment is limited. Moreover, most of these studies on the aquatic ecotoxicity of ILs have dealt with phytoplankton. Nonetheless, benthic organisms are just as important if we wish to fully understand the prospective fate and effects of ILs in this environment. In the case of benthic organisms, it is not only water but also sedimentary matter that is a significant source of chemicals previously sorbed in this environmental compartment. An attempt to evaluate the impact of ILs on the marine environment has been made: Latała et al., (2005, 2009 a,b) investigated the toxic effect of imidazolium and pyridinium ILs on various algal taxa: green algae (*C. vulgaris* and *O. submarina*) characteristic of freshwater and brackish environments as well as marine diatoms (*C. meneghiniana, S. marinoi* and *B. paxillifer*) and a cyanobacterium (*G. amphibium)*. The results indicated that sensitivity to ILs decreased in the following order: cyanobacterium > diatoms > green algae. The authors suggested that the most important determinant of the observed difference was the cell wall structure – the peptidoglycan cell wall in the cyanobacterium, the silica based cell wall in diatoms and the cellulose wall in the case of green algae. If IL cations are of concern, it is highly likely that the much more effective electrostatic biosorption on the diatom surface than on green algae will determine the extent of toxic action in this taxonomic group. Differences in the sensitivity of algae exposed to ILs were also observed by Kulacki and Lamberti (2008); the EC_{50} they obtained for *S. quadricauda* (cellulose-based cell walls) was several orders of magnitude lower than that obtained for the same ILs for *C. reinhardtii* (with a cell wall composed primarily of glycoproteins).

PHYTOTOXICITY

The impact of ILs on the aquatic environment has also been studied with the macrophyte *Lemna minor* (duckweed). This common aquatic vascular plant is recommended for toxicity tests because of its small size, rapid growth, clonal reproduction and widespread distribution. In general, 1-alkyl-3-methylimidazolium ILs with longer alkyl chains were more toxic to *L. minor* than those with short alkyl chains: the EC_{50} of IM14 Br was approximately 10 times greater than that of IM18 Br in respect of both frond and root growth. The root growth EC_{50} was 8.56 ppm when a butyl chain was present but was 0.25 ppm when an octyl chain was substituted (Larson et al., 2008). The side chain length effect was confirmed by Matzke et al., (2007). Moreover the EC_{50} of 8 μM obtained for IM18 in the duckweed growth inhibition test was within the range of the toxic references. Cation class was also an important factor determining toxicity. Butyl-substituted 3-methylpyridinium (root growth EC_{50} = 7.49 ppm) and 3-methylimidazolium cations had similar toxicities, whereas the tetrabutylammonium cation was considerably less toxic (root growth EC_{50} = 32.71 ppm). In addition to quantitative measurements of growth, frequent discoloration of fronds and occasional gibbosity was observed at high IL concentrations. Stolte et al., (2007) investigated the influence of different head groups, functionalized side chains and anions of ILs on *L. minor.* Most of the compounds (pyridinium, imidazolium, 1-methylpiperidinium, 1-methylpyrrolidinium) displayed a significant toxicity with EC_{50}s from 145 to 295 μM. 1-butyl-4-(dimethylamino)pyridinium and the quaternary ammonium compound (*N,N*-dimethylethylammonium) were highly toxic (EC_{50} < 10 μM). In contrast, the morpholinium compound exhibited only a moderate effect on *L. minor* (EC_{50} > 1000 μM). Although the toxicity towards *Vibrio fischeri* and *Scenedesmus vacuolatus* increased in the case of all the compounds combined with the $(CF_3SO_2)_2N^-$, the anion had no influence on the observed effects of ILs on *L. minor*. The significant anion effect was observed in duckweed growth inhibition test for Tris(pentafluoroethyl)trifluorophosphate, EC_{50} obtained for IM12

$(C_2F_5)_3PF_3$ was one order of magnitude lower than the values for IM12 $(CN)_2N$, IM12 $C(CN)_3$, IM12 $B(CN)_3$ and IM12 $(CF_3SO_2)_2N$ (Steudte et al., 2012).

Lemna minor was found to be the most sensitive species in case of the toxicity test for the group of compounds based on substituted amines as cations and organic acids with different numbers of carbon atoms as anions. The most toxic compounds were 2-hydroxydiethanolamine butanoate, 2-hydroxyethanolamine butanoate and 2-hydroxydiethanolamine isobutanoate with EC_{50} values of 33, 59 and 79 mg/l, respectively (Peric et al., 2013). The acute toxicity of the new family of ILs was below the levels observed for "classic" ILs (substituted imidazolium and pyridinium chlorides) therefore it was concluded that they could be consider as environmentally safer alternatives for more toxic ILs and organic solvents.

The design and use of ILs and other new chemicals should incorporate not only standard toxicity tests but also information on how such chemicals will interact with other components of aquatic ecosystems. When the influence of DOM (dissolved organic matter) on the toxicity of the cations was tested, no effect of a low-molecular-weight organic acid or commercial humic matter was observed. In contrast, natural DOM reduced the toxicity only at low concentrations of imidazolium ILs but had no effect on the toxicity of pyridinium and ammonium ILs (Larson et al., 2008).

The release of ILs into the aquatic environment is still most likely *via* waste water processes, but accidental spills, waste disposal or leakage from landfill sites may cause the terrestrial environment to be exposed as well. Thus it is important to assess the prospective hazard assessment of ILs in the environment. One of the first studies on IL toxicity to plants in terrestrial compartments was by Jastorff et al., (2005), who investigated the phytotoxicity of IM14 BF_4 and IM18 BF_4 to the terrestrial plant *Lepidium sativum* (cress). The number of seedlings of *L. sativum* decreased significantly in test samples with IM18 BF_4 and IM14 BF_4 at respective concentrations of 100 mg/kg and 1000 mg/kg. The low toxicity of ILs (1-ethyl-3-methylimidazolium, 1-butyl-3-methylimidazolium, and 1-hexyl-3-methylimidazolium chlorides) towards seedlings and primary root growth in *L. sativum* was observed in both solutions and soils artificially contaminated with those compounds. With the aquatic solutions the root growth EC_{50}s were 0.1046 mg/ml for IM12, 0.0510 mg/ml for IM14 and 0.0122 mg/ml for IM16. The hazardous nature of these compounds depends strongly on their hydrophobicity: the more hydrophobic the ionic liquid, the fewer the number of seeds that germinate. The influence of the matrix in which the test was performed was demonstrated. EC_{50}s ranged from 0.001 to > 6 mg/g for IM12, from 0.08 to > 6 mg/g for IM14, and from 0.001 to 2.3 mg/g for IM16, depending on the soil type. The relationship between the soil organic matter content and the toxicity of IL to cress was observed, the phytotoxicity increased with decreasing organic matter content in the soil. It is probably a consequence of sorption: soil with more organic carbon sorbed IL cations more extensively than soil with little or no organic matter. This sorption was, however, reversible (Studzińska et al., 2008). Moreover, the more hydrophobic ILs like IM14 or IM16 were more strongly bound by the matrix than IM12. Ions can enter the plant in many different ways: it has been suggested that a small IM12 can easily cross a plant cell wall. For these two reasons, the toxic effects of IM12 may be more serious than those of IM14: it is present in the soil solution in larger quantities and can probably migrate through plant cells more easily and quickly. The visible effects of IL toxic activity on garden cress were similar to those elicited by a nutrient deficit in plants: the highest IL concentration caused chromatosis in every soil type. This may suggest that ILs block macroelement uptake by plants or are preferentially sorbed by cress. HPLC was used to determine the amount of IL taken up by cress, and from this the bioavailability (IL concentration in solution vs. the amount of the cation adsorbed by cress) of the ILs was estimated: for IM12 this was equal to 0.2595, for IM14 it was 0.2751 and for IM16 0.2974. The relationship between the decrease in root germination and the amount of ionic liquid taken up by cress was almost linear (Studzińska and Buszewski, 2009).

The influence of chiral ILs (-)-1-[(1*R*)-Nopyl]-3-methylimidazolium chloride and nitrate introduced to the soil on germination and early growth stages of the terrestrial higher plants spring barley (*Hordeum vulgare*) and common radish (*Raphanus sativus*) was investigated. The toxicity of these ILs was dependent on their concentration. The lowest concentration causing a distinct reduction in plant germination/growth and permanent damage to the plants was 100 mg/kg of soil d.w. Spring barley tolerated the ILs (200 mg/kg) better than common radish (100 mg/kg). Seed germination was inhibited only in common radish exposed to chloride salt (Bałczewski et al., 2007). It was also shown that wheat (*Triticum aestivum*) germination was reduced to 38% in the presence of 4.4 mM IM14 BF_4. Similarly, the root and shoot length of wheat seedlings decreased with increasing concentrations of IM14 BF_4. The activity of amylase increased in shoots and roots, but decreased significantly in germinating seeds when the IM14 BF_4 concentration exceeded 1.8 mM. The activity of peroxidase, a key enzyme in plants subjected to environmental stress, as well as the soluble protein content in shoots treated with IM14 BF_4 changed in a similar manner. Increasing concentrations of IM14 BF_4 reduced the chlorophyll content, thereby impairing photosynthesis and plant growth. Thus, at concentrations of 0.9 mM or more, IM14 BF_4 was toxic to wheat seedlings (Wang et al., 2009).

Of six different anions (Cl^-, BF_4^-, $(CF_3SO_2)_2N^-$, $(CF_3)_2N^-$, $8OSO_3^-$ and bis(1,2-benzenediolato)borate) studied, an evident anion effect on terrestrial organisms was observed with $(CF_3SO_2)_2N^-$. In wheat and cress both growth inhibition and phenotypic abnormal appearance occurred, e.g., dwarfism, pseudo-succulence and a loss of gravitropism. Possible explanations for these deformations are interferences with the phytohormone system (Matzke et al., 2007).

The influence of soils of different composition – varying contents of the clay minerals smectite and kaolinite – on the toxicity of different imidazolium-based ILs combined with four different anions – $(CF_3SO_2)_2N^-$, BF_4^- and the more biocompatible Cl^- and HSO_4^- – was studied with respect to growth inhibition in wheat. The influence of the anion on toxicity depending on the clay mineral added was evident. Growth inhibition was significant with 1000 mg/kg dry weight soil of IM14 Cl, IM14 BF_4 and IM14 HSO_4. The IM14$(CF_3SO_2)_2N$ was one order of magnitude more toxic than the other three IM14 compounds, and in this case the addition of clay minerals increased toxicity. With IM14 HSO_4 the presence of 10% smectite completely eliminated the toxic effect. The authors suggested that ILs whose toxic action is due to the anionic moiety are especially hazardous to soils, in particular those with high clay contents (Matzke et al., 2009). Soils in general are negatively charged matrices, so smaller interactions between the anionic moiety and the matrices are expected, resulting in potentially more serious effects in soil organisms (Ranke et al., 2007).

Toxicity to Animals

Aquatic snails are important aquatic populations to maintain a balanced ecosystem. *Physa* spp. are key members of freshwater food webs, because they graze algae and are themselves important prey for fish and invertebrate predators. Alterations of *Physa* habitat use and behaviour can affect the species composition, standing crop and productivity of attached algae. Changes to locomotion induced by toxicants can make individual snails more or less susceptible to predation. Thus, IL-induced changes in *Physa* behaviour may have substantial indirect effects on freshwater food webs. The acute toxicity of several ILs with imidazolium, pyridinium, quaternary ammonium and quaternary phosphonium cations combined with Br^- and PF_6^- anions to the freshwater pulmonate snail, *Physa acuta* was studied. The 96h-LC_{50}s (median lethal concentrations) of imidazolium- and pyridinium-based ILs ranged from 1 to 325 mg/l. The most toxic IL to *P. acuta* was Py8-3Me Br (LC_{50} = 1.0 mg/l) and the least toxic was tetrabutylammonium bromide (LC_{50} = 580.2 mg/l). Tetrabutylphosphonium displayed moderate toxicity with LC_{50} = 208 mg/l. ILs with eight-carbon substituents were more toxic than those with six- and four-carbon substituents. Toxicity did not consistently depend on whether the IL cation was imidazolium- or pyridinium-based. Sublethal concentrations of imidazolium- and pyridinium-based ILs affected the behaviour (movement and feeding rates) of this snail. Compared to the controls, the snails moved more slowly when exposed to butyl- and hexyl-cation ILs at 1 to 3% of LC_{50} concentrations but were not affected at higher IL concentrations (up to 10% of LC_{50}). Snail movement was not affected by ILs with octyl groups. Grazing patterns, however, indicated that snails grazed less at higher IL concentrations. *P. acuta* egestion rates were reduced in the presence of ILs at 3 to 10% of LC_{50} concentrations. Thus, nonlethal IL concentrations affected *P. acuta* behaviour, potentially impacting on individual fitness and food web interactions (Bernot et al., 2005b). The embryonic and postembryonic developmental toxicity of imidazolium-based ILs to the snail *P. acuta* was reported. The exposure to low concentrations (1.5 and 2.1 mg/l) of IM18 Br inhibited hatching rate and increased mortality of snail embryos. In addition snail embryonic malformations induced by the IL were observed at concentrations higher that 4.16 mg/l. The differences in hatching rates between three developmental stages of *P. acuta* embryos indicated that the veliger stage was more sensitive to IM18 Br exposure than the blastula and gastrula stages. The 96h-LC_{50} values determined for juvenile, subadult, and adult snails (70.83, 97.59 and 109.3 mg/l respectively) revealed developmental stage specific sensitivity to the IL (Li et al., 2012b). Subchronic toxicity test results showed that IM18 Br exposure significantly affected growth of the juvenile snails and their reproductive ability. Moreover acute exposure induced changes in the activities of antioxidant enzymes (catalase (CAT), superoxide dismutase (SOD) and glutathione S-transferase (GST)) and levels of the antioxidant glutathione (GSH) and the lipid peroxidation by-product malondialdehyde (MDA). These results suggest that oxidative stress and lipid peroxidation may be involved in the mechanism of toxicity for ILs (Ma et al., 2014).

In many freshwater ecosystems, bivalves represent the largest biomass of filter-feeding organisms, which makes them important components of aquatic food webs. The effect of imidazolium and pyridinium ILs on mortality and feeding behaviour of a sessile bivalve, the zebra mussel (*Dreissena polymorpha*), was studied. The ILs tested caused acute mortality over a wide range of concentrations, with the lowest LC_{50} of 21.4 mg/l obtained for Py8-3Me

Br and highest LC_{50} of 1290 mg/l for IM14 Br. For each reduction in alkyl chain length by two carbons, a lowering of toxicity of approximately one order of magnitude was observed. However, the toxicity was not influenced by the heterocyclic base, and the toxicity of pyridinium- and imidazolium-based ILs was similar. Comparison of the sensitivity to the same IL across organisms shows that zebra mussels exhibit some of the highest LC_{50}s and would be among the most resistant aquatic organisms in the event of an IL release. The filtering behaviour of individual mussels fed with *Chlamydomonas reinhardtii* and exposed to sublethal concentrations of imidazolium-based ILs was examined. Short-term exposure to any IL reduced zebra mussel feeding. The lowest observed effect concentrations (LOECs) for IM14 Br, IM16 Br and IM18 Br of 1200, 100 and 6 mg/l respectively indicated that disruption of zebra mussel filtering was also influenced by IL structure (Costello et al., 2009).

Earthworms play a key role in the soil ecosystem, controlling as they do the cycling of substances and the energy transformations within a terrestrial ecosystem. Because of its sensitivity to environmental pollutants the earthworm *Eisenia foetida* has been recommended as a model species for monitoring environmental soil pollution; it has also been used to assess the toxic effect of 1-methyl-3-octylimidazolium bromide. The results of an acute toxicity test performed in artificial soil showed that the 7 d-LC_{50} and the 14 d-LC_{50} of IM18 Br were 206.8 mg/kg and 159.4 mg/kg respectively. It was suggested that exposure to the IL induced the formation of reactive oxygen species and stimulated antioxidant defence systems. Activity changes of CAT, SOD and GST and elevated levels of GSH and MDA were observed in *E. foetida* exposed to different concentrations (10 - 160 mg/kg) of IM18 Br for 14 days (Li et al., 2010). Moreover subchronic exposure to IM18 Br resulted in either inhibition or induction of AChE and cellulase activities, depending on the IL concentration or the exposure time; the consequences were disruptions of the nervous function in the earthworms and alterations to their cellulosic metabolism (Luo et al., 2009). The same research group indicated that high concentrations of IM18 Br significantly inhibited the growth of the earthworms. These effects were enhanced with prolonged exposure to increasing concentrations of the IL. The reproductive ability of *E. foetida* was significantly inhibited after 42 d of subchronic exposure to IM18 Br at concentrations > 5 mg/kg soil (dry weight). During acute exposure, a high concentration (> 40 mg/kg) inhibited the activities of $Na^{+}K^{+}$ATPase and Mg^{++}ATPase after 3 and 7 d. After 42 d of subchronic exposure, the activity of both ATPases was inhibited in groups exposed to > 5 mg/kg (Luo et al., 2009).

Daphnia magna is one of the most widely distributed planktonic crustaceans worldwide. It is also an important link in the food web between microbial and higher trophic levels in freshwater ecosystems, and has been the international standard model animal for toxicity testing studies. Notably, *D. magna* was much more sensitive than *Vibrio fischeri* to the toxicity of 1-butyl-3-methylimidazolium salts (Garcia et al., 2005). Therefore, *D. magna* was used as a model organism to evaluate the toxicity of ILs and the biological responses they induced. IL toxicity to *D. magna* ranged from the highly toxic 3-methyl-1-octylimidazolium bromide (LC_{50} = 0.04 μM) to the less toxic 1-butyl-3,5-dimethylpyridinium bromide (LC_{50} = 97 μM). In general, the toxicity of ILs with longer alkyl chain substituents was comparable to that of phenol, while ILs with shorter substituents (e.g., 1-butyl-3-methylimidazolium bromide) were more toxic to *D. magna* than benzene or methanol (Couling et al., 2005). The toxicities of choline saccharinate (EC_{50} = 1219 mg/l) and choline acesulfamate (EC_{50} = 1378 mg/l) to *D. magna* were low (Nockemann et al., 2007). The acute effects of imidazolium-based ILs on the survival of the crustacean *D. magna* and their chronic effects on the reproduction of the species was studied by Bernot et al., (2005a). Lethal concentrations of IM14 with various anions (e.g., Cl^-, Br^-, PF_6^- and BF_4^-) ranged from 8.03 to 19.91 mg/l. Thus, toxicity appeared to be related to the imidazolium cation and not to the anion. A 21-d chronic test of individual *D. magna* exposed to nonlethal IL concentrations showed significant reductions in reproductive output (total neonates, first brood neonates and average neonates) after exposure to low concentrations (0.3 mg/l) of imidazolium ILs. The impact of ILs on the development of *D. magna* was investigated in detail for IM18 Br. The respective 24h- and 48h-LC_{50}s were 1.99 and 0.95 mg/l. A series of multigenerational toxicity tests showed that this IL exerts a toxic effect on the development of *D. magna* (Luo et al., 2008). Significantly higher activities of antioxidant defence enzymes (CAT, SOD, glutathione peroxidise (GPX) and GST) as well as elevated levels of GSH and MDA, found after exposure to IM18 Br at concentrations between 0.06 - 0.95 mg/l, suggested that oxidative stress might be involved in the mechanism of IL-induced toxicity in animals (Yu et al., 2009b). Samori et al., (2007) compared the acute toxicity of IM14 BF_4 to that of oxygenated chain–substituted ionic liquids: 1-methoxyethyl-3-methylimidazolium tetrafluoroborate and dicyanamide to the *D. magna*. The EC_{50}s of the two ILs were very similar, ranging from 209 to 222 mg/l, and were higher by two orders of magnitude than the EC_{50} of IM14 BF_4 (5.15 mg/l). The results indicate that the introduction of an oxygenated side chain into the imidazolium cation can greatly reduce toxicity of ILs. Pretti et al., (2009) tested the impact of different head groups (imidazolium, pyridinium, pyrrolidinium, ammonium, morpholinium, thiophenium and sulfonium) on the toxicity of ILs to *D. magna.* The more toxic compounds were the quaternary ammonium-based ILs AMMOENG 100 and 130 with respective EC_{50}s of 1.57 mg/l (slightly toxic) and 0.55 mg/l (highly toxic). Toxicity to *D. magna* was strongly affected by the cationic head group, decreasing from aromatic heterocyclic nitrogen-containing pyridinium (Py4 $(CF_3SO_2)_2N$, EC_{50} = 1.7 mg/l) and imidazolium (IM14 $(CF_3SO_2)_2N$, EC_{50} = 18.9 mg/l) compounds to non-aromatic cyclic and acyclic pyrrolidinium (Pyr14 $(CF_3SO_2)_2N$, EC_{50} = 37.2 mg/l), ammonium and morpholinium (EC_{50} > 100 mg/l) compounds. The toxicity of sulfonium salts was also low; investigated sulfonium- and thiophenium-based ILs were practically harmless. It was demonstrated that in the case of imidazolium ILs, substitution of one or two carbon atoms in a longer alkyl chain with a more electronegative atom (chlorine or oxygen) or the substitution of an alkyl chain by a hydrogen atom reduced acute toxicity to *D. magna.* For several ammonium cations combined with $1OSO_3^-$, $1SO_3^-$ and $(CF_3SO_2)_2N^-$ as counter anions EC_{50} values lower than 100 mg/l were determined in the test with *D. magna.* Especially ILs with N1114 and N1112O1 cations seemed to cause toxic effects regardless of the anion structure. N1222 $1OSO_3$ and the above-mentioned compounds with EC_{50} values < 100 mg/l but > 10 mg/l can be classified as "harmful to aquatic organisms" (European Commission, 2011). N1112O1 $1SO_3$ with the EC_{50} value of 8 mg/l can even be considered as "toxic to aquatic organisms". Pyr14 $1OSO_3$, N1222 and choline-based ILs were generally above the threshold of 100 mg/l, indicating their low acute toxicity to aquatic environment. *D. magna* turned out the most sensitive to the studied compounds as no effect was observed in acute toxicity test with IPC-81 cell line, marine bacteria *V. fischeri* and green algae *S. capricornutum* (Stolte et al., 2012). In the study performed by Steudte et al., (2012) for the most common ionic liquids anions like $(CN)_2N^-$, $C(CN)_3^-$, $B(CN)_3^-$, $(CF_3SO_2)_2N^-$ and $(C_2F_5)_3PF_3^-$ only the latter elicited strong toxic effect in the test with *D. magna* with EC_{50} value of 0.42 mg/l.

When evaluating the impact of ILs on an aquatic ecosystem, it is necessary to determine their toxicity to fish. Located at the top of the aquatic food chain, they are important consumers, playing an important role in maintaining the balance of freshwater and marine ecosystems. Once they have accidentally leaked into an aquatic ecosystem, ILs can endanger the survival, growth and development of fish populations. Pretti et al., (2009) screened 18 ILs with different anions and cations for acute toxicities towards zebrafish (*Danio rerio*). This species plays an important role as a model vertebrate in ecotoxicological studies. Since the LC_{50}s of imidazolium, pyridinium, pyrrolidinium, morpholium and choline ILs were > 100 mg/l, they can be regarded as practically harmless towards zebrafish. By contrast, the ammonium salts AMMOENG 130 and AMMOENG 100 (EC_{50} = 5.9 and 5.2 mg/l respectively) were as toxic as common organic solvents, if not more so. Behavioural changes such as reduction in general activity, loss of equilibrium and erratic swimming and remaining motionless at mid-water level for prolonged periods were observed in treated fish exposed to both ILs at concentrations of 10 mg/l or more. Histological evaluation of dead zebrafish showed skin and gill lamellae alterations (Pretti et al., 2006). The LC_{50} of 244 mg/l for IM18 Br, obtained for another fish species, the goldfish (*Carassius auratus*), after 24 hours of exposure also indicated moderate or low IL toxicity. Further histological observations with a scanning electron microscope revealed that acute exposure to IM18 Br induced obvious superficial damage to the skin, gill filaments and intestines, which were suggested as being the first direct targets of the ionic liquid in this fish. Further damage to the goldfish's kidney and hepatopancreas together with a significant increase in the malondialdehyde level in the hepatopancreas were observed (Li et al., 2011). Acute exposure to 100 – 300 mg/l of IM18 Br for 7 days disrupted the immune system function in the brocaded carp (*Cyprinus carpio*) (Li et al., 2012a). The results showed that exposure to the highest tested concentration of IM18 Br inhibited the activity of specific and nonspecific immune systems, mainly the IgM level, lysozyme activity and the complement C3 content, while 100 mg/1 of IM18 Br stimulated the fish immune system and hence activated lysozyme in the earlier times of exposure (2 – 5 days). This result indicates that IM18 Br is immunotoxic to brocaded carp. Moreover, histological examination revealed considerable damage to the hepatopancreas, kidney and spleen of these fish after 7 days of 300 mg/l IM18 Br treatment. Wang et al., (2010) examined the toxic effect of the same ionic liquid on the early embryonic development of the goldfish *C. auratus* and evaluated the future consequences and environmental risk of the presence of ILs in the aquatic ecosystem. The results of acute toxicity tests showed that the 72h-LC_{50}s at the early gastrula, blastopore closure and heart beating stages of development were 187, 245 and 298 mg/l respectively, indicating that this toxicity was development stage-specific. In the subchronic tests, IM18 Br exposure prolonged the duration of embryo dechorionation and decreased the hatching rates of the treated embryos compared to the control. The results of the study showed a remarkable dose-response relationship between embryonic malformation and mortality, and the IL exposure concentration.

The developmental toxicity of IM18 Br was also confirmed in the study with the frog *Rana nigromaculata* (Li et al., 2009). Amphibians are often used to test the risk of exposure to contaminants in aquatic systems, largely because their larvae live in water. The developmental toxicity of IM18 Br was stage sensitive: 94h-LC_{50}s at the early cleavage, early gastrula and neutral plate stages were 85.1, 43.4 and 42.4 mg/l respectively. In embryos exposed to IM18 Br, the duration of embryonic dechorionation was prolonged in the early cleavage and neural plate but not the early gastrula stages of development. In addition, the mortality of embryos exposed to IM18 Br during their early development was found to be concentration-dependent; embryos in the neural plate developmental stage suffered the highest mortality. Morphological malformations were also observed in IM18 Br treated embryos. The authors suggested that membrane effects may be the main mechanism leading to IM18 Br-induced damage of frog embryos (Li et al., 2009).

The effects of exposure of mice and rats to ILs were studied in view of potential human exposure as a result of water or soil contamination from industrial effluents or accidental spills. Landry et al., (2005) demonstrated that ILs can be absorbed after dermal exposure. The topical application of IM14 Cl (2000 mg/kg) resulted in dermal irritation in rats and severe systemic toxicity in mice. However, the extent and rate of absorption were greatly influenced by the vehicle used for application, toxic effects were observed when it was applied in DMF but not in water. The absorption, distribution, metabolism and excretion of IM14 Cl were studied after oral (50 mg/kg) and intravenous (5 mg/kg) administration in male F-344 rats and B6C3F1 female mice. After oral administration IM14 Cl was extensively absorbed into the systemic circulation, with the maximum concentration achieved after one hour. Systemic bioavailability was determined to be 62% of a 50 mg/kg dose in rats; no evidence was obtained for the metabolism of systemically available IL. Urinary excretion of the parent compound was the major route of elimination in rats, regardless of gender, species, dose, route or number of exposures. These studies illustrate that the systemic bioavailability of IM14 Cl is high, tissue disposition and metabolism are negligible, and the absorbed compound is extensively extracted by the kidney and eliminated in the urine as the parent compound (Sipes et al., 2008). After prenatal exposure of mice to 1-butyl-3-methylimidazolium chloride, foetal weight was significantly lower and malformations, suggesting that IM14 Cl may be teratogenic, occurred at the two highest dosages (169 and 225 mg/kg/d). A dose-dependent increase in maternal morbidity and mortality during the course of treatment showed maternal toxicity. On the basis of these results, therefore, IM14 Cl appears to be developmentally toxic at maternally toxic dosages (Bailey et al., 2008). The acute and subchronic oral toxicity of another IL didecyldimethylammonium saccharinate [DDA][Sac] was investigated in rats. This compound was classified in the 4th toxicity class with LD_{50} = 500 mg/kg. Pathological changes in an acute experiment were observed in the surface layer of the colon and in lung parenchyma. In a subchronic experiment rats were administered 10, 30 and 100 mg/kg/day [DDA][Sac] for 28 days. Reduced body weight gain and slightly reduced food consumption were observed, particularly in high-dose rats. Slight haematological changes and significant changes in clinical chemistry parameters included increases in serum enzyme ALT, SDH, ALP and GGT activities, and in glucose, blood urea nitrogen and creatinine concentrations. No treatment-related microscopic changes were observed in the subchronic experiment. Under the conditions of this 28-day study and based on the toxicological endpoints evaluated, the LOAEL (the lowest observed adverse effect level) of [DDA][Sac] for males and females was determined to be 10 mg/kg b.w. (Jodynis-Liebert et al., 2010). It was demonstrated that exposure (intraperitoneal injection) to 1-octyl-3-methylimidazolium bromide induced oxidative stress and pathological changes in the mouse liver. A significant increase in hepatosomatic index, a decrease in liver protein content and a decrease in the antioxidant enzymes (CAT and GPX) were found in a group treated with 35.7 mg/kg (24h-LD_{50}). An increase in the activity of glutathione-S-transferase – an enzyme involved in the detoxification of many toxicants – together with pathological alterations in the liver were also observed (Yu et al., 2009a). Surprisingly, there was no effect of IM18 Br on lipid peroxidation. In contrast, levels of malondialdehyde in goldfish hepatopancreases after treatment with IM18 Br (Li et al., 2011) and in daphnias and earthworms after exposure to ILs were significantly higher (Yu et al., 2009b; Li et al., 2010).

The available data clearly demonstrate considerable variation in organism sensitivity to ILs both across and within taxonomic groups. Algae appear to be more sensitive than snails to IM14 Br, because the EC_{50} of IM14 Br for periphyton chlorophyll *a* (68.6 mg/l) was substantially lower than LC_{50} found for the freshwater gastropod *Physa acuta* (229 mg/l) (Bernot et al., 2005b). On the other hand the algal EC_{50} was higher than that for a freshwater zooplankter, *Daphnia magna* (48h-LC_{50} = 8 mg/l) (Bernot et al., 2005a), and a marine bacterium, *Vibrio fischeri* (EC_{50} = 0.67 mg/l) (Ranke et al., 2004). The green algae were approximately two orders of magnitude more sensitive to the mixture scenarios than wheat and proved to be a good reference test system for the evaluation of measured effects within a prospective hazard assessment of ionic liquids (Matzke et al., 2008). Hence the generalization and extrapolation of toxicity results (limited to a single endpoint) obtained with a model organism should be avoided at all costs.

The negative impact of IM14 Br on freshwater periphyton was observed in mesocosm experiments at concentrations as low as 1 to 10 mg/l, which were approximately 6- to 68-fold lower than estimated the EC_{50}s of the 96-h acute toxicity test, probably because of the longer exposure times (Evans-White and Lamberti, 2009). Thus one should remain aware of the fact that in the natural environment where living organisms will be chronically exposed, lower concentrations of ILs will be sufficient to exert toxic effects.

Standardized single-species bioassays are a rapid, and cost-effective way of determining the toxicity of novel chemicals and of comparing the relative toxicities of different compounds. However, standard bioassays do not normally mimic the conditions to which an organism is exposed in the environment. Interactions among organisms, seasonality, influence of abiotic conditions and the fact that other pollutants are already present in the environment are often overlooked. It has been shown that the aquatic toxicity of ILs can be influenced by such environmental factors as dissolved organic matter (Larson et al., 2008), nutrients (Kulacki and Lamberti, 2008) or salinity (Latała et al., 2005, 2010).

It was observed that exposure to IM14 Cl in the soil–ecological system inhibited the growth of soil microorganisms (bacteria and actinomycetes) and altered the soil's physico-chemical properties. There was an increase in organic mass and soluble salts in the soil as well as a change in its pH with increasing IM14 Cl concentration (Tong et al., 2012). Indirect influence of the altered physico-chemical conditions on living soil organisms could be assumed.

Laboratory experiments showed that soils and sediments represent a greater interaction potential with ILs and that bioavailability can be reduced, e.g., as a result of sorption. The rate at which compounds become bioavailable from sediments depends mainly on the strength and reversibility of sorption. Latała et al., (2009b) suggested that the greater toxicities of long-alkyl ILs would most probably not endanger benthic organisms as they would be rapidly and very strongly adsorbed to sedimentary matter. On the other hand, the less toxic but bioavailable short-chain compounds may pose a significant risk to organisms inhabiting the benthic environment.

The results obtained for the zebra mussel *Dreissena polymorpha* (Costello et al., 2009) and the snail *Physa acuta* (Bernot et al., 2005b) showed that a complete chemical hazard assessment cannot be obtained simply by measuring a single endpoint. As the zebra mussel can tolerate relatively high concentrations of ILs, the total freshwater bivalve biomass may not decline at environmentally relevant concentrations, but trophic interactions may be altered through behavioural changes. Reduced feeding by zebra mussels in the presence of ILs could have substantial effects on other trophic levels, which could lead to changes in the affected ecosystems that would not have been predicted solely from standard mortality bioassays. In such cases, environmentally more realistic experiments that include multiple species are required. It has been demonstrated that sublethal concentrations of ILs reduced snail grazing rates and movement. Alterations in snail behaviour could affect the species composition of algae, and the size of the snail population could be affected as a result of the greater or lesser susceptibility of an individual snail to predation.

Contaminants can exert direct as well as indirect effects that would not be detected by single species toxicity tests. The results obtained by Evans-White and Lamberti (2009) in an aquatic mesocosm experiment illustrated the direct and indirect effects of exposure to sublethal concentrations of contaminants on ecosystems. The indirect impact of a potential aquatic contaminant, an ionic liquid, on benthic algal biomass and primary production through direct effects on herbivorous snails was demonstrated. When the influence of the IL alone was studied, reduced chlorophyll *a* and algal biovolume, a shift in the relative composition of major algal taxa (from more sensitive (diatoms) to more tolerant ones (filamentous greens)), altered feeding behaviour, a decline in reproduction and growth rates of snails were observed. In the presence of both IM14 Br and snails, the chlorophyll *a* abundance and algal biovolume were greater than would be expected if both factors acted additively. In the presence of IM14 Br snails were less efficient grazers, thereby releasing the algae from grazer control. Snails stimulated periphyton primary production in the absence, but not in the presence, of IM14 Br, suggesting that IM14 Br also can indirectly alter ecosystem functioning. These findings suggest that sublethal contaminant levels can negatively impact on communities and ecosystem processes *via* complex interactions; they also provide baseline information regarding the potential effects of an emergent industrial chemical on aquatic systems.

Contaminants are present in the natural environment as mixtures and may interact antagonistically or synergistically and be more or less toxic than individually. On entering the environment, mixtures of ILs or of ILs and other pollutants are likely to occur, so the impact of mixture toxicity should be taken into account. The effects of three differently composed mixtures containing ILs as well as ILs and the heavy metal cadmium on limnic green algae and wheat was investigated by Matzke et al., (2008). The experimental data they obtained did not fit in with the predictions of the concentration addition (CA) and independent action (IA) concepts, however. Both concepts underestimated the effect of a mixture consisting exclusively of ILs, leading to the conclusion that interactions among the mixture's components or between the mixture's components and the environmental matrix must have occurred.

Every assessment of the environmental risk of ILs should take into account not just their toxicity but also their biodegradability. Anions like methylsulfate, octylsulfate and methylsulfonate are recommendable with respect to their biodegradability as well as from an (eco)toxicological point of view. On the other hand, the fluorine-containing anion $(CF_3SO_2)_2N^-$ should be avoided when designing inherently safer ILs because of its toxic potential, proven in the various toxicological systems discussed above, and because of its resistance to biological and abiotic degradation, which makes it extremely persistent. In contrast, the 8OSO$_3^-$ anion, despite being highly toxic as indicated by the *Vibrio fischeri* assay, can be recommended because of its ready biodegradability: environmental exposure times are short, thus lessening the hazard posed by toxic concentrations being reached (Matzke et al., 2007).

Some authors found that the toxicity of ILs in the above tests was comparable to or even greater than that of the commonly used solvents (e.g., methanol, acetone, toluene, dimethyl sulphoxide, propanol, ethanol, acetonitrile) they are intended to replace (Stepnowski et al., 2004, Bernot et al., 2005b; Pretti et al., 2006; Ranke et al., 2007; Cho et al., 2008). On the other hand the results suggest the possibility of tailoring an ionic liquid by adding a suitable functional group to reduce its toxicity and thus make it environmentally friendlier.

REFERENCES

Arning, J.; Stolte, S.; Böschen, A.; Stock, F.; Pitner, W.-R.; Welz-Biermann, U.; Jastorff, B.; Ranke, J.; 2008. Qualitative and quantitative structure activity relationships for the inhibitory effects of cationic head groups, functionalized side chains and anions of ionic liquids on acetylcholinesterase. *Green Chem.*10, 47–58.

Bailey, M.M.; Townsend, M.B.; Jernigan, P.L.; Sturdivant, J.; Hough-Troutman, W.L.; Rasco, J.F.; Swatloski, R.P., Rogers, R.D.; Hood, R.D.; 2008. Developmental toxicity assessment of the ionic liquid 1-butyl-3-methylimidazolium chloride in CD-1 mice. *Green Chem.* 10, 1213-1217.

Bałczewski, P.; Bachowska, B.; Białas, T.; Biczak, R.; Wieczorek, W.M.; Balińska, A.; 2007. Synthesis and phytotoxicity of new ionic liquids incorporating chiral cations and/or chiral anions. *J. Agric. Food Chem.* 55, 1881–1892.

Bernot, R.J.; Brueseke, M.A.; Evans-White, M.A.; Lamberti, G.A.; 2005a. Acute and chronic toxicity of imidazolium-based ionic liquids on Daphnia magna. *Environ. Toxicol. Chem.* 24(1), 87-92.

Bernot, R.J.; Kennedy, E.E.; Lamberti, G.A.; 2005b. Effects of ionic liquids on the survival, movement, and feeding behavior of the freshwater snail, Physa acuta. *Environ. Toxicol. Chem.* 24, 1759–1765.

Brycki, B.; Dego-Szafran, Z.; Mirska, I.; 2010. Synthesis and antimicrobial activities of some quaternary morpholinium chlorides. *Polish J. Microbiology* 59, 49-53.

Busetti, A.; Crawford, D.E.; Earle, M.J.; Gilea, M.A.; Gilmore, B.F.; Gorman, S.P.; Laverty, G.; Lowry, A.F.; McLaughlin, M.; Seddon, K.R.; 2010. Antimicrobial and antibiofilm activities of 1-alkylquinolinium bromide ionic liquids. *Green Chem.* 12, 420-425.

Carson, L.; Chau, P.K.W.; Earle, M.J.; Gilea, M.A.; Gilmore, B.F.; Gorman, S.P.; McCann, M.T.; Seddon, K.R.; 2009. Antibiofilm activities of 1-alkyl-3-methylimidazolium chloride ionic liquids. *Green Chem.* 11, 492–497.

Chefson, A.; Auclair, K.; 2007. CYP3A4 activity in the presence of organic cosolvents, ionic liquids, or water-immiscible organic solvents. *ChemBioChem.* 8(10), 1189-1197.

Cho, C.-W.; Pham, T.P.T.; Jeon, Y.C.; Vijayaraghavan, K.; Choe, W.S.; Yun, Y.-S.; 2007. Toxicity of imidazolium salt with anion bromide to a phytoplankton Selenastrum capricornutum: Effect of alkyl-chain length. *Chemosphere* 69(6), 1003-1007.

Cho, C.-W.; Jeon, Y.C.; Pham, T.P.T.; Vijayaraghavan, K.; Yun, Y.-S.; 2008a. The ecotoxicity of ionic liquids and traditional organic solvents on microalga Selenastrum capricornutum. *Ecotoxicol. Environ. Saf.* 71(1), 166-171.

Cho, C.-W.; Pham, T.P.T.; Jeon, Y.C.; Yun, Y.-S.; 2008b. Influence of anions on the toxic effects of ionic liquids to a phytoplankton Selenastrum capricornutum. *Green Chem.* 10, 67-72.

Cho, C.-W.; Pham, T.P.T.; Jeon, Y.C.; Min, J.H.; Jung, H.Y.; Lee, D.S.; Yun, Y.-S.; 2008c. Microalgal photosynthetic activity measurement system for rapid toxicity assessment. *Ecotoxicology* 17(6), 455-463.

Cieniecka-Roslonkiewicz, A.; Pernak, J.; Kubis-Feder, J.; Ramani, A.; Robertsond, A.J.; Seddon, K.R.; 2005. Synthesis, anti-microbial activities and antielectrostatic properties of phosphonium-based ionic liquids. *Green Chem.* 7, 855-862.

Costello, D.M.; Brown, L.M.; Lamberti, G.A.; 2009. Acute toxic effects of ionic liquids on zebra mussel (Dreissena polymorpha) survival and feeding. *Green Chem.* 11, 548 -553.

Couling, D.J.; Bernot, R.J.; Docherty, K.M.; Dixon, J.K.; Maginn, E.J.; 2006. Assessing the factors responsible for ionic liquid toxicity to aquatic organisms via quantitative structure-property relationship modeling. *Green Chem.* 8, 82-90.

Cybulski, J.; Wiśniewska, A.; Kulig-Adamiak, A.; Lewicka, L.; Cieniecka-Roslonkiewicz, A.; Kita, K.; Fojutowski, A.; Nawrot, J.; Materna, K.; Pernak, J.; 2008. Long-alkyl-chain quaternary ammonium lactate based ionic liquids. *Chem. Eur. J.* 14, 9305-9311.

Docherty, K.; Kulpa, C.; 2005. Toxicity and antimicrobial activity of imidazolium and pyridinium ionic liquids. *Green Chem.* 7, 185-189.

Docherty, K.M.; Hebbeler, S.Z.; Kulpa, C.F., 2006. An assessment of ionic liquid mutagenicity using the Ames Test. *Green Chem.* 8, 560–567.

Evans-White, M.A.; Lamberti, G.A.; 2009. Direct and indirect effects of a potential aquatic contaminant on grazer-algae interactions. *Environ. Toxicol. Chem.* 28(2), 418-426.

Fent, K.; 2001. Fish cell lines as versatile tools in ecotoxicology: assessment of cytotoxicity, cytochrome P4501A induction potential and estrogenic activity of chemicals and environmental samples. *Toxicol. in Vitro* 15, 477-488.

Frade, R.F.M.; Matias, A.; Branco, L.C.; Afonso, C.A.M.; Duarte, C.M.M.; 2007. Effect of ionic liquids on human colon carcinoma HT-29 and CaCo-2 cell lines. *Green Chem.* 9, 873-877.

Garcia, M.T.; Gathergood, N.; Scammells, P.J.; 2005. Biodegradable ionic liquids-Part II. Effect of the anion and toxicology. *Green Chem.* 7, 9-14.

Garcia, H.; Ferreira, R.; Petkovic, M.; Ferguson, J.; Leitão, M.C.; Gunaratne, N.; Seddon K.R.; Rebelo, L.P.N.; Silva Pereira, C.; 2010. Dissolution of cork biopolymers in biocompatible ionic liquids. *Green Chem.* 12, 367-369.

Guilhermino, L.; Barros, P.; Silva, M.C.; Soares, A.M.V.M.; 1998. Should the use of inhibition of cholinesterases as a specific biomarker for organophosphate and carbamate pesticides be questioned? *Biomarkers* 3, 157-163.

Herbert, A.; Guilhermino, L.; Da Silva de Assis, H.C.; Hansen, P.D.; 1995. Acetylcholinesterase activity in aquatic organismsas pollution biomarker. *Angew. Zool.* 3, 1-15.

Hou, X.D.; Liu, Q.P.; Smith, T.J.; Li, N.; Zong, M.H.; 2013. Evaluation of toxicity and biodegradability of cholinium amino acids ionic liquids. *PLoS One.* 8(3), e59145.

Hough-Troutman, W.L.; Smiglak, M.; Griffin, S.; Reichert, M.W.; Mirska, I.; Jodynis-Liebert, J.; Adamska, T.; Nawrot, J.; Stasiewicz, M.; Rogers, R.D.; Pernak, J.; 2009. Ionic liquids with dual biological function: sweet and antimicrobial, hydrophobic quaternary ammonium-based salts. *New J. Chem.* 33, 26-33.

Hugget, R.J.; Kimerle, R.A.; Mehrle Jr, P.M; Bergman, H.L.; 1992. Biomarkers. Biochemical, physiological, and histological markers of antropogenic stress. In: SETAC Special Publications Series. Edited by C.H. Ward, B.T. Walton & T.W. LaPoint. pp. 347. Lewis Publish., Boca Raton.

Jastorff, B.; Mölter, K.; Behrend, P.; Bottin-Weber, U.; Filser, J.; Heimers, A.; Ondruschka, B.; Ranke, J.; Schaefer, M.; Schröder, H.; Stark, A.; Stepnowski, P.; Stock, F.; Störmann, R.; Stolte, S.; Welz-Biermann, U.; Ziegert, S.; Thöming, J.; 2005. Progress in evaluation of risk potential of ionic liquids-basis for an eco-design of sustainable products. Gr*een Chem.* 7, 362–372.

Jodynis-Liebert, J.; Nowicki, M.; Murias, M.; Adamska, T.; Ewertowska, M.; Kujawska, M.; Piotrowska, H.; Konwerska, A.; Ostalska-Nowicka, D.; Pernak, J.; 2010. Cytotoxicity, acute and subchronic toxicity of ionic liquid, didecyldimethylammonium saccharinate, in rats. *Regul. Toxicol. Pharm.* 57, 266–273.

Kulacki, K.J.; Lamberti, G.A.; 2008. Toxicity of imidazolium ionic liquids to freshwater algae. *Green Chem.* 10, 104-110.

Kumar, R.A.; Papaiconomou, N.; Lee, J.M.; Salminen, J.; Clark, D.S.; Prausnitz, J.M.; 2009. In vitro cytotoxicities of ionic liquids: effect of cation rings, functional groups, and anions. *Environ. Toxicol.* 24, 388-395.

Larson, J.H.; Frost, P.C.; Lamberti, G.A.; 2008. Variable toxicity of ionic liquidforming chemicals to Lemna minor and the influence of dissolved organic matter. *Environ. Toxicol. Chem.* 27(3), 676-681.

Latała, A.; Stepnowski, P.; Nędzi, M.; Mrozik, W.; 2005. Marine toxicity assessment of imidazolium ionic liquids: Acute effects on the Baltic algae Oocystis submarina and Cyclotella meneghiniana. *Aquat. Toxicol.* 73, 91- 98.

Latała, A.; Nędzi, M.; Stepnowski, P.; 2009a. Toxicity of imidazolium and pyridinium based ionic liquids towards algae. Chlorella vulgaris, Oocystis submarina (green algae) and Cyclotella meneghiniana, Skeletonema marinoi (diatoms). *Green Chem.* 11, 580-588.

Latała, A.; Nędzi, M.; Stepnowski, P.; 2009b. Toxicity of imidazolium and pyridinium based ionic liquids towards algae. Bacillaria paxillifer (a microphytobenthic diatom) and Geitlerinema amphibium (a microphytobenthic blue green alga). *Green. Chem.* 11, 1371-1376.

Latała, A.; Nędzi, M.; Stepnowski, P.; 2010. Toxicity of imidazolium ionic liquids towards algae. Influence of salinity variations. *Green Chem.*12, 60-64.

Li, X.-Y.; Zhou, J.; Yu, M.; Wang, J.J.; Pei, Y.C.; 2009. Toxic effects of 1-methyl-3-octylimidazolium bromide on the early embryonic development of the frog Rana nigromaculata. *Ecotoxicol. Environ. Saf.* 72(2), 552-556.

Li, X.-Y.; Luo, Y.-R.; Yun, M.-X.; Wang, J.; Wange, J.-J.; 2010. Effects of 1-methyl-3-octylimidazolium bromide on the anti-oxidant system of earthworm. *Chemosphere* 78, 853–858.

Li X.-Y., Zeng S.-H., Zhang, W.-H.; Liu,L.; Ma, S.; Wang J.-J.; 2011. Acute toxicity and superficial damage to goldfish from the ionic liquid 1-methyl-3-octylimidazolium bromide. *Environ. Toxicol.* 28(4), 207-214.

Li, X.-Y.; Miao, X.-Q.; Zhang, L.-F.; Wang, J.-J.; 2012a. Immunotoxicity of 1-methyl-3-octylimidazolium bromide on brocarded carp (Cyprinus carpio L.). *Ecotoxicol. Environ. Saf.* 75, 180–186.

Li, X.-Y.; Dong, X.-Y.; Bai, X.; Liu, L.; Wang, J.-J.; 2012b The embryonic and postembryonic developmental toxicity of imidazolium-based ionic liquids on Physa acuta. *Environ. Toxicol.*

Luo, Y.-R.; Li, X.-Y.; Chen, X.-X.; Zhang, B.-J.; Sun, Z.-J.; Wang, J.-J.; 2008. The developmental toxicity of 1-methyl-3-octylimidazolium bromide on Daphnia magna. *Environ. Toxicol.* 23(6), 736-744.

Luo, Y.-R.; Wang, S.-H.; Yun, M.-X.; Li, X.-Y.; Wange, J.-J.; Sun, Z.-J.; 2009. The toxic effects of ionic liquids on the activities of acetylcholinesterase and cellulase in earthworms. *Chemosphere* 77, 313–318.

Łuczak, J.; Jungnickel, C.; Lacka, I.; Stolte, S.; Hupka, J.; 2010. Antimicrobial and surface activity of 1-alkyl-3-methylimidazolium derivatives. *Green Chem.* 12, 593-601.

Ma, J.-M.; Cai, L.-L.; Zhang, B.-J.; Hu, L.-W.; Li, X.-Y.; Wang J.-J.; 2010. Acute toxicity and effects of 1-alkyl-3-methylimidazolium bromide ionic liquids on green algae. *Ecotoxicol. Environ. Saf.* 73(6):1465-1469.

Ma, J.-M.; Dong, X.-Y.; Fang, Q; Li, X.-Y., Wang, J.-J.; 2014. Toxicity of imidazolium-based ionic liquids on Physa acuta and the snail antioxidant stress response *J. Biochem. Mol. Toxicol.* 28(2), 69-75.

Matzke, M.; Stolte, S.; Thiele, K.; Juffernholz, T.; Arning, J.; Ranke, J.; Welz-Biermann, U.; Jastorff, B.; 2007. The influence of anion species on the toxicity of 1-alkyl-3-methylimidazolium ionic liquids observed in an (eco)toxicological test battery. *Green Chem.* 9, 1198–1207.

Matzke, M.; Stolte, S.; Böschen, A.; Filser, J.; 2008. Mixture effects and predictability of combination effects of imidazolium based ionic liquids as well as imidazolium based ionic liquids and cadmium on terrestrial plants (Triticum aestivum) and limnic green algae (Scenedesmus vacuolatus). *Green Chem.* 10, 784-792.

Matzke, M.; Stolte, S.; Arning, J.; Uebers, U.; Filser, J.; 2009. Ionic liquids in soils: effects of different anion species of imidazolium based ionic liquids on wheat (Triticum aestivum) as affected by different clay minerals and clay concentrations. *Ecotoxicology* 18, 197–203.

McLaughlin, M.; Earle, M.J.; Gîlea, M.A.; Gilmore, B.F.; Gorman, S.P.; Seddon, K.R.; 2011. Cytotoxicity of 1-alkylquinolinium bromide ionic liquids in murine fibroblast NIH 3T3 cells. *Green Chem.* 13, 2794-2800.

Morrissey, S.; Pegot, B.; Coleman, D.; Garcia, M.T.; Ferguson, D.; Quilty, B.; Gathergodd, N.; 2009. Biodegradable, non-bactericidal oxygen-functionalised imidazolium esters: A step towards 'greener' ionic liquids. *Green Chem.* 11, 475-483.

Mulkiewicz, E.; Jastorff, B.; Składanowski, A.C.; Kleszczyński, K.; Stepnowski, P.; 2007. Evaluation of the acute toxicity of perfluorinated carboxylic acids using eukaryotic cell lines, bacteria and enzymatic assays. *Environ. Toxicol. Pharmacol.* 23, 279-285.

Nockemann, P.; Thijs, B.; Driesen, K.; Janssen, C.R.; Van Hecke, K.; Van Meervelt, L.; Kossmann, S.; Kirchner, B.; Binnemans, K.; 2007. Choline saccharinate and choline acesulfamate: Ionic liquids with low toxicities. *J. Phys. Chem. B.* 111(19), 5254-5263.

Peric, B.; Sierra, J.; Martí, E.; Cruañas, R.; Garau, M.A.; Arning, J.; Bottin-Weber, U.; Stolte S.; 2013. (Eco)toxicity and biodegradability of selected protic and aprotic ionic liquids. *J. Hazard. Mater.* 261, 99-105.

Pernak, J.; Kalewska, J.; Ksycifiska, H.; Cybulski, J.; 2001a. Synthesis and antimicrobial activities of some pyridinium salts with alkoxymethyl hydrophobic group. *Eur. J. Med. Chem.* 36, 899-907.

Pernak, J.; Rogoza, J.; Mirska, I.; 2001b. Synthesis and antimicrobial activities of new pyridinium and benzimidazolium chlorides. *Eur. J. Med. Chem.* 36, 313-320.

Pernak, J.; Branicka, M.; 2003. The properties of 1- alkoxymethyl-3-hydroxypirydinium and 1-alkoxymethyl-3-dimethylaminopirysdinium chlorides. *J. Surfactants Deterg.* 6, 119-123.

Pernak, J.; Chwała, P.; 2003. Synhesis and antimicrobial activities of choline-like quaternary ammonium chlorides. *Eur. J. Med. Chem.* 38, 1035-1042.

Pernak, J.; Sobaszkiewicz, K.; Mirska, I.; 2003. Anti-microbial activities of ionic liquids. *Green Chem.* 5, 52-56.

Pernak, J.; Goc, I.; Mirska, I.; 2004a. Anti-microbial activities of protic ionic liquids with lactate anion. *Green Chem.* 6, 323-329.

Pernak, J.; Sobaszkiewicz, K.; Foksowicz-Flaczyk, J.; 2004b. Ionic liquids with symmetrical dialkoxymethyl-substituted imidazolium cations. *Chem. Eur. J.* 10, 3479–3485.

Pernak, J.; Feder-Kubis, J.; 2005. Synthesis and properties of chiral ammonium-based ionic liquids. *Chem. Eur. J.* 11, 4441-4449.

Pernak, J.; Feder-Kubis, J.; 2006. Chiral pyridinium-based ionic liquids containing the (1R,2S,5R)-(-)-menthyl group. *Tetrahedron: Asymmetr.*17, 1728-1737.

Pernak, J.; Smiglak, M.; Griffin, S.T.; Hough, W.L.; Wilson, T.B.; Pernak, A.; Zabielska-Matejuk, J.; Fojutowski, A.; Kitad, K.; Rogers, R.D.; 2006. Long alkyl chain quaternary ammonium-based ionic liquids and potential applications. *Green Chem.* 8, 798-806.

Pernak, J.; Feder-Kubis, J.; Cieniecka-Roslonkiewicz, A.; Fischmeister, C.; Griffin, S.T.; Rogers, R.D.; 2007a. Synthesis and properties of chiral imidazolium ionic liquids with a (1R,2S,5R)-(-)-menthoxymethyl substituent. *New J. Chem.* 31, 879-892.

Pernak, J.; Skrzypczak, A.; Lota, G.; Frackowiak, E.; 2007b. Synthesis and properties of trigeminal tricationic ionic liquids. *Chem. Eur. J.* 13, 3106-3112.

Pernak, J.; Syguda, A.; Mirska, I.; Pernak, A.; Nawrot, J.; Prądzyńska, A.; Griffin, S.T.; Rogers, R.D.; 2007c. Choline-derivative-based ionic liquids. *Chem. Eur. J.* 13, 6817-6827.

Pernak, J.; Borucka, A.; Walkiewicz, F.; Markiewicz, B.; Fochtman, P.; Stolte, S.; Steudte, S.; Stepnowski, P.; 2011. Synthesis, toxicity, biodegradability and physicochemical properties of 4-benzyl-4-methylmorpholinium-based ionic liquids. *Green Chem.* 13, 2901-2911.

Petkovic, M.; Ferguson, J.; Bohn, A.; Trindade, J.; Martins, I.; Carvalho, M.B.; Leitão, M.C.; Rodrigues, C.; Garcia, H.; Ferreira, R.; Seddon, K.R; Rebelo, L.P.N.; Silva Pereira, C.; 2009. Exploring fungal activity in the presence of ionic liquids. *Green Chem.* 11, 889-894.

Petkovic, M.; Ferguson, J.; Gunaratne, H.Q.N.; Ferreira, R.; Leitão, M.C.; Seddon, K.R; Rebelo, L.P.N.; Silva Pereira, C.; 2010. Novel biocompatible cholinium-based ionic liquids—toxicity and biodegradability. *Green Chem.* 4, 643-649.

Petkovic, M.; Seddon, K.R.; Rebeloa, L.P.N.; Pereira, C.S.; 2011. Ionic liquids: a pathway to environmental acceptability. *Chem. Soc. Rev.* 40, 1383-1403.

Pham, T.P.T.; Cho, C.-W.; Vijayaraghavan, K.; Min, J. H.; Yun, Y.-S.; 2008a. Effect of imidazolium-based ionic liquids on the photosynthetic activity and growth rate of Selenastrum capricornutum. *Environ. Toxicol. Chem.* 27(7), 1583-1589.

Pham, T.P.T.; Cho, C.-W.; Min, J.; Yun, Y.-S.; 2008b. Alkyl-chain length effects of imidazolium and pyridinium ionic liquids on photosynthetic response of Pseudokirchneriella subcapitata. *J. Biosci. Bioeng.* 105(4), 425-428.

Pinto, P.C.; Costa, A.D.; Lima, J.L.; Saraiva, M.L.; 2011. Automated evaluation of the effect of ionic liquids on catalase activity. *Chemosphere* 82(11), 1620-1628.

Pretti, C.; Chiappe, C.; Pieraccini, D.; Gregori, M.; Abramo, F.; Monni, G.; Intorre, L.; 2006. Acute toxicity of ionic liquids to the zebrafish (Danio rerio). *Green Chem.* 8, 238 -240.

Pretti, C.; Chiappe, C.; Baldetti, I.; Brunini, S.; Monni, G.; Intorre, L.; 2009. Acute toxicity of ionic liquids for three freshwater organisms: Pseudokirchneriella subcapitata, Daphnia magna and Danio rerio. *Ecotoxicol. Environ. Saf.* 72(4), 1170-1176.

Ranke, J.; Mölter, K.; Stock, F.; Bottin-Weber, U.; Poczobutt, J.; Hoffmann, J.; Ondruschka, B.; Filser, J.; Jastorff, B.; 2004. Biological effects of imidazolium ionic liquids with varying chain length in acute Vibrio fischeri and WST-1 cell viability assays. *Ecotox. Environ. Saf.* 58, 396–404.

Ranke, J.; Cox, M.; Müller, A.; Schmidt, C.; Beyersmann, D.; 2006. Sorption, cellular distribution and toxicity of imidazolium ionic liquids in mammalian cells - Influence of lipophilicity. *Toxicol. Environ. Chem.* 88 (2), 273-285.

Ranke, J.; Stolte, S.; Störmann, R.; Arning, J.; Jastorff, B.; 2007. Design of sustainable chemical products – the example of ionic liquids. *Chem. Rev.* 107, 2183–2206.

Raves, M.L.; Harel, M.; Pang, Y.P.; Silman, I.; Kozikowski, A.P.; Sussman, J.L.; 1997. Structure of acetylcholinesterase complexed with the nootropic alkaloid, (-)-huperzine A. *Nat. Struct. Biol.,* 4, 57-63.

Saadeh, S.M.; Yasseen, Z.; Sharif, F.; Shawish, H.M.A.; 2009 New room temperature ionic liquids with interesting ecotoxicological and antimicrobial properties. *Ecotox. Environ. Safe.* 72, 1805-1809.

Samori, C.; Pasteris, A.; Galletti, P.; Tagliavini, E.; 2007. Acute toxicity of oxygenated and nonoxygenated imidazolium-based ionic liquids to Daphnia magna and Vibrio fischeri. *Environ. Toxicol. Chem.* 26(11), 2379-2382.

Samori, C.; 2011. Ionic liquids and their biological effects towards microorganismas. *Curr. Org. Chem.* 15, 1888-1904.

Schaffran, T.; Justus, E.; Elfert, M.; Chen, T.; Gabel, D.; 2009. Toxicity of *N,N,N* trialkylammoniododecaborates as new anions of ionic liquids in cellular, liposomal and enzymatic test systems. *Green Chem.* 11, 1458-1464.

Sheldon, R.; 2001. Catalytic reactions in ionic liquids. *Chem. Commun.* 23, 2399–2407.

Sipes, I.G.; Knudsen, G.A.; Kuester, R.K.; 2008. The effects of dose and route on the toxicokinetics and disposition of 1-butyl-3-methylimidazolium chloride in male F-344 rats and female B6C3F1 mice. *Drug Metab. Dispos.* 36, 284–293.

Składanowski, A.C.; Stepnowski, P.; Kleszczyński, K.; Dmochowska, B.; 2005. AMP deaminase in vitro inhibition by xenobiotics. A potential molecular method for risk assessment of synthetic nitro- and polycyclicmusks, imidazoliumionic liquids and N-glucopyranosyl ammoniumsalts. *Environ. Toxicol. Phar.* 19, 291–296.

Stasiewicz, M.; Mulkiewicz, E.; Tomczak-Wandzel, R.; Kumirska, J.; Siedlecka, E.M.; Golebiowski, M.; Gajdus, J.; Czerwicka, M.; Stepnowski, P.; 2008. Assessing toxicity and biodegradation of novel, environmentally benign ionic liquids (1-alkoxymethyl-3-hydroxypyridinium chloride, saccharinate and acesulfamates) on cellular and molecular level. *Ecotox. Environ. Safe.* 71(1), 157-165.

Stepnowski, P.; Składanowski, A.C.; Ludwiczak, A.; Laczynska, E.; 2004. Evaluating the cytotoxicity of ionic liquids using human cell line HeLa. *Hum. Exp. Toxicol.* 23(11), 513-517.

Steudte, S.; Stepnowski, P.; Cho, C.-W., Thöming, J.; Stolte, S.; 2012. (Eco)toxicity of fluoro-organic and cyano-based ionic liquid anions. *Chem. Commun.* 48, 9382-9384.

Stock, F.; Hoffmann, J.; Ranke, J.; Störmann, R.; Ondruschka, B.; Jastorff, B.; 2004. Effects of ionic liquids on the acetycholinesterase – a structure - activity relationship consideration. *Green Chem.* 6, 286–290.

Stolte, S.; Arning, J.; Bottin-Weber, U.; Matzke, M.; Stock, F.; Thiele, K.; Uerdingen, M.; Welz-Biermann, U.; Jastorff, B.; Ranke, J.; 2006. Anion effects on the cytotoxicity of ionic liquids. *Green Chem.* 8, 621-629.

Stolte, S.; Matzke, M.; Arning, J.; Böschen, A.; Pitner, W.-R.; Welz-Biermann, U.; Jastorff, B.; Ranke, J.; 2007. Effects of different head groups and functionalized side chains on the aquatic toxicity of ionic liquids. *Green Chem.* 9, 1170–1179.

Stolte, S.; Steudte, S.; Areitioaurtena, O.; Pagano, F.; Thöming, J.; Stepnowski, P.; Igartua, A.; 2012. Ionic liquids as lubricants or lubrication additives: an ecotoxicity and biodegradability assessment. *Chemosphere* 89(9), 1135-1141.

Studzińska, S.; Buszewski, B.; 2009. Study of toxicity of imidazolium ionic liquids to watercress (Lepidium sativum L.). *Anal. Bioanal. Chem.* 393, 983–990.

Studzińska, S.; Sprynskyy, M.; Buszewski, B.; 2008. Study of sorption kinetics of some ionic liquids on different soil types. *Chemosphere* 71(11), 2121–2128.

Swatloski, R.P.; Holbrey, J.D.; Rogers, R.D.; 2003. Ionic liquids are not always green: hydrolysis of 1-butyl-3-methylimidazolium hexafluorophosphate. *Green Chem.* 5, 361-63.

Tee, K.L.; Roccatano, D.; Stolte, S.; Arning, J.; Jastorff, B.; Schwaneberg U.; 2008. Ionic liquid effects on the activity of monooxygenase P450BM-3. *Green Chem.* 10, 117-123.

Tong, Y.; Wang, Q.; Bi, Y.; Lei, M.; Lv, Y.; Liu, Y.J.; Liu, J.; Lu, L.; Ma, Y.; Wu, Y.; Zhu, S.; 2012. Influence of ionic liquid 1- butyl-3-methylimidazolium chloride on the soil micro-ecological system. *Open Biotech. J.* 6, 1-4.

Wang, X.; Ohlin, C.A.; Lu, Q.; Fei, Z.; Hu, J.; Dyson, P.J.; 2007. Cytotoxicity of ionic liquids and precursor compounds towards human cell line HeLa. *Green Chem.* 9, 1191-1197.

Wang, L.-S.; Wang, L.; Wang, L.; Wang, G.; Li, Z.-H.; Wang, J.-J.; 2009. Effect of 1-butyl-3-methylimidazolium tetrafluoroborate on the wheat (Triticum aestivum L.) seedlings. *Environ. Toxicol.* 24, 296–303.

Wang, S.-H.; Huang, P.-P; Li, X.-Y.; Wang, C.-Y.; Zhang, W.-H.; Wang, J.-J., 2010. Embryonic and developmental toxicity of the ionic liquid 1-methyl-3-octylimidazolium bromide on goldfish. *Environ. Toxicol.* 25(3), 243-250.

Weaver, K.D.; Kim, H.J.; Sun, J.; MacFarlane, D.R.; Elliott, G.D.; 2010. Cyto-toxicity and biocompatibility of a family of choline phosphate ionic liquids designed for pharmaceutical applications *Green Chem.* 12, 507-513.

Wells, A.S.; Coombe, V.T.; 2006. On the freshwater ecotoxicity and biodegradation properties of some common ionic liquids. *Org. Process Res. Dev.* 10(4), 794-798.

Xie, Y.N.; Wang, S.F.; Zhang, Z.L.; Pang, D.W.; 2008. Interaction between room temperature ionic liquid [bmim]BF4 and DNA investigated by electrochemical micromethod. *J. Phys. Chem. B* 112 (32), 9864-9868.

Yu, M.; Li, S.-M.; Li, X.-Y.; Zhang, B.-J.; Wang, J.-J.; 2009a. Acute effects of 1-octyl-3-methylimidazolium bromide ionic liquid on the antioxidant enzyme system of mouse liver. *Ecotox. Environ. Saf.* 71, 903–906.

Yu, M.; Wang, S.H.; Luo, Y.R.; Han, Y.W.; Li, X.Y.; Zhang, B.J.; Wang, J.J.; 2009b. Effects of the 1-alkyl-3-methylimidazolium bromide ionic liquids on the antioxidant defense system of Daphnia magna. *Ecotoxicol. Environ. Saf.* 72(6), 1798-1804.

In: Ionic Liquids: Synthesis, Characterization and Applications ISBN: 978-1-63117-716-3
Editor: Anson Brooks

Chapter 9

ASSESSING THE ROLE OF TEMPERATURE, IRRADIATION AND OXIDATION ON ILs STABILITY

E. Siedlecka**, *M. Czerwicka and P. Stepnowski
Faculty of Chemistry, University of Gdańsk, Gdańsk, Poland

ABSTRACT

ILs show unique properties, which are useful for their application in industry. They are very promising solvents and reagents, however it has to be kept in mind that in practice ILs should be stable under application conditions. After used the ionic liquids should be recovered and reused. Therefore the knowledge of their resistant to temperature, irradiation, oxidizing and reducing agents and activity lose during application is necessary. The attention must also be paid to the purity of ILs used to chemical, photochemical and electrochemical processes, due to their lower stability in presence of small amount of water and mineral acids.

In the review biomedical, industrial and scientific applications of ILs were presented. The influence of operating conditions such as: temperature, the type of irradiation, the redox potential, electrical current and purity of IL on its stability during application are discussed. Additionally the effect of chemical structure of ILs on the degradation pathway were considered.

Sooner or later ILs applied in industry will be waste, and efficient degradation process should be used to their disposal. Most of ILs are poorly biodegradable and do not inert to water organisms. Advanced Oxidation Processes appear to be promising in this respect. The progress in wastewater treatment and disposal of used ILs in oxidative conditions, were also discussed.

INTRODUCTION

When in 1914 Walden synthesized the first room-temperature ionic liquid, he didn't expect that it was the first step to create such a fascinating group of compounds [1]. Although

* Corresponding author: ewa.siedlecka@ug.edu.pl.

almost one hundred years have passed from that moment, the best for ionic liquids were last 20 years. Today all chemists know what ionic liquids (ILs) are, and that typical ionic liquid consists of usually bulky organic cation and inorganic or organic anions. A really great number of cations and anions combinations is the reason of the structural diversity of ILs. Because the chemical structure determines the properties of each compound, it is also same responsible for their applicability. In the beginning ILs were treated as alternatives for volatile organic solvents. After some time, these extraordinary Green solvents became much more popular. They are very stable under the extreme hard conditions, and therefore they are used as electrolytes, catalysts or solvents in different environment such as high temperature, oxidation/reduction potential, irradiation etc. In spite of the high stability, under such conditions applied for a long time, they could be destroyed in small or even large degree [2]. The modern ionic liquids are tunable and it is possible to design unique compounds for specific purposes, and obtain the human health and environmentally friendly ILs. Although ILs properties create favourable conditions for their regeneration and recovery sometimes it is impossible, and then the well-judged degradation system is needed. Some selected applications of ILs and their stability in different thermal, electrochemical, photochemical and chemical processes will be presented [Figure 1]. Additionally the degradation methods of wasted ILs will be discussed in this chapter.

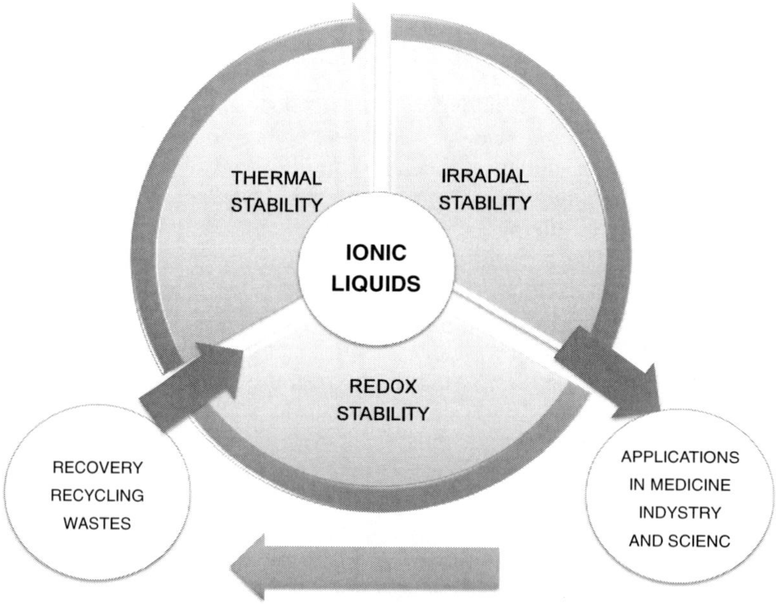

Figure 1. Ionic liquids circle of life.

Biomedical, Industrial and Scientific Applications of ILs

Stability of ionic liquids is definitely a desirable feature from their application point of view and it provokes representatives of different scientific disciplines like chemistry, physics, biology, medicine, pharmacy, etc. to think how to take advantage of them. If it is possible the used ionic liquids should be regenerated or recovered, but sometimes their recycling is not sufficient or impossible to apply and ILs appear in wastes. In this case specific degradation methods are needed.

Although many ILs are present in industry some of them are used in scientific laboratories. They have an inestimable importance for separations in analytical chemistry [3-5]. They are used in different extraction methods such as classical liquid-liquid extraction, liquid-phase microextraction (LPME) or solid-phase microextraction (SPME) [4-6]. Thank to them isolation of various metal ions [7-9], organic contaminants [3,10], bioactive compounds from plants [11,12] or even microorganisms [13] is reached with the simultaneous reduced consumption of classical organic solvents. The chromatographic techniques are next separations procedures where ILs are used. Ionic liquids stationary phases are successfully applied to gas chromatography [14,15] as well as liquid chromatography especially HPLC (RP and NP) [16,17]. The low volatility, high thermal stability and wide range of polarity decide about this type of applicability. They are also used as mobile phases additives [4,16,18]. Apart from gas and liquid chromatography, also capillary electrophoresis takes advantage of ILs. The modification of capillary surface by some ILs (they are covalently bonded or dynamically coated) reverse unwanted process of electroosmotic flow (EOF) [4]. It is worth to mention that ionic liquids are used for separation of the specific substances for example CO_2 or H_2S from gaseous samples [3,19]. The specific properties of ILs make them useful agents for mass spectrometry mainly with soft ionisation techniques. They are applied as liquid matrices for matrix-assisted laser desorption/ionization mass spectrometry (MALDI-MS) and as additives in ESI-MS for the better anions detection [4,20]. While ILs are used as solvents for many reactions and catalysis, they can interact with dissolved substances, what has an effect on reactivity and selectivity in various important reaction transformations [21,22].

One of the most important consumers of ionic liquids is industry. In this area their unique stability plays a key role. It has been proved that some of them have excellent tribological properties comparing with reference lubricants. Their long-term thermo-oxidative stability, non-flammability, non-corrosive behavior decide about their usage as lubricating fluids and nanofluids in vacuum machines (especially working in specific conditions: in space applications, where the fire risk is significant, in clean rooms etc.) and also as lubricants for magnetic media [23-26]. Ionic liquids perform significant function as a component of energy related products. It had been found that some energetic ionic liquids (EILs) characterized by low melting point, low viscosity, low surface tension and long-term thermal stability were proposed for space propulsion [27]. Another ILs application is electrolyte membrane for batteries [28-31]. Thanks to thermo-oxidative stability they protect electrodes from the decomposition simultaneously improving the safety of battery cells. Ionic liquids are used as electrolytes also in dye-sensitized solar cells (DSSC) [31-33] and as heat-transfer and storage fluids for conversion of solar power [34].

Room temperature ionic liquids, have been shown recently to function as efficient solvent systems for the oxidative leaching of gold and silver from a complex sulfidic ore and electronic waste. The [IM14] [HSO_4]/oxidant system in particular gave a significantly enhanced recovery of silver compared with aqueous leaching, in addition to excellent gold recovery [35]. It is known that a lot of metal compounds associated with mining or metallurgy are likely to pose a risk for an environment. Application of ionic liquids in extraction and separation of rare earth metals is widely studied. Recent results now show a great promise in the treatment of wastewater from hydrometallurgical processes [36,37].

The conversion of cellulose into ionic liquids to deal with the depletion of fossil fuel has got much attention. Completing the hydrolysis of cellulose under mild conditions is the key step, which is possible in ILs. They act as both acid catalyst to promote the hydrolysis of cellulose (SO_3H-functionalized acidic ionic liquids), and the solvent of reaction (1-butyl-3-methylimidazolium chloride) [38,39]. Ionic liquids seem to be a promising alternative in the process of lignocelluloses biomass pre-treatment [40,41]. It was proved the similar efficiency of ILs based technologies compared to the conventional pre-treatment methods [40]. As a result, cellulose is more accessible for hydrolysis and other reactions [41,42].

Ionic liquids are beginning to be used as a new kind of binder to make an ionic liquid modified electrode. Because they have many specific electrochemical properties such as wide electrochemical window, high ionic conductivity and good solubility, they have been recognized as a useful non-aqueous media for various electrochemical processes [43-46]. IL modified electrodes have been used to human DNA [47], hydrogen peroxide [46] or morphine [48] detection.

Utilization of ionic liquids in medicine seems to be impossible. ILs are very stable, with rather poor biodegradability, and potentially toxic what excludes them from biomedical usage. But on the other hand they have been pharmaceutically acceptable salts for a long time. A lot of pharmaceutically active compounds have a liquid salt form (for example antiseptic and disinfectant known from the 1937 - cetylpyridinium chloride or drug for functional gastrointestinal disorders known from the 1902 – methylscopolamine bromide) [49,50]. ILs play a significant role in syntheses or/and analyses of many drugs [51]. They are also involved in preparation of biopolymer-based materials useful for tissue engineering, drug delivery system, wound treatment etc. [51,52]. They can help in modification of some polymers for biomedical applications such as chitin (the dopamine-mediated sclerotization of regenerated chitin) [53], chitosan (the production of blended hydrogels) [54] or PVC (the production of biomedical-grade suspension of poly(vinyl chloride) [55]. Some of ionic liquids reveal activity against Gram-positive and Gram-negative bacteria [51,56], fungi [51,57], and tumor cells [51,58]. Nevertheless, the biomedical applications of ionic liquids have been not so popular yet, the role of new designed non-toxic ILs in medicine looks promising for the coming years.

Thermal Stability

Taking applications of ionic liquids with high temperature environments into consideration, thermal stability and decomposition are significant properties. In the context of structural diversity of ILs the following factors could have direct impact on these physicochemical properties: type of cation and anion, the alkyl chain length, different substituent groups in the alkyl chain or in the ring, and confined geometry [59-61]. There are some investigations reporting that imidazolium based ionic liquids seem to be more stable in comparison with phosphonium, pyridinium, ammonium, pyrrolidinium, and piperidinium. Nevertheless, the obtained results are not sufficient to create a definite rank [59,60]. According to the studies of Cassidy et al., ILs based upon tetraarylphosphonium cations characterize superior long-term thermal stability compared to the analogical salts based upon imidazolium, quaternary ammonium, or tetraalkylphosphonium cations. [62]. The alkyl chain length effect is strongly observed for ionic liquids with shorter chains only. More stable in higher temperature are ILs with shorter alkyl chains. The longer the substituted alkyl chain is the less remarkable differences between degradation temperatures (T_g) eg. for [Pip13][(CF3SO2)2N], [Pip14][(CF3SO2)2N], [Pip18][(CF3SO2)2N] are observed. In this case the Tg decreases in the following order 341°C>320°C>315°C [63]. The same relation was observed for imidazolium, pyrrolidinium, pyridinium, piperidinium and ammonium ionic liquids [60].

The recent results obtained by Villaneuva et al., are in a good agreement with these observations [64]. They investigated the thermal stability of three ionic (1-ethyl-3 methylimidazolium bis(trifluoromethylsulfonyl) imide ([IM12] [(CF3SO2)2N]), 1-propyl-3-methylimidazolium bis(trifluoromethylsulfonyl)imide ([IM13] [(CF3SO2)2N]), and 1-buthyl-3-methylimidazolium bis(trifluoromethylsulfonyl)imide ([IM14] [(CF3SO2)2N]) liquids for their application as new absorbents. The most stable was ([IM12] [(CF3SO2)2N]). The greater influence on the thermal degradation is related to presence of different substituent groups or chemical modifications. The most examples known for imidazolium based ionic liquids are listed below:

- the presence of ether group in the alkyl chain decreases Tg [65];
- comparing to one with carbonyl group the presence of hydroxyl group in the alkyl chain makes the IL more stable [66];
- in 1,3-disubstituted imidazolium tetrachloroferrates (MILs - magnetic ILs) the presence of β–cyanoethyl group (instead of methyl group) decreases thermal stability [67];
- the substitution of hydrogen of the 1,3-disubstituted imidazolium ring by alkyl group increases the thermal stability [68].

Basing on present studies it is possible to say that the most critical role in the thermal properties of ionic liquids plays the anion type. It has been found that the relative anion stability follows the order:

- [(CF3SO2)2N]$^-$ > [(CF3SO2)3C]$^-$ ~ [(C4F9)SO3]$^-$ ~ [(C2F5SO2)2N]$^-$ > [CF3SO3]$^-$ > [N(CN)2]$^-$ >[PhCOO]$^-$ >[Br]$^-$ > [I]$^-$ ~ [Cl]$^-$ > [1COO]$^-$ > [CF3COO]$^-$ ([BF4]$^-$ and [PF6]$^-$ were not included because of lack of unambiguous data - the temperature difference was even 100C°) for 1-butyl-3-methylimidazolium based ILs [60];
- [(C2F5SO2)2N]$^-$ > [AsF6]$^-$ > [(CF3SO2)2N]$^-$ ~ [(C4F9)SO3]$^-$ ~ [CF3SO3]$^-$ > [BF4]$^-$ >> [N(CN)2]$^-$ > [BF3(Ph)]$^-$ > [B(Ph)4]$^-$ for 1-octyl-4-methylpyridinium based ILs [69];
- [(CF3SO2)2N]$^-$ ~ [AsF6]$^-$ > [BF4]$^-$ for 1-methyl-1-butylpiperidinium based ILs [63];
- [(CF3SO2)2N]$^-$ > [(C2F5SO2)2N]$^-$ > [(CF3SO2)3C]$^-$ > [PF6]$^-$ > [BF4]$^-$ > [Cl]$^-$ for tetraethylammonium based ILs [70];
- [(CF3SO2)2N]$^-$ > [C(CN)3]$^-$ > [N(CN)2]$^-$ for N-butyl,N-methylpyrrlidinium based ILs [71];
- [(CF3SO2)2N]$^-$ > [(C2F5)3PF3]$^-$ > [N(CN)2]$^-$ > [Cl]$^-$ for triheksyl-tetradecylphosphonium based ILs [72].

Summarising above data and the rest of published results ILs halides were the least resistant to high temperatures. The reverse dependence between the thermal stability and the coordinating properties of the anion, hydrophilicity and anion size was proved. In agreement with this trend anions such as hexafluorophosphate and tetrafluoroborate or organic anions such as bis(pentafluoroethylsulfonyl)imide and bis(trifluoromethylsulfonyl)imide were found to be the most thermally stable the inorganic [60].

The thermal stability of 1-butyl-3-methylimidazolium hexaflurophosphate in confined geometry was analysed by Singh et al. It was found that the decomposition temperature of unconfined form is higher than in case of the same ionic liquid in a confined geometry [61].

The thermo-oxidative stability of ionic liquids for industrial use is a complex problem. From the one hand there are their own properties but from the other hand one cannot forget about other substances present in ionic liquids containing products, and of course about possible impurities [73]. Glenn et al., investigated the thermal stability of [IM14] [BF4] in the presence of following nucleophiles: thiophenol, sodium phenylthiolate, sodium benzoate, aniline, and sodium phenoxide [74]. They proved that in the presence of nucleophiles the thermal stability of imidazolium based ILs decreases. It was revealed by Meine et al., that the same effect occurred in the presence of water [75]. They studied long-term stability of [IM14][Cl] under anhydrous conditions and in the addition of 2 wt% of water. The degradation resistance of anhydrous sample was four times higher after heating for 10 days at 100°C and over six times higher at 140°C. Van Valkenburg et al., investigated the influence of metal ions, chloride, and water on the degradation temperature of [IM12][BF4], [IM14][BF4], and 1,2-dimethyl-3-propylimidazolium bis(trifluorosulfonyl)imide [34]. They found the negligible impact of all tested impurities including water. Therefore, considering the temperature limit values the time of exposure and the isocratic or gradient temperature conditions depending on the given ILs destination have to be considered. Connections between imidazolium, pyrrolidinium, piperidinium or quaternary ammonium cations and bis(trifluoromethylsulfonyl)imide anion, or tetrafluoroborate are the most readily applied ILs as batteries electrolytes. Due to the thermal properties of ILs the evaporation of electrolyte and flammability at increased temperatures are reduced when comparing to classical organic solvents [29,76].

The visual effects of thermal degradation are the easiest to detect, but the color changes concern only several ILs. Basing on the literature data it was found that the preferable and commonly used method to observe degradation temperature is thermogravimetric measurements [60]. However there is a report concerning the potentiometric titration which seems to be quick, precise, and low-cost analytical method to trace thermal transformation. It was proved that in case of some imidazolium ionic liquids the thermal decomposition starts at lower temperature when comparing with the values obtained using TGA (*Thermal Gravimetric Analysis*) [75]. Additionally this fact was confirmed by the results of ESI-MS and GC-MS analyses. There were identified several degradation products of [IM14][Cl] what provided the conclusions about the preferential reaction type as follows:

- retroalkylation (1-chloromethane, 1-chlorobutane, 1-methylimidazole and 1-butylimidazole;
- elimination of the n-butyl side-chain (1-methylimidazolium species, 1-butene);

- trans-alkylation (1,3-dibutylimidazolium species, 1,3-dimethylimidazolium) – this process is invisible for TGA because is not related with weight loss [60].

Many mechanisms of thermal decomposition of different types of ionic liquids have been known. Summarizing all of obtained results it is known that the S_N2 mechanism of nucleophilic substitution is characteristic for 1,3-alkylimidazolium and allyl-imidazolium halides followed by forming *N*-alkylimidazoles and 1-alkylhalides [59,77-79]. This mechanism explains the mentioned above reduced stability of ILs in the presence of other nucleophiles [74]. The same degradation pathways were observed for the 1-alkylpyridinium, dialkylpyrrolidinium, and tetraalkylammonium-based ILs [70]. Another type of nucleophilic reaction typical for quaternary salts is β-elimination resulting in production of alkenes, amines, and trialkylammonium cations in case of quaternary ammonium ILs, and alkenes, 1-methylimidazolium species in case of 1,3-alkylimidazolium halides [59,67]. It was stated that for disubstituted imidazolium halides, dicyanamide, hexafluorophosphate and tetrafluoroborate the dealkylation by C-N bond cleavage was preferred [60,70,78]. For $[BF4]^-$ and $[PF6]^-$ containing dialkylimidazolium ILs the most readily formed products are alkenes and haloalkanes while in case of halides there were haloalkanes and 1-alkylimidazoles [78]. Some trialkylimidazolium-based ILs or dialkylimidazolium bis(trifluoromethylsulfonyl)-imides are decomposed also via the S_N1 mechanism [60,70]. Moreover, the polymerization process for cyano containing anions and cyclic quaternary ammonium cation and volatile products formed for phosphonium ILs was indicated [71]. The knowledge about all of these mechanisms is very helpful for designing new ionic liquids for industry purpose as well as for waste treatment.

Irradiation Stability

In this part the degradation of ionic liquids under the influence of different type of irradiation will be discussed. Brands et al., studied the sensitivity of [IM14][I], [IM16][Cl], [Pyr14][(CF3SO2)2N] and [P666-14][(CF3SO2)2N] for UV photolysis [80]. There were found di-, and trihalides ions as a result of continuous irradiation, and suppression of this product formation when the excitation pulse for a short time was used. Also the slowly darkening with increasing irradiation time was observed. Katoh&Takashi compared the UV-laser irradiation effect on 1-methyl-3-buthylimidazolium bis(trifluoromethanesulfonyl)amide [IM14][(CF3SO2)2N] (220nm light pulse excitation) and iodide solution of [IM14][(CF3SO2)2N] (280nmexcitation) [81]. They observed the relative stability of neutral radical $[IM14]^{\bullet}$ and effective degradation in case of the excited state of [IM14]*. The ILs stability for UV photolysis is very important for these ionic liquids which are potential components of products converting solar energy into chemical one.

Hydrophobic room-temperature ionic liquids radiation chemistry is fundamental for treatment of high-level nuclear waste (as possible replacements for molecular diluents for nuclear separations, as well as the basis of new separations processes) [82-84]. Because the anion types influence for many physicochemical properties, their effect on radiochemical stability of room-temperature ionic liquids was carefully studied [85-87]. Le Rouzo et al., investigated the stability of 1-butyl-3-methylimidazolium ($[IM14]^+$) salts ($[(CF3SO2)2N]^-$, $[CF3SO3]^-$, $[PF6]^-$ and $[BF4]^-$) under gamma irradiation for high irradiation doses (up to 2.0

MGy) [85]. For analysis of radiolysis products they used HPLC, ESI-MS, ^{19}F NMR, UV-visible spectroscopy, and GC/MS. By the means of the above techniques they found a large number of radiolysis products. They were derived from reactions of primary generated species of cation and anion by indirect radiolysis which could later react together throughout the irradiation by indirect radiolysis. It was proved that relative degradation of the $[IM14]^+$ cation depending on the anion in the following order: $[(CF3SO2)2N]^- \leq [CF3SO3]^- < [PF6]^- \leq [BF4]^-$ (in a slight degree). In spite of small amount of many radiolysis products, all tested ILs occurred to be suitable for applications in the nuclear fuel cycle. The latest experience of Shkrob scientific group concerned the radiation stability of ILs based on such "sweet" sulfonyl imide anions (phthalimide, saccharinate, o-benzenedisulfonimide, and acesulfamate) [87]. They found all of them practically useful upon exposure to ionizing radiation, but saccharinate and o-benzenedisulfonimide remarkably stable to radiation-induced fragmentation. Previously, Shkrob et al., tested radiochemical stability of different ammonium, phosphonium, and imidazolium based ILs [88]. They identified the fragmentation pathways for constituent anions (among others $[Br]^-$, $[NO3]^-$, $[BF4]^-$, $[PF6]^-$, $[CF3SO3]^-$, $[N(CN)2]^-$) using electron paramagnetic resonance (EPR) spectroscopy. They observed following consequences of radiolytically induced redox reactions:

- for fluorinated anions, F^- loss following the electron attachment are quite prominent,
- extensive bond scission was observed $[NO3]^-$ and $[B(CN)4]^-$ anions and indirectly implicated for $[BF4]^-$ and $[PF6]^-$ anions.

Although the relative stability of constituent anions followed in the order carboxylates < dialkyl phosphates <alkyl sulfates < alkanesulfonates < imides (including dicyanamide). There were not general principles for choosing radiation-resistant anions. In summary they showed among larger anions, the derivatives of benzoate and imide anions were found to be relatively stable, while for ensemble of small anions the most stable were $[CF3SO3]^-$ and $[N(CN)2]^-$. Obviously, the next step was to study the radiochemical stability of different cation types [88-92]. For this purpose Shkrob et al., followed the radiolytic degradation of 1,3-dialkyl substituted imidazolium cations, and for comparison also 1,3-dimethoxy- and 2-methylsubstituted imidazolium and 1-butyl-4-methylpyridinium cations [88]. To identify the result species they used electron paramagnetic resonance spectroscopy (EPR), ^1H and ^{13}C nuclear magnetic resonance (NMR) and tandem electrospray ionization mass spectrometry. They indicated the readily attacked places. One of them was C(2) position in unsubstituted imidazolium cations, the second one was the aliphatic chain. Taking into consideration all results obtained by several scientific groups, it is clear that radiation chemistry of ionic liquids is very complex [82-88]. The reaction cascade includes the following steps: the initial interactions of αβγ radiation with chemical systems by ionization and excitation, geminate recombination of formed cations to give more excited states or ion dissociation, and additional recombinations and hydrogen abstraction processes [82]. In consequence of irradiation primary neutral species followed by their fragmentation, protonation, deprotonation, structural rearrangements, and the subsequent (secondary) transformations including forming radicals and molecular fragments recombination, disproportionation, and reaction with stable radiolytic products [82-84,88] are generated. Shkrob et al., investigated radiation stability of derivatives of 5-membered ring heterocycles [89], 1-benzylpyridinium [90], and guanidinium cations [91]. The analytical techniques were similar to those used during 1,3-dialkyl substituted imidazolium cations studies (EPR, NMR, ESI MS). They confirmed a partial or complete loss of aliphatic side chain as a result of irradiation of 1-alkyl-3-methylimidazolium, 1-alkyl-3-methyl-1,3,4-triazolium, and 1-alkyl-5-methylthiazolium cations (extensive C−N and C−C bond fragmentation) [89]. In case of 1-benzyl derivatives of imidazolium and 1,3,4-triazolium cations the loss of benzyl radicals was observed. Looking for ionic liquids suitable for nuclear extractions they studied the stability of 1-benzylpyridinium cations [90]. The quite good resistance for radiation damage was explained by charge delocalization in both their reduced and oxidized forms in the ILs. There was also observed the reduced yield of fragmentation because the carbocation stabilization. The last mentioned example concerned the guanidinium cations investigations [91]. The cyclic guanidinium cations were found resistant to radiation induced oxidation and reduction, but much less resistant to fragmentation connected with multiple C−C and C−N bond dissociation.

The radiation resistance of ILs could be achieved by using the effective additives. Huang et al., studied the influence of $FeCl_3$ on the degree of radiolysis of [IM14][Cl] [93]. They proved the better protection effect with increasing of Fe(III) concentration. They observed that in excess of [IM14][Cl], new ionic liquid [IM14][FeCl4] was formed.

Ionic liquids are applied as a media for many reactions during which the assistance of ultrasound or microwave irradiation is needed. Oxley et al., investigated the degradation effects during the sonication of the following ILs: butylmethylimidazolium chloride [IM14][Cl], butylmethylimidazolium tetrafluoroborate [IM14][BF4], butylmethylimidazolium hexafluorophosphate [IM14][PF6], urea ammonium nitrate [CO(NH2)2][NH4NO3] (UAN), and decylmethylimidazolium tetraphenylborate [IM1-10][B(Ph)4] [94]. In consequence of acoustic cavitation a number of decomposition products was detected by GC-MS. In the head-gas over sonicated ionic liquids they found: light hydrocarbons, nitriles, and alkyl halides in case of imidazolium ILs (darkening to amber color was also observed), chlorobutane and chloromethane in case of [IM14][Cl], trace amounts of several cyclic products including benzene in case of [IM1-10][BPh4], and NH_3 and CO_2 in case of UAN. During the same analyses of reference (untreated with ultrasound) samples no effect was observed what additionally confirmed the sonochemical decomposition of ILs. Except for ultrasound, also microwaves activate many processes carried out in ionic liquids [22]. In this case ILs play a role of a very good medium for absorbing microwaves, what decides about the high heating rate and increasing the rate and yield of reaction. Unfortunately the high temperature can favor the degradation of ILs what was mentioned above [74]. This has forced us to use only thermal stable ILs for microwave assisted reactions.

Electrochemical Stability

Considering ionic liquids as electrolytes, it is important to know their electrochemical stability during an electrolysis. Their stability could differ depending on particular electrodes. For this purpose in the individual set-up the electrochemical window has been studied. It has been found that, ammonium-, pyrrolidinium-, and piperidinium-based ionic liquids with triflate or triflyl anions show the largest electrochemical stability [95-98]. The electrochemical window determined at a glassy carbon (GC) or platinum electrode is within a range of ca. 6 V. These results cannot be surprising, because quaternary ammonium salts are usually used as a supporting electrolyte in the conventional electrolyte which has wide electrochemical windows. The branching of the alkyl group in pyrrolidinium bis(trifluoromethylsulfonyl)imide ionic liquids leads to a moderate increase of the electrochemical stability in changing from n-butyl to iso-butyl while a dramatic decrease was observed from the sec-butyl group. The progressive increase of the cathode limit potential was also observed along the increase of the side chain length [98]. Recent data presents low-cost protic salts which are a sub-class of ionic liquids (ILs) containing azepanium units as cation, and formate, trifluoroacetate, or hydrogen sulfate groups as anions with the stability of ca. 4V [99]. Imidazolium and morpholinium salts also show stability of ca. 4 V, while quaternary phosphonium ionic liquids exhibit the wide electrochemical window of approximately 6 V [100-102]. A shift to a lower potential of the anode limit with an increase in the length of the side-chain in tri-n-butylalkylphosphonium fluorohydrate (alkyl = methyl, butyl or octyl) was observed [102].

The cathodic stability of ILs is mainly determined by the potential at which the reduction of the cations takes place. The cathodic reduction of imidazolium-based ILs has been thought to yield neutral [IM1A]• radicals (A = alkyl) that subsequently undergo disproportionation or recombination. Shkrob and Wishart imply the formation of a dimer product $(IM1A)_2$ in electrochemical decomposition of $[IM14^+]$ [103]. Their study suggests that these decomposition products are more likely to be formed via deprotonation of the dimer radical cation than by the reaction of these unstable [IM1A]• radicals [103].

The intermediates, which occurred at the cathodic limit of the pyrrolidinium- and imidazolium ionic liquids were also predicted by quantum chemical calculations and verified by experiments by Kroon et al., [104]. It was showed that $[IM14^+]$ electrochemically decomposed on dimer of 1-butyl-3-methylimidazolium formed in the coupling reaction of two 1-butyl-3-methylimidazolium radicals, and 1-butyl-2-hydro-3-methylimidazole formed in a disproportionation reaction. In the case of $[Pyr14]^+$ all predicted decomposition reaction products were found: 1-methylpyrrolidine, octanes, octenes and 2-butanol, which indicates that at the first stage the unstable [Pyr14]• radical (in contrast to imidazolium radical) decomposes into 1-methylpyrrolidine and a butyl radical [104].

The reduction process of protic ionic liquids (1-methylpyrrolidinium, 1-methylpiperidiniuim, andazepanium), consisting of Brönsted acids and bases is expected to mainly involve the positively charged nitrogen in the pyrrolidinium, piperidiniuim, and azepanium rings, the reduction of cation proceeds by concerted-exchange mechanism. First, the cation is deprotonated, and then the proton is reduced to H_2. The results of the cathodic limiting potentials in pyrrolidinium, piperidiniuim, andazepanium ring show no significant relationship with cation size and none with methyl-substituted and non-methyl-substituted cations. These cations show a cathodic reduction at a very negative potential, −2.5 V [99].

However, if the cathodic stability of the anion is poor the type of counter ion may influence the cathodic stability limit [105].

The anodic stability of the materials is mainly determined by the potential at which the oxidation of the anions takes place. Halide counter ions such as $[F]^-$ or $[Br]^-$, which undergo anodic oxidation at relatively low potentials, are responsible for the narrow stability of ionic liquids of the order of ca. 2–3 V. With regard to the anodic limit, the carboxylate anion formate and trifluoroacetate is oxidized at a relatively low anodic potential, +0.7 V. The Kolbe reaction involves the formation of dimers, whereas the oxidation of the hydrogen radical and trifluoro-radical higher anodic potential [106]. The hydrogen sulfate anion is oxidized at an appropriate anodic potential, +1.5 V. Hydrogen sulfate-based ILs have a higher anodic limit potential than those of formate- and trifluoroacetate-based ILs, this can be attributed to the formation of persulfate ion when the hydrogen sulfate anions are oxidized. Accordingly, hydrogen sulfate-based ILs have a larger electrochemical window than those of formate- and trifluoroacetate-based ILs [99]. The amide anion on the contrary, is oxidized at relatively high anodic potentials, which implies the broad stability of ionic liquids based on this anion [102,107]. However, the anodic limit potential of 1-etyl-3-methylimidazolium and p-dimethylamino-N-butylpyridinium pairing with very stable anions is determined not by the oxidation of the anions, but the cation, since the anodic limiting potential of these system doesn't depend on the anions [108].

Quantum chemical calculations used to predict the electrochemical stability of ionic liquids [109] showed that reductive and oxidative stability can be correlated respectively to the lowest unoccupied molecular orbital (LUMO) energy level of the cation and to the highest occupied molecular orbital (HOMO) energy level of the anion pairing in IL [109].

It is hard to make a comparison of the electrochemical window of different ILs because of the influence of the measurement conditions on the results. Among them there is the quality of ILs and kind of working and references electrode [108,110]. The presence of a small amount of water or oxygen, which dissolves very well in ILs, (also from air), can significantly change the electrochemical window, because cyclic voltammogram technique is very sensitive to these impurities. The effect of water on the electrochemical window of imidazolium and pyrimidinium salts has been studied [108]. The addition of more than 3 wt% of water to 1-butyl-3-methylimidazolium tetrafluoroborate or hexafluorophosphate caused a decrease in the width of electrochemical window of over 2.0 V [108]. However the influence of water significantly depended on the kind of working electrode. For glassy carbon (GC) and tungsten (W) electrodes, no significant effects were observed regardless of the addition of water for N-butylpyridinium chloroaluminate [111]. For a platinum working electrode, a reduction current of $[Cl]^-$ was observed at a potential much more positive than that of the cathodic limiting potential of this system. Stability windows for ILs obtained at Pt and GC electrodes are usually comparable, while the ionic liquid/tungsten interface shows more enhanced stability in the range of 6.5–7 V. Recently the unexpected decomposition of the bis(trifluoromethylsulfonyl)amide anion under quite mild electrochemical conditions has been reported. The results show clearly that the highly stable bis(trifluoromethylsulfonyl)amide anion at GC anode can easily be decomposed during anodic oxidation of copper in the ionic liquid - 1-butyl-1-methylpyrrolidinium bis(trifluoromethylsulfonyl)amide - at 70 °C leading to the formation of CuF_2. At room temperature, however, no significant decomposition was obtained [112].

Despite of difficulties comparisons of presented in literature electrochemical windows data among the various salts (imidazolium, ammonium, phosphonium), provide important information to design electrochemically stable ILs.

Hydrolysis of Ionic Liquids

In general, most pure ionic liquids can be considered as not susceptible to hydrolysis. Maybe, a higher water sensitivity could be observed for some functionalized side chains containing e.g., ester groups but this has not been investigated so far. However even water-immiscible ILs are hygroscopic and usually contain some amount of water. Knowledge of ionic liquids behavior in the presence of water is thus indispensable for any potential application. Many ILs, especially those containing halide, acetate, formate, and phosphate anions, have shown high capacity for cellulose or chitosan dissolution at a temperature of 100 °C, and this dissolution can be significantly increased by microwave heating [113,114]. Therefore ILs could be good solvents for their hydrolysis. It was also demonstrated that good depolymerization level could be achieved for chitosan in ILs in the presence of mineral acids. Despite the drastic experimental conditions in these investigations the authors have not consider the possibly of IL degradation. However the high temperature, irradiation by microwave and the presence of water, mineral acids or trace amount of metal ions influence on the stability of ILs, as it was discussed above.

The hydrolysis instability of $[PF6]^-$ or $[BF4]^-$. $[PF6]^-$ or $[BF4]^-$ containing ionic liquids has often been noted in the literature [115,116]. For both anions hydrolytic processes were observed and a higher susceptibility for the $[BF4^-]$ anion was found. In case of [IM14][PF6] a hydrolysis was found during the extraction process under acid conditions [117]. The authors supposed that HNO_3 might act as a catalyst, which promoted the degradation of $[PF6]^-$ to $[PO4]^{3-}$. The catalyzed formation of $[PO4]^{3-}$ was also accelerated in the presence of SiO_2. Moreover SiF_4 and H^+ were also formed, which further led to the production of H_3PO_4. Additionally, hydrolysis of $[PF6]^-$ forms a mixture of products, including HF, POF_3, PO_2F_2, which are very toxic and corrosive [118]. Recently it has been shown that actinide hexachloro complexes ($An(IV)Cl_6$) are unstable with regard to hydrolysis in [IM14][PF6]. $An(IV)Cl_6$ complexes, where An(IV) is Th(IV), Np(IV) or Pu(IV), are able to accelerate $[PF6]^-$ anion hydrolysis in hydrated [IM14][PF6], followed by the formation of soluble and insoluble species of An(IV) [119].

The hydrolytic stability of [IM14][CH3OSO3] and [IM14][C2H5OSO3] also depends on the anions. In the presence of water, methyl sulfate and ethyl sulfate anions form the corresponding alcohol at elevated temperatures. For alkyl sulfates with longer alkyl chains (from 8 to 18 carbon atoms), the sensitivity to hydrolysis is much lower [120].

OXIDATIVE REMOVAL OF ILS FROM WATER

Chemical Degradation

Advanced oxidation processes (AOPs) based on the generation of a very reactive oxidizing agent (•OH), are very promising methods for industrial waste water treatment especially the removal of refractory and hazardous organic compounds from contaminated water. In literature there are several oxidative processes, which have been used successfully for ILs removal [121-133]. The results clearly show that the resistant of ILs to decay in water solutions is connected with ILs composition as the kind and quantity of oxidizing entities in AOP system.

In 2005, they were used for the first time by Stepnowski and Zalewska for degradation of four imidazolium ionic liquids: 1-butyl-, 1-hexyl-, 1-octyl-3-methylimidazolium chlorides, and 1-ethyl-3-ethylimidazolium tetrafluoroborate H_2O_2/UV and TiO_2/UV systems [121]. In this study the H_2O_2/UV system was found to be more effective than a photocatalytic (TiO_2/UV) system. Additionally the stability of studied ILs correlates directly with the length of the N-1 *n*-alkyl side chain [121].

The reaction of ozone with a pyridinium IL containing 1-alkoxymethyl and 1-alkylthiomethyl substituents and imidazolium ILs with lactate anions was investigated by Pernak et al., [122,123]. Both types of salts undergo the decomposition very quickly and were quantitatively removed. Additionally, the investigations showed that ozonation was strongly dependent on the kind and the positions of the substituents on the pyridinium ring. The most favorable were the 3-position on the ring and the substituents including hydroxyl or dimethylamino groups [122,123].

Further results presented by Siedlecka and co-workers reported that [IM14][Cl], was very effectively degraded in other advanced oxidation processes: chemical (Fe^{3+}/H_2O_2) [124] and electrochemical system [125-127,132,133]. However the electrochemical process are the most promising method because mineralization of organic matter is the most advanced.

Electrochemical depletion of the 1-butyl-3-methylimidazolium chloride dissolved in water was studied using different electrodes such as commonly used in industry dimensionally stable anodes (IrPt, IrO_2, PbO_2) and promising new boron-doped diamond (BDD) thin film electrode. Ionic liquid oxidation at the conventional electrodes (IrPt, IrO_2) were ineffective for the removal of IM14Cl. As the IrO_2 anode surface was deactivated by organic intermediates generated in this process. In contrast to this at PbO_2 and BDD oxidation of IL were successful.

The study also demonstrated that the electrolysis at conventional electrode materials led to the different distribution of products then at BDD, due to different mechanism of IL degradation. Electrochemical oxidation at diamond electrode mainly occurred indirectly via •OH radicals and by other chemical entities such as active chlorine species [132]. More, the results obtained in single and double compartment reactor with BDD indicated that the removal of $IM14^+$ was more rapid in the reactor with separate anode and cathode space and the elimination of imidazolium cation was obtained only by oxidation [133].

To better understanding the mechanism of ILs oxidation and the influence of their structure on the stability in AOPs conditions the experiments with different ILs were carried out.

The degradation by Fenton's reagent of the imidazolium ionic liquids containing in structure from four to eight carbon atoms in N-alkyl chains showed that the most stable compound in this system was 1-octyl-3-methylimidazolium chloride, and the least stable - the 1-butyl entity [127]. Replacing the imidazolium head group with pyridinium ([Py4-3Me][Cl]) no significant change in the resistance of ionic liquids to degradation was observed.

In the vigorous Fenton-like system where the different radicals such as •OH, HO_2•, O_2•, HO_2^- are generated simultaneously with •OH radicals, and [Py4-3Me][Cl] was degraded to a lesser extent than [IM14][Cl]. This surprising fact of enhancing resistance of [Py4-3Me][Cl] with increasing H_2O_2 concentration in Fenton-like reagent, was concluded by the authors probably as a change in the degradation mechanism [127].

The influence of anions pairing in of ILs has been also investigated. A comparative study of 1-butyl-3-methylimidazolium salts with different anions as a counter ion in Fenton-like system [128] and electrochemical oxidation process [127] indicated that the stability of ionic liquids depends on the reactivity of anion. In Fenton-like system degradation of IL decreases when the counter ion competes with the cation for hydroxyl radicals. Therefore the resistance to oxidation decreased in the following order: [IM14][Cl] > [IM14][C(CN)3] > [IM14][CF3SO3]. Additionally, the anions were undergoing significantly complex reactions with ferric ions, inhibiting the generation of •OH radicals [130]. In electrochemical oxidation process at boron doped diamond electrode (BDD) the IM14X composition influenced the decay rate of the $[IM14]^+$ cation in the following order: [IM14][Cl] > [IM14][Br] ≈ [IM14][BF4]≈ [IM14][PF6] ≈ [IM14][CF3SO3] > [IM14][CH3C6H4SO3]. The $[Cl]^-$ counter ion enhanced IL degradation, probably by the anodic shift of the water decomposition at surface of anode, whereas $[CH3C6H4SO3]^-$competed with the cation for anodically produced hydroxyl radicals [127]. It was observed that the anions presented in electrolyte or as a background in matrix also had a significant influence on the efficiency of degradation process [125, 128].

However, ultrasonic chemical oxidative degradation of 1-alkyl-3 methylimidazolium ionic liquids showed that altering the nature of the side chains and the type of anions used, surprisingly, does not affect the degradation process [129]. The authors consider that, under ultrasound, the reactions in the ionic liquid mixture $/H_2O_2/CH_3COOH$ took place rapidly between the radicals and specific sites of the molecule as H_2O_2 is in excess while the new radicals were generated continuously with the aid of ultrasonic treatment leading to exhaustive degradation [129].

In AOPs, organic compounds can react in many different ways: they can be oxidized, or attacked by nucleophiles in relative high temperature.

Therefore we can assume that the rate and mechanism of degradation of ionic liquids by different AOPs depend on the experimental conditions. The identification of intermediates in degradation mixtures confirms this thesis.

The identification of ionic liquid by-products in H_2O_2/UV and Fe^{3+}/H_2O_2 [130, 131], showed that these AOP gave a large number of degradation intermediates. In spite of the different methods used for their identification: LC-MS in H_2O_2/UV and GC-MS in Fenton-like system, they indicated a similar by-products of degradation. During electrochemical oxidation and application of ultrasound/H_2O_2/CH_3COOH system, oxidation of ILs was more advanced and only a few by-products were found. Their identification showed that they were also presented in previous mentioned oxidation processes.

Generally the degradation of [IM14][Cl] at the first stage of AOP leads to:

- elimination of the n-butyl side-chain (1-alkylimidazole) from parent compound;
- substitution of hydroxyl group or groups to the alkyl side chain in different positions or substitution of carbonyl group or groups to imidazolium ring in C4 and C5 position. These intermediates are expected to be unstable under further oxidation, and they were decomposed by oxidative cleavage of imidazolium ring
- substitution of O_2• radicals (they are generated in vigorous Fenton-like system and during electrolysis) to C2 position in imidazolium ring, forming imidazolones (Figure 2).

These studies offer preliminary insights into the oxidation processes of imidazolium ionic liquids in water solution [131]. However, the most important factor for the wastewater treatment process seems to be that the intermediates of ionic liquids generated by AOPs are more biodegradable than the parent compounds.

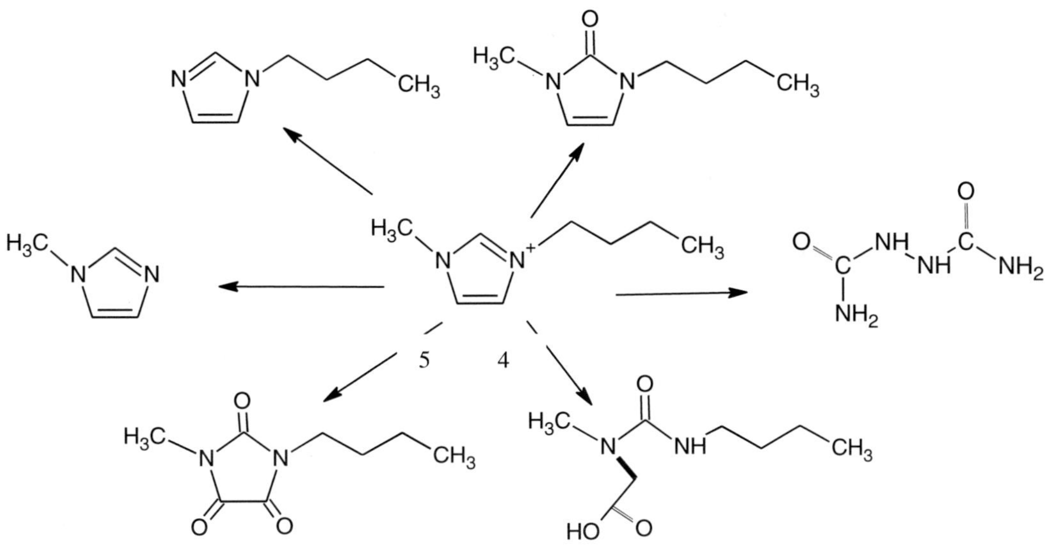

Figure 2. Major abundant by-products of the [IM14]$^+$ degradation by advanced oxidation processes.

In contrast to pyridinium, *N*-substituted imidazolium head group is known to be poorly biodegradable [134]. The introduction of hydroxyl/carbonyl groups associated with a ring-opening reaction or the removal of aromaticity by AOPs along with complete pollutants mineralization are very promising attempts to increase their biodegradability. In general, it is known that the introduction of hydroxyl groups into the side chain of ionic liquid cations results in a significantly decreased acute toxicity [135,136] comparing to the unsubstituted alkyl compounds. However, it has to be kept in mind that the reactive intermediates formed during degradation may still cause toxic effects on the acute or chronic scale, which have to be investigated in more detail.

Biological Degradation

A number of research groups have performed biodegradation tests with alkyl substituted imidazolium cations using activated sewage sludge [137-139]. In most cases ILs were used as a sole source of organic carbon and organic nitrogen. The use of living organisms, especially microorganisms, for ILs treatment seem to be a natural and obvious and friendlier to the environment approach. The investigations showed that the more toxic ILs with different head groups (e.g., imidazolium, pyridinium) with longer side-chains the more susceptible to biodegradation they were. The factor responsible for the increasing tendency towards toxicity and biodegradability is probably the lipophilicity, but this fact creates a conflict of interests in looking for biological treatment ILs [136-139]. Romero et al., [140] discussed the biodegradability of imidazolium ILs in the presence of additional carbon source. It was found that the ILs tested were not biodegradable when D-glucose was available. However, ILs with no additional carbon were also not degraded (2–10%), which is in contrast to other research where, e.g., complete primary biodegradation of 1-methyl-3-octylimidazolium chloride [IM18][Cl] was shown [136, 141].

To more accurately predict biodegradation under real conditions it is beneficial to take adaptation into account especially if biodegradation requires induction of specific metabolic pathways, e.g., aromatic ring break-down [142]. Docherty et al., observed complete biodegradation of 1-hexyl-3-methylimidazolium bromide after extending duration of the test and concluded that though IL could not be classified as readily biodegradable it is not expected to persist in the environment [142].

In the context of biodegradation in activated sludge alkylammonium salts were also tested. The problem is that many quaternary ammonium salts possess biocidic properties. Nishiyama et al., found that alkyltrimethylammonium salts with long alkyl side-chain were ultimately biodegradable in the presence of activated sludge. Quaternery ammonium salt were degraded to tertiary amine by N-dealkylation at the first stage of the biodegradation pathway and then the tertiary amine rapidly disappeared [143]. For choline and methoxy-choline ionic liquids ready biodegradability was observed within 5 or 10 d, respectively [144].

Garcia et al., investigated the influence of imidazolium-based IL anions on their biodegradation. It was shown that perfluorinated IL ions were more susceptible to biodegradation than halides [145] and the octyl sulphate anion was better biodegradable than hexafluorophosphate, bromide, tetrafluoroborate, and bis[(trifluoromethyl)sulphonyl]amide [138]. This was in agreement with the results of Gathergood et al. [139] but contradicted the results obtained for pyridinium ILs in the same research group by Hartjani et al.. They did not find any significant influence of the octyl sulphate anion on IL biodegradability [146]. It could be concluded that the anion on its own influences the biodegradability of ILs in a minimal degree, but if combined with some other cations may alter the susceptibility to biodegradation [147].

The inhibition of activated sludge (from different domestic and industrial Sources) respiration according to OECD guideline 209 by a selection of imidazolium, pyridinium, piperidinium, ammonium ionic liquids were studied. The structure activity relationships of IL toxicity towards activated sludge are in good agreement with those found for other organisms and test systems. The inhibitory potential of tested ILs substituted with short alkyl chains (<4) and polar anions was low, whereas more significant toxic effects were determined for highly hydrophobic ionic cations and anions with low - some < 50 mM (<10 mg $L^{-}1$). Recently it was found that the source of the inoculum has a strong influence on the biodegradation rates of tested ILs, but in terms of toxicity no significant differences in sensitivity/tolerance of the activated sludge could be determined. The study demonstrated that the EC_{50} values from *Vibrio fischeri* can be used for a reliable assessment of the sludge inhibition potential of tested ILs [148].

The results of comprehensive IL biodegradability studies are collated in several valuable papers and reviews [138, 139, 147-149].

Conclusion

Evidently ILs show very attractive properties for their application in separation techniques. Many applications of ILs are very attractive, however it has to be kept in mind that in practice they also should be stable under application conditions. Therefore the knowledge of their resistant to temperature, oxidizing and reducing agents and activity lose during application and recycling is necessary. However the researchers often present new applications without examining the stability of ionic liquids under the applied conditions.

Most attention must also be paid to the purity of ILs used to chemical, photochemical and electrochemical processes, due to their lower stability in presence of small amount of water and mineral acids.

Along with sustainable processes regeneration and recycling after usage in various applications and operations extend chemical substances whole life cycle. At the same time, there should be a progress in wastewater treatment and disposal of used ILs, because most of them are poorly biodegradable and do not inert to water organisms. AOPs appear to be promising in this respect.

CATIONS			
ABBREVIATION	NAME	ABBREVIATION	NAME
$[IM12]^+$	**1-ethyl-3-methylimidazolium**	$[P666\text{-}14]^+$	trihexyltetradecylphosphonium
$[IM13]^+$	**1-propyl-3-methylimidazolium**	$[Pip13]^+$	**1-methyl-1-propylpiperidinium**
$[IM14]^+$	**1-butyl-3-methylimidazolium**	$[Pip14]^+$	**1-methyl-1-butylpiperidinium**
$[IM16]^+$	**1-hexyl-3-methylimidazolium**	$[Pip18]^+$	**1-methyl-1-octylpiperidinium**
$[IM18]^+$	**1-octyl-3-methylimidazolium**	$[Py4\text{-}3Me]^+$	**1-butyl-3-methylpyridinium**
$[IM1\text{-}10]^-$	decylmethylimidazolium	$[CO(NH2)2]^+$	**urea**
$[IM1A]^+$	**1-alkyl-3-methylimidazolium**	$[Pyr14]^+$	**1-methyl-1-butylpyrrolidinium**
ANIONS			
ABBREVIATION	NAME	ABBREVIATION	NAME
$[C(CN)3]^-$	tricyanomethide	$[CF3COO]^-$	trifluoroacetate
$[BF4]^-$	tetrafluoroborate	$[1COO]^-$	acetate
$[PF6]^-$	hexafluorophosphate	$[B(CN)4]^-$	tetracyanoborate
$[B(Ph)4]^-$	tetraphenylborate	$[PhCOO]^-$	benzoate
$[FeCl4]^-$	tetrachloroferrate	$[(CF3SO2)3C]^-$	tris(trifluoromethylsulfonyl) methide
$[(CF3SO2)2N]^-$	bis(trifluoromethylsulfonyl) imide	$[Cl]^-$	chloride
$[N(CN)2]^-$	dicyanamide	$[I]^-$	iodide
$[CF3SO3]^-$	triflate	$[Br]^-$	bromide
$[BF3(Ph)]^-$	trifluorophenylborate	$[PO4]^{3-}$	phosphate
$[(C4F9)SO3]^-$	nonaflate	$[NO3]^-$	nitrate
$[AsF6]^-$	hexafluoroarsenate	$[NH4NO3]^-$	ammonium nitrate
$[(C2F5SO2)2N]^-$	bis(pentafluoroethylsulfo-nyl)imide	$[CH3OSO3]^-$	methylsulphate
$[(C2F5)3PF3]^-$	tris(pentafluoroethyl)trifluoro-phosphate	$[CH3C6H4SO3]^-$	toluenesulfonate
$[C2H5OSO3]^-$	ethylsulphate	$[HSO4]^-$	hydrogen sulfate

ACKNOWLEDGMENTS

Financial support was provided by Polish Ministry of Research and High Education under grant DS/530 8150 D199-14F

REFERENCES

[1] Castner, E.W., Wishart, J.F., Spotlight on ionic liquids. *J. Chem. Phys.* 132 (2010) 120901; doi: 10.1063/1.3373178.

[2] Feng, R., Zhao, D., Guo, Y., Revisiting Characteristics of Ionic Liquids: A Review for Further Application Development. *J. Environ. Prot.* 1 (2010) 95-104.

[3] Han D., Row, K.H., Recent Applications of Ionic Liquids in Separation Technology, *Molecules* 15 (2010) 2405-2426; doi:10.3390/molecules15042405.

[4] Sun, P., Armstrong, D.W., Ionic liquids in analytical chemistry. *Anal. Chim. Acta* 661 (2010) 1–16.

[5] Zaijun, L., Xiulan, S., Junkang, L., Ionic Liquid as Novel Solvent for Extraction and Separation in Analytical Chemistry, Chapter 8 in "Ionic Liquids: Applications and Perspectives", Edited by Kokorin, A., InTech. (2011) 153-180.

[6] Ren, Q., Yang, Q., Yan, Y., Xing, H., Bao, Z., Su, B., Yang, Y., Ionic Liquid-Mediated Liquid-Liquid Extraction, Chapter 16 in „Applications of Ionic Liquids in Science and Technology", Edited by Handy, S., InTech (2011) 343-360.

[7] Heitzman, H., Young, B.A., Rausch, D.J., Rickert, P., Stepinski, D.C., Dietz, M.L., Fluorous ionic liquids as solvents for the liquid–liquid extraction of metal ions by macrocyclic polyethers. *Talanta* 69 (2006) 527–531.

[8] Manzoori, J.L., Amjadi, M., Abulhassani, J., Ionic liquid-based single drop microextraction combined with electrothermal atomic absorption spectrometry for the determination of manganese in water samples. *Talanta* 77 (2009) 1539-1544.

[9] Germani, R., Mancini, M.V., Spreti, N., Di Profio, P., Savelli, G., Cu(II) Extraction in Ionic Liquids and Chlorinated Solvents: Temperature Effect. *Green and Sustainable Chemistry* 1 (2011) 155-164.

[10] Zaijun, L., Peipei, C., Cuiping, Y., Yinjun, F., Zhongyun, W., Ming, L., Haixia, S., A Novel Temperature-Controlled Ionic Liquid Dispersive Liquid Phase Microextraction for Determination of Dicofol and DDT in Environmental Water Samples Prior to Gas Chromatography Mass Spectrometry. *Curr. Anal. Chem.* 5 (2009) 324-329.

[11] Chowdhury, S.A., Vijayaraghavan, R., MacFarlane, D.R., Distillable ionic liquid extraction of tannins from plant materials. *Green Chem.* 12 (2010) 1023-1028.

[12] Kim, Y.H., Choi, Y.K., Park, J., Lee, S., Yang, Y.H., Kim, H.J., Park, T.J., Hwan Kim, Y., Lee, S.H., Ionic liquid-mediated extraction of lipids from algal biomass. *Bioresour Technol.* 109 (2012) 312-315.

[13] Mester, P., Wagner, M., Rossmanith, P., Use of Ionic Liquid-Based Extraction for Recovery of *Salmonella* Typhimurium and *Listeria monocytogenes* from Food Matrices. *J. Food Protect.* 73 (2010) 680-687.

[14] Poole, C.F., Poole, S.K., Ionic liquid stationary phases for gas chromatography, *J. Sep. Sci.* 34 (2011) 888–900.

[15] Payagala, T., Zhang, Y., Wanigasekara, E., Huang, K., Breitbach, Z.S., Sharma, P.S., Sidisky, L.M., Armstrong, D.W., Trigonal tricationic ionic liquids: A generation of gas chromatographic stationary phases. *Anal. Chem.* 81 (2009) 160-173.

[16] Flieger, J., Application of Ionic Liquids in Liquid Chromatography, Chapter 12 in "Ionic Liquids: Applications and Perspectives", Edited by Kokorin, A., InTech. (2011) 243-272.

[17] Van Meter, D.S., Stuart, O.D., Carle, A.B., Stalcup, A.M., Characterization of a Novel Pyridinium Bromide Surface Confined Ionic Liquid Stationary Phase for High Performance Liquid Chromatography under Normal Phase Conditions via Linear Solvation Energy Relationships. *J. Chromatogr. A.* 1191 (2008) 67–71.

[18] Hu, X., Peng, J., Huang, Y., Yin, D., Liu, J., Ionic liquids as mobile phase additives for high-performance liquid chromatography separation of phenoxy acid herbicides and phenols. *J. Sep. Sci.* 32 (2009) 1–7.

[19] Shiflett, M.B., Niehaus, A.M.S., Yokozeki, A., Separation of CO_2 and H_2S Using Room-Temperature Ionic Liquid [bmim][$MeSO_4$]. *J. Chem. Eng. Data* 55 (2010) 4785–4793.

[20] Fukuyama, Y., Liquid Matrices in MALDI-MS, Chapter 17 in „Applications of Ionic Liquids in Science and Technology", Edited by Handy, S., InTech (2011) 361-374.

[21] Hallett, J.P., Welton, T., Room-Temperature Ionic Liquids: Solvents for Synthesis and Catalysis. 2. *Chem. Rev.* 111 (2011) 3508–3576.

[22] Sarda, S.R., Pathan, M.Y., Paike, V.V., Pachmase, P.R., Jadhav, W.N., Pawar, R.P., A facile synthesis of flavones using recyclable ionic liquid under microwave irradiation. *Arcivoc* xvi (2006) 43-48.

[23] Minami, I., Kamimura, H., Mori, S., Thermo-oxidative stability of ionic liquids as lubricating fluids. *J. Synthetic Lubrication* 24 (2007) 135–147.

[24] Pisarova L., Gabler Ch., Dorr N., Pittenauer E., Allmaier G., Thermo-oxidative stability and corrosion properties of ammonium based ionic liquids. *Tribol. Int.* 46 (2012) 73–83.

[25] Wang, B., Wang, X., Lou, W., Hao, J., Gold-ionic liquid nanofluids with preferably tribological properties and thermal conductivity. *Nanoscale Res. Lett.* 6(1) (2011) 1-10.

[26] Kondo, H., Ionic Liquid Lubricant with Ammonium Salts for Magnetic Media, Chapter 20 in "Applications of Ionic Liquids in Science and Technology", Edited by Handy, S., InTech (2011) 421-446.

[27] Kamal, F., Yann, B., Rachid, B., Charles, K., Application of Ionic Liquids to Space Propulsion, Chapter 21 in "Applications of Ionic Liquids in Science and Technology", Edited by Handy, S., InTech (2011) 447-466.

[28] Eunhee Cha, Junyoung Mun, E rang Cho, Taeeun Yim, Young Gyu Kim, Seung M. Oh, Soo A Lim, and Jea Wook Lim, The Corrosion Study of Al Current Collector in Phosphonium Ionic Liquid as Solvent for Lithium Ion Battery. *J. Korean Electrochem. Soc.* 14 (2011) 152-156.

[29] Yun, Y.S., Kim, J.H., Lee, S.-Y., Shim, E.-G., Kim, D.-W., Cycling performance and thermal stability of lithium polymer cells assembled with ionic liquid-containing gel polymer electrolytes. *J. Power Sources* 196 (2011) 6750–6755.

[30] An, Y., Zuo, P., Cheng, X., Liao, L., Yin, G., Preparation and Properties of Ionic-Liquid Mixed Solutions as a Safety Electrolyte for Lithium Ion Batteries. *Int. J. Electrochem. Sci.* 6 (2011) 2398 – 2410.

[31] MacFarlane, D.R., Tachikawa, N., Forsyth, M., Pringle, J.M., Howlett, P.C., Elliott, G.D., Davis, Jr., J.H., Watanabe, M., Simon, P., Angell, C.A., Energy applications of ionic liquids. *Energy Environ. Sci.* 7 (2014) 232-250.

[32] Lee, C.-P., Chen, P.-Y., Ho, K.-C., Ionic Liquid Based Electrolytes for Dye-Sensitized Solar Cells, Chapter 25 in Ionic Liquids: Theory, Properties, New Approaches, Edited by Kokorin, A., InTech (2011) 631-656.

[33] Wei, D., Andrew, P., Ryhänen, T., Electrochemical photovoltaic cells – review of recent development. *J. Chem. Technol. Biotechnol,* 85 (2010) 1547–1552.

[34] Van Valkenburg, M.E., Vaughn, R.L., Wilkes, J.S., Thermochemistry of ionic liquid heat-transfer fluids. *Thermochim. Acta* 425 (2005) 181-188.

[35] Whitehead J.A., Zhang J., McCluskey A., Lawrance G.A., Comparative leaching of a sulfidic gold ore in ionic liquid and aqueous acid with thiourea and halides using Fe(III) or HSO5 – oxidant. *Hydrometallurgy* 98 (2009) 276–280.

[36] Tian, G., Li, J., Hua, Y., Application of ionic liquids in hydrometallurgy of nonferrous metals, Trans. Nonferrous, *Met. Soc. China* 20 (2010) 513−520.

[37] Liu, Y., Chen, J., Li, D., Application and Perspective of Ionic Liquids on Rare Earths Green Separation, in Special Issue: Ionic Liquids: Challenging the Frontiers of Separations Processes Separation Science and Technology 47 (2) (2012) 223-232.

[38] Liu Y., Xiao W., Xia S., Ma P., SO_3H-functionalized acidic ionic liquids as catalysts for the hydrolysis of cellulose. *Carbohyd. Polym.* 92 (2013) 218– 222.

[39] Tao, F., Song, H., Chou, L., Dehydration of fructose into 5- hydroxymethylfurfural in acidic ionic liquids. *RSC Advances* 1 (2011) 672–676.

[40] da Costa Lopes, A.M., João, K.G., Morais, A.R.C., Bogel-Łukasik, E., Bogel-Łukasik, R., Ionic liquids as a tool for lignocellulosic biomass fractionation. *Sustainable Chemical Processes* (2013) 1:3.

[41] Holm, H., Lassi, U., Ionic Liquids in the Pretreatment of Lignocellulosic Biomass, Chapter 24 in "Ionic Liquids: Applications and Perspectives", Edited by Kokorin, A., InTech. (2011) 545-560.

[42] Sun, N., Liu, H., Sathitsuksanoh, N., Stavila, V., Sawant, M., Bonito, A., Tran, K., George, A., Sale, K.L., Singh, S., Simmons, B.A., Holmes, B.M., Production and extraction of sugars from switchgrass hydrolyzed in ionic liquids. *Biotechnology for Biofuels* (2013) 6:39.

[43] S. Salmanpour, T. Tavana, A. Pahlavan, M.A. Khalilzadeh, A.A. Ensafi, H. Karimi-Maleh, H. Beitollahi, E. Kowsari, D. Zareyee, Voltammetric determination of norepinephrine in the presence of acetaminophen using a novel ionic liquid/multiwall carbon nanotubes paste electrode. *Mater. Sci. Eng.* C 32 (2012) 1912–1918.

[44] M. Bijad, H. Karimi-Maleh, M.A. Khalilzadeh, Application of ZnO/CNTs nanocomposite ionic liquid paste electrode as a sensitive voltammetric sensor for determination of ascorbic acid in food samples. *Food Anal. Method* (2013), http://dx.doi.org/10.1007/s12161-013-9585-9.

[45] J. Vahedi, H. Karimi-Maleh, M. Baghayeri, A.L. Sanati, M.A. Khalilzadeh, M. Bahrami,A fast and sensitive nanosensor based onMgO nanoparticle room-temperature ionic liquid carbon paste electrode for determination of methyldopa in pharmaceutical and patient human urine samples. *Ionics* (2013) http://dx.doi.org/10.1007/ s11581-013-0940-z.

[46] Cheng H., Chen H.T., Hsien K.-Ch., Lu Ch.-Y., Chen P.-Y., New copper complexes incorporated with the one-step preparation of ionic liquid carbon paste electrode for highly selectively reducing hydrogen peroxide. *Electrochem. Commun.* 40 (2014) 38–41.

[47] C. Serpi L. Kovatsi S. Girousi, Electroanalytical quantification of total dsDNA extracted from human sample using, an ionic liquid modified, carbon nanotubes paste electrode. *Anal. Chim. Acta, (in press)* http://dx.doi.org/doi:10.1016/j.aca.2013.12.031.

[48] Sanati A.L., Karimi-Maleh H., Badiei A., Biparva P., Ensafi A.A., A voltammetric sensor based on NiO/CNTs ionic liquid carbon paste electrode for determination of morphine in the presence of diclofenac. *Mater. Sci. Eng.* C 35 (2014) 379–385.

[49] Kumar, V., Malhotra, S.V., Ionic Liquids as Pharmaceutical Salts: A Historical Perspective, Chapter 1 In in „Ionic Liquid Applications: Pharmaceuticals, Therapeutics, and Biotechnology", Edited by: Malhotra, S.V., Volume 1038, American Chemical Society (2010) 1-12.

[50] Bica, K., Rodríguez, H., Gurau, G., Cojocaru, O.A., Riisager, A., Fehrmann, R., Rogers, R.D., Pharmaceutically active ionic liquids with solids handling, enhanced thermal stability, and fast release. *Chem. Commun.* 48 (2012) 5422–5424.

[51] Sekhon, B.S., Ionic liquids: Pharmaceutical and Biotechnological Applications. *Asian J. Pharm. Biol. Res.* 1 (2011) 395-411.

[52] Lee, S.H., Miyauchi, M., Dordick, J.S., Linhardt, R.J., Preparation of Biopolymer-Based Materials Using Ionic Liquids for the Biomedical Application, Chapter 10 in „Ionic Liquid Applications: Pharmaceuticals, Therapeutics, and Biotechnology", Edited by: Malhotra, S.V., Volume 1038, American Chemical Society,1038 (2010) 115–134.

[53] Oh, D.X., Shin, S., Lim, c., Hwang, D.S.H., Dopamine-Mediated Sclerotization of Regenerated Chitin in Ionic Liquid. *Materials* 6 (2013) 3826-3839.

[54] Silva, S.S., Santos, T.C., Cerqueira, M.T., Marques, A.P., Reys, L.L., Silva, T.H., Caridade, S.G., Mano, J.F., Reis, R.L.., The use of ionic liquids in the processing of chitosan/silk hydrogels for biomedical applications. Green Chem. 14 (2012) 1463-1470.

[55] Dias, A.M.A., Marceneiro, S., Braga, M.E.M., Coelho, J.F.J., Ferreira, A.G.M., Simőes, P.N., Veiga, H.I.M., Tomé, L.C., Marrucho, I.M., Esperança, J.M.S.S., Matias, A.A., Duarte, C.M.M., Rebelo, L.P.N., de Sousa, H.C., Phosphonium-based ionic liquids as modifiers for biomedical grade poly(vinyl chloride). *Acta Biomater.* 8 (2012) 1366–1379.

[56] Gilmore, B.F., Antimicrobial Ionic Liquids, Chapter 26 in "Ionic Liquids: Applications and Perspectives", Edited by Kokorin, A., InTech. (2011) 587-604.

[57] Petkovic, M., Ferguson, J., Bohn, A., Trindade, J. R., Martins, I., Leitão, C., Carvalho, M. B., Rodrigues, C., Garcia, H., Ferreira, R., Seddon, K. R., Rebelo, L. P. N., Silva Pereira, C., On the Merge of Fungal Activity with Ionic Liquids towards the Development of New Biotechnological Processes, Chapter 16 in „Ionic Liquid Applications: Pharmaceuticals, Therapeutics, and Biotechnology", Edited by: Malhotra, S.V., American Chemical Society 1038 (2010) 197–207.

[58] Kumar, V., Malhotra, S.V., Antitumor Activity of Ionic Liquids on Human Tumor Cell Lines, Chapter 8 in „Ionic Liquid Applications: Pharmaceuticals, Therapeutics, and Biotechnology", Edited by: Malhotra, S.V., American Chemical Society 1038 (2010) 91-102.

[59] Sowmiah, S., Srinivasadesikan, V., Tseng, M.-C., Chu, Y.-H., On the chemical stabilities of ionic liquids. *Molecules* 14 (2009) 3780-37813.

[60] 60) Siedlecka, E.M., Czerwicka, M., Stolte, S., Stepnowski, P., Stability of Ionic Liquids in Application Conditions. *Curr. Org. Chem.* 15 (2011) 1974-1991.

[61] Cassity, C.G., Mirjafari, A., Mobarrez, N., Strickland, K.J., O'Briena, R.A., Davis, Jr., J.H., Ionic liquids of superior thermal stability. *Chem. Commun.* 49 (2013) 7590-7592.

[62] Singh, M.P., Singh, R.K., Chandra, S., Thermal stability of ionic liquid in confined geometry. *J. Phys. D: Appl. Phys.* 43 (2010) 92001-92004.

[63] Salminen, J., Papaiconomou, N., Kumar, R.A., Jong-Min, L., Newman, J., Prausnitz, M., Physicochemical properties and toxicities of hydrophobic piperidinium and pyrrolidinium ionic liquids. *Fluid Phase Equilib.* 261 (2007) 421-426.

[64] Villanueva, M., Coronas,A., García, J., Salgado, J., Thermal Stability of Ionic Liquids for Their Application as New Absorbents. *Ind. Eng. Chem. Res.* 52 (2013) 15718−15727.

[65] Schrekker, H.S., Silva, D.O., Gelesky, M.A., Stracke, M.P., Schrekker, C.M.L., Gonçalves, R.S., Dupont, J., Preparation, cation-anion interactions and physicochemical properties of ether-functionalized ionic liquids. *J. Braz. Chem. Soc.* 19 (2008) 426-433.

[66] Bösmann, A., Schultz, P.S., Wasserscheid, P., Enhancing Task Specific Ionic Liquids' Thermal stability by Structural Modification. *Monatsh. Chem.* 138 (2007) 1159-1161.

[67] Zhuravlev, O.E., Verolainen, N.V., Voronc, L.I., Thermal Stability of 1,3-Disubstituted Imidazolium Tetrachloroferrates, Magnetic Ionic Liquids. *Russ. J. Appl. Chem.* 84 (2011) 1158−1164.

[68] Ngo H.L., LeCompte K., Hargens L., McEwen A.B., Thermal properties of imidazolium ionic liquids. *Thermochim. Acta* 357-358 (2000) 97-102.

[69] Papaiconomou, N., Salminen, J., Jong-Min, L., Prausnitz, M., Physicochemical properties of hydrophobic ionic liquids containing 1-octylpyridinium, 1-octyl-2-methylpyridinium, or 1-octyl-4-methylpyridinium cations. *J. Chem. Eng. Data* 52 (2007) 883-840.

[70] Kroon M.C., Buijs W., Peters C.J., Witkamp G.-J., Quantum chemical aided prediction of the thermal decomposition mechanisms and temperatures of ionic liquids. *Thermochim. Acta* 465 (2007) 40-47.

[71] Wooster, T.J., Johanson, K.M., Fraser, K.J., MacFarlane, D.R., Scott, J.L., Thermal degradation of cyano containing ionic liquids. *Green Chem.* 8 (2006) 691-696.

[72] Ferreira, A.F., Simoes, P.N., Ferreira, A.G.M., Quaternary phosphonium-based ionic liquids: Thermal stability and heat capacity of the liquid chase, *J. Chem. Thermodynamics* 45 (2012) 16–27.

[73] Maton, C., De Vos, N., Stevens, C.V., Ionic liquid thermal stabilities: decomposition mechanisms and analysis tools. *Chem. Soc. Rev.,* 42 (2013) 5963-5977.

[74] Glenn A.G., Jones P.B., Thermal stability of ionic liquid BMI(BF4) in the presence of nucleophiles. *Tetrahedron Lett.* 45 (2004) 6967-6969.

[75] Meine, N., Benedito, F., Rinaldi, R., Thermal stability of ionic liquids assessed by potentiometric titration. *Green Chem.* 12 (2010) 1711–1714.

[76] Lewandowski, A., Świderska-Mocek, A., Ionic liquids as electrolytes for Li-ion batteries—An overview of electrochemical studies. *J. Power Source*, 194 (2009) 601–609.

[77] Chambreau, S.D., Boatz, J.A., Vaghjiani, G.L., Koh, C., Kostko, O., Golan, A., Leone, S.R., Thermal Decomposition Mechanism of 1-Ethyl-3-methylimidazolium Bromide Ionic Liquid. *J. Phys. Chem. A*, 116 (2012) 5867–5876.

[78] Ohtani, H., Ishimura, S., Kumai, M., Thermal decomposition behaviors of imidazolium-type ionic liquids studied by pyrolysis-gas chromatography. *Anal. Sci.* 24 (2008) 1335-1340.

[79] Hao, Y., Peng, J., Hu, S., Li, J., Zhai, M., Thermal decomposition of allyl-imidazolium-based ionic liquid studied by TGA–MS analysis and DFT calculations. *Thermochim. Acta* 501 (2010) 78–83.

[80] Brands, H., Chandrasekhar, N., Unterreiner, A.-N., Ultrafast Dynamics of Room Temperature Ionic Liquids after Ultraviolet Femtosecond Excitation. *J. Phys. Chem. B* 111 (2007) 4830–4836.

[81] 81) Katoh, R., Takahashi, K., Photo-degradation of imidazolium ionic liquids. *Radiat. Phys. Chem.* 78 (2009) 1126–1128.

[82] Allen, D., Baston, G., Bradley, A.E., Gorman, T., Haile, A., Hamblett, I., Hatter, J.E., Healey, M.J.F., Hodgson, B., Lewin, R., Lovell, K.V., Newton, B., Pitner, W.R., Rooney, D.W., Sanders, D., Seddon, K.R., Sims, H.E., Thied, R.C., An investigation of the radiochemical stability of ionic liquids. *Green Chem.* 4 (2002) 152–158.

[83] Shkrob, I.A., Chemerisov, S.D., EPR Study of Radicals in Irradiated Ionic Liquids and Implications for the Radiation Stability of Ionic Liquid-Based Extraction Systems (2007) arXiv:physics/0703027v1 [physics.chem-ph].

[84] Wishart, J.F., Ionic Liquids and Ionizing Radiation: Reactivity of Highly Energetic Species. *J. Phys. Chem. Lett.* 1 (2010) 3225–3231.

[85] Le Rouzo, G., Lamouroux, C., Dauvois, V., Dannoux, A., Legand, S., Durand, D., Moisy, P., Moutiers, G., Anion effect on radiochemical stability of room-temperature ionic liquids under gamma irradiation. *Dalton Trans.* 31 (2009) 6175–6184.

[86] Shkrob, I.A., Marin, T.W., Chemerisov, S.D., Hatcher, J.L., Wishart, J.F., Radiation Induced Redox Reactions and Fragmentation of Constituent

[87] Ions in Ionic Liquids. 1. Anions. *J. Phys. Chem. B* 115 (2011) 3872–3888.

[88] Shkrob, I.A., Marin, T.W., Chemerisov, S.D., Hatcher, J.L., Wishart, J.F., Toward Radiation-Resistant Ionic Liquids. Radiation Stability of Sulfonyl Imide Anions. *J. Phys. Chem. B* 116 (2012) 9043–9055.

[89] Shkrob, I.A., Marin, T.W., Chemerisov, S.D., Hatcher, J.L., Wishart, J.F., Radiation Induced Redox Reactions and Fragmentation of Constituent Ions in Ionic Liquids. 2. Imidazolium Cations. *J. Phys. Chem. B* 115 (2011) 3889–3902.

[90] Shkrob, I.A., Marin, T.W., Luo, H., Dai, S., Radiation Stability of Cations in Ionic Liquids. 1. Alkyl and Benzyl Derivatives of 5-Membered Ring Heterocycles. *J. Phys. Chem. B* 117 (2013) 14372−14384.

[91] Shkrob, I.A., Marin, T.W., Hatcher, J.L., Cook, A.R., Szreder, T., Wishart, J.F., Radiation Stability of Cations in Ionic Liquids. 2. Improved Radiation Resistance through Charge Delocalization in 1-Benzylpyridinium. *J. Phys. Chem. B* 117 (2013) 14385−14399.

[92] Shkrob, I.A., Marin, T.W., Bell, J.R., Luo, H., Dai, S., Radiation Stability of Cations in Ionic Liquids. 3. Guanidinium Cations. *J. Phys. Chem. B* 117 (2013) 14400−14407.

[93] Shkrob, I.A., Marin, Radiation Stability of Cations in Ionic Liquids. 4. Task-Specific Antioxidant Cations for Nuclear Separations and Photolithography. *J. Phys. Chem. B* 117 (2013) 14797−14807.

[94] Huang, L., Huang, W., Fu, H.Y, WU, G.Z, GUO, Z.J., WU, W.S., CHEN, S.M., Influence of FeCl3 on radiation stability of ionic liquid BmimCl. *Chin. Sci. Bull.* 58 (2013) 1150-115.

[95] Oxley, J.D., Prozorov, T., Suslick, K.S., Sonochemistry and Sonoluminescence of Room-Temperature Ionic Liquids. *J. Am. Chem. Soc.* 125 (2003) 11138-11139.

[96] Sun, J., Forsyth, M., MacFarlane, D.R., Room-temperature molten salts based on the quaternary ammonium ion. *J. Phys. Chem. B* 102 (1998) 8858-8864.

[97] MacFarlane, D.R., Meakin, P., Sun, J., Amini, N., Forsyth, M.J. Pyrrolidiniumimides: A new family of molten salts and conductive plastic cristalphases. *J. Phys. Chem. B* 103 (1999) 4164-4170.

[98] Kim, K.S., Choi, S., Demberelnyamba, D., Lee, H., Oh, J., Lee, B.B., Mun,S.J., Ionic liquids based on *N*-alkyl-*N*-methylmorpholinium salts as potentialelectrolytes. *Chem. Commun.* 7 (2004) 828-829.

[99] Appetecchi, G.B., Montanino, M., Zane, D., Carewska, M., Alessandrini, F., Passerini, S., Effect of the alkyl group on the synthesis and the electrochemicalproperties of N-alkyl-N-methyl-pyrrolidiniumbis(trifluoro-methanesulfonyl) imide ionic liquids. *Electrochim. Acta* 54 (2009) 1325-1332.

[100] Wu, T-Y., Su, S-G., Lin, Y-Ch., Wang, H. P., Lin, M-W, Gung, S-T, Sun I.-W., Electrochemical and physicochemical properties of cyclic amine-based Brönsted acidic ionic liquids. *Electrochim. Acta* 56 (2010) 853–862.

[101] Kanematsu, S., Matsumoto, K., Hagiwara, R., Electrochemically stable fluorohydrogenated ionic liquids based on quaternary phosphoniumcation. *Electrochem. Commun.* 11 (2009) 1312-1315.

[102] Tsunashima, K., Sugiya, M., Physical and electrochemical properties of low viscosity phosphonium ionic liquids as potential electrolytes. *Electrochem Commun.* 9 (2007) 2353-2358.

[103] Tsunashima, K. Kawabata, A. Matsumiya, M. Kodama, S. Enomoto, R. Sugiya, M. Kunugi, Y., Low viscous and highly conductive phosphonium ionic liquids based on bis(fluorosulfonyl)amide anion as potential electrolytes. *Electrochem. Commun.* 13 (2011) 178–181.

[104] Shkrob, I.A., Wishart, J.F., Charge trapping in imidazolium ionic liquids. *J. Phys. Chem. B* 113 (2009) 5582-5592.

[105] Kroon, M.C., Buij, W., Peters, C.J., Witkamp, G.J., Decomposition of ionic liquids in electrochemical processing. *Green Chem.* 8 (2006) 241-245.

[106] Sakaebe, H., Matsumoto, H., Tatsumi, K., Discharge-charge properties of Li/$LiCoO_2$ cell using room temperature ionic liquids based on quaternary ammonium cation - Effect of the structure. *J. Power Sources* 146 (2005) 693-697.

[107] Andrieux, C.P. Gonzalez, F. Savéant, J.-M., Homolytic and heterolytic radical cleavage in the Kolbe reaction: Electrochemical oxidation of arylmethyl carboxylate ions. *J. Electroanal. Chem.* 498 (2001) 171-180.

[108] Galiński, M., Lewandowski, A., Stępniak, I., Ionic liquids as electrolytes. *Electrochim. Acta* 51 (2006) 5567-5580.

[109] Motsumoto, H., Electrochemical windows of room-temperature ionic liquids, in book Electrochemical aspects of ionic liquids, edited by H. Ohro, Wiley & son Inc. (2005) 35-54.

[110] Tian, G.C., Zhou, X.J., Li, J., Hua, Y.X., Quantum chemical aided molecular design of ionic liquids as green electrolytes for electrodeposition of active metals. *Trans. Nonferrous Met. Soc. China* 19 (2009) 1639-1644.

[111] Kurzweil, P., Chwistek, M., Electrochemical stability of organic electrolytes in supercapacitors: Spectroscopy and gas analysis of decomposition products. *J. Power Sources* 176 (2008) 555-567.

[112] Fuchigami, T., Electrolytic reactions, in book Electrochemical aspects of ionic liquids, edited by H. Ohro, Wiley & son Inc. (2005) 89-109.

[113] Ismail, A.S., Zein El Abedin, S., Höfft, O., Endres, F., Unexpected decomposition of the bis (trifluoromethylsulfonyl) amide anion during electrochemical copper oxidation in an ionic liquid. *Electrochem. Commun.* 12 (2010) 909–911.

[114] Zhang, Z., Li, C., Wang, Q., Zhao, Z.K., Efficient hydrolysis of chitosan in ionic liquids. *Carbohyd. Polym.* 78 (2009) 685-689.

[115] Su, C-H., Chung, M-H., Hsieh, H-J., Chang, J-K., Ding, J-C., Wu, H-M., Enzymatic hydrolysis of lignocellulosic biomass in ionic liquid media for fermentable sugar production. *J. Taiwan Inst. Chem. E.* 43 (2012) 573-577.

[116] Villagran, C., Deetlefs, M., Pitner, W. R., Hardacre, C., Quantification of halide in ionic liquids using ion chromatography. *Anal. Chem.* 76 (2004) 2118-2123.

[117] Baker, G. A., Baker, S. N., A simple colorimetric assay of ionic liquid hydrolytic stability. *Aust. J. Chem.* 58 (2005) 174-177.

[118] Visser, A.E., Swatlowski, R.P., Reichert, W.M., Griffin, S.T., Rogers, R.D., Traditional extractants in nontraditional solvents: groups 1 and 2 extraction by crown ethers in room-temperature ionic liquids. *Ind. Eng. Chem. Res.* 39 (2000) 3596-3604.

[119] Swatlowski, R.P., Holbrey, J.D., Rogers, R.D., Ionic liquids are not alwaysgreen: hydrolysis of 1-butyl-3-methylimidazolium hexafluorophosphate. *Green Chem.* 5 (2003) 361-363.

[120] Nikitenko, S.I., Berthon, C., Moisy, P., Instability of actinide(IV) hexachloro complexes in room-temperature ionic liquid [BuMeIm]PF_6 due to hydrolysis of the hexafluorophosphate anion. *Comptes Rendus Chimie* 10 (2007) 1122-1127.

[121] Wasserscheid, P., van Hal, R., Bosmann, A., 1-n-butyl-3-methylimidazolium([bmim]) octylsulfate - an even "greener" ionic liquid. *Green Chem.* 4 (2002) 400-404.

[122] Stepnowski, P., Zaleska, A., Comparison of different advanced oxidationprocesses for degradation of room temperature ionic liquids. *Photochem. Photobiol.* 170 (2005) 45-50.

[123] Pernak, J., Goc, I., Mirska, I., Anti-microbial activities of protic ionic liquidswith lactate anion. *Green Chem.* 6 (2004) 323-329.

[124] Pernak, J., Branicka, M., Synthesis and aqueous ozonation of some pyridinium salts with alkoxymethyl and alkylthiomethyl hydrophobic groups. *Ind. Eng. Chem. Res.* 43 (2004) 1966-1974.

[125] Siedlecka, E.M., Mrozik, W., Kaczyński, Z., Stepnowski, P., Degradation of1-butyl-3-methylimidazolium chloride ionic liquid in a Fenton-like system. *J. Hazard. Mater.* 154 (2008) 893-900.

[126] Fabiańska, A., Ossowski, T., Stepnowski, P., Stolte, S., Thöming, J., Siedlecka, E.M., Electrochemical oxidation of imidazolium-based ionic liquids: the influence ofanions. *Chem. Eng. J.* 198–199 (2012) 338–345.

[127] Siedlecka, E. M., Stolte, S., Gołębiowski, M., Nienstedt, A., Stepnowski, P., Thöming J., Advanced oxidation process for the removal of ionic liquids from wastewater: the influence of functionalized side chains on the electrochemical degradability of imidazolium cations, Separation and Purification Technology – in press.

[128] Siedlecka, E.M., Stepnowski, P., The effect of alkyl chain length on thedegradation of alkylimidazolium- and pyridinium-type ionic liquids in a Fenton-like system. *Environ. Sci. Pollut. Res.* 16 (2009) 453-458.

[129] Siedlecka, E.M., Gołębiowski, M., Kaczyński, Z., Czupryniak, J., Ossowski, T., Stepnowski, P., Degradation of ionic liquids by Fenton reaction; the effect of anions as counter and background ions. *Appl. Catal., B* 91 (2009) 573- 579.

[130] Li, X., Zhao, J., Li, Q., Wang, L., Tsang, S.C., Ultrasonic chemical oxidative degradation of 1,3-dialkylimidazolium ionic liquids and their mechanisticelucidation. *Dalton Trans.* (2007) 1875-1880.

[131] Czerwicka, M., Stolte, S., Muller, A., Siedlecka, E.M., Gołębiowski, M., Kumirska, J., Stepnowski, P., Identification of ionic liquid breakdown productsin an advanced oxidation system. *J. Hazard. Mater.* 171 (2009) 478- 483.

[132] Siedlecka, E.M., Gołębiowski, M., Kumirska, J., Stepnowski, P., Identification of 1-butyl-3-methylimidazolium chloride degradation products formed in Fe(III)/H2O2 oxidation system. *Chem. Anal. (Warsaw)* 53 (2008) 943- 951.

[133] Ewa M. Siedlecka, Aleksandra Fabianska, Stefan Stolte, Anne Nienstedt, Tadeusz Ossowski, Piotr Stepnowski, Jorg Thoming, Electrocatalytic Oxidation of 1-Butyl-3-Methylimidazolium Chloride: Effect of the Electrode Material. *Int. J. Electrochem. Sci.* 8 (2013) 5560-5574.

[134] Czupryniak J., Fabianska A., Stepnowski P., Ossowski T., Bogdanowicz R., Gnyba M., Siedlecka E. M., Application of BDD thin film electrode for electrochemical decomposition of heterogeneous aromatic compounds. *Cent. Eur. J. Phys.* DOI: 10.2478/s11534-012-0058-3.

[135] Philipp, B., Hoff, M., Germa, F., Schink, B., Beimborn, D., Mersch-Sundermann, V., Biochemical interpretation of quantitative structure-activity relationships (QSAR) for biodegradation of N-heterocycles: a complementary approach to predict biodegradability. *Environ. Sci. Technol.* 41 (2007) 1390-1398.

[136] Jastorff, B., Mölter, K., Berendt, P., Bottin-Weber, U., Filser, J., Heimers, A., Ondruschka, B., Ranke, J., Schaefer, M., Schroder, H., Stark, A., Stepnowski, P., Stock, F., Störmann, R., Stolte, S., Welz-Biermann, U., Ziegert, S., Thöming, J., Progress in evaluation of risk potential of ionic liquids-basisfor an eco-design of suitable products. *Green Chem.* 7 (2005) 362-372.

[137] Stolte, S., Matzke, M., Arning, J., Böschen, A., Pitner, W.-R., Welz-Biermann, U., Jastorff, B., Ranke, J., Effects of different head groups and functionalised side chains on the aquatic toxicity of ionic liquids. *Green Chem.* 9 (2007) 1170-1179.

[138] Stolte, S., Abdulkarim, S., Arning, J., Blomeyer-Nienstedt, A.-K., Bottin-Weber, U., Matzke, M., Ranke, J., Jastorff, B., Thoming, J., Primary biodegradation of ionic liquid cations, identification of degradation products of 1-methyl-3-octylimidazolium chloride

and electrochemical waste water treatment of poorlybiodegradable compounds. *Green Chem.* 10 (2008) 214–224.

[139] Garcia, M.T., Gathergood, N., Scammells, P.J., Biodegradable ionic liquids. Part II. Effect of the anion and toxicology. *Green Chem.* 7 (2005) 9–14.

[140] Gathergood, N., Scammells, P.J., Garcia, M.T., Biodegradable ionic liquids. Part III. The first readily biodegradable ionic liquids. *Green Chem.* 8 (2006) 156–160.

[141] Romero, A., Santos, A., Tojo, J., Rodrıguez, A., Toxicity and biodegradability of imidazolium ionic liquids. *J. Hazard. Mater.* 151 (2008) 268–273.

[142] Markiewicz, M., Jungnickel, C., Markowska, A., Szczepaniak, U., Hupka, J., Paszkiewicz, M., 1-Methyl-3-octylimidazolium chloride-sorption and primary biodegradation analysis in activated sewage sludge. *Molecules* 14 (2009) 4396–4405.

[143] Docherty, K.M., Dixon, J.K., Kulpa Jr., C.F., Biodegradability of imidazolium and pyridinium ionic liquids by an activated sludge microbial community. *Biodegradation* 18 (2007) 481–493.

[144] Nishiyama, N., Toshima, Y., Ikeda, Y., Biodegradation of alkyltrimethylammonium salts in activated sludge. *Chemosphere* 30 (1995) 593-603.

[145] Stolte, S., Steudte, S., Areitioaurtena, O., Pagano, F., Thöming, J., Stepnowski, P., Igartua, A., Ionic liquids as lubricants or lubrication additives: An ecotoxicity and biodegradability assessment. *Chemosphere* 89 (2012) 1135-1141.

[146] Gathergood, N., Scammells, P.J., Design and Preparation of Room-Temperature Ionic Liquids Containing Biodegradable Side Chains. *Aust. J. Chem.* 55 (2002) 557-560.

[147] Harjani, J.R., Singer, R.D., Garcia, M.T., Scammels, P.J., The design and synthesis of biodegradable pyridinium ionic liquids. *Green Chem.* 10 (2008) 436–438.

[148] Stolte S., Steudte S., Igartua A., Stepnowski P., The biodegradation of ionic liquids - the view from a chemical structure perspective. *Curr. Org. Chem.* 15 (2011) 1946-1973.

[149] Markiewicz M., Piszora M., Caicedo N, Jungnickel Ch, Stolte S., Toxicity of ionic liquid cations and anions towards activated sewage sludge organisms from different sources - Consequences for biodegradation testing and wastewater treatment plant operation. *Water Res.* 47 (2013) 2921-2928.

[150] Coleman, D., Gathergood N., Biodegradation studies of ionic liquids. *Chem. Soc. Rev.* 39 (2010) 600–637.

In: Ionic Liquids: Synthesis, Characterization and Applications ISBN: 978-1-63117-716-3
Editor: Anson Brooks

Chapter 10

Analytical Methods for Determination of ILs Constituents

Monika Paszkiewicz* and Piotr Stepnowski
Institute for Environmental and Human Health Protection,
Faculty of Chemistry, University of Gdansk, Gdansk, Poland

Abstract

Room-temperature ionic liquids (RTILs) salts with melting points below 100 °C have been intensively studied in recent years. Unique physiochemical properties and broad application possibilities of ionic liquids make their industrial use a matter of the nearest future. Therefore, analytical methods are required to determine their purity and identification and quantitative determination of their components in various matrices. The primary aim of this review is to outline current developments in chromatographic and electrophoretic methods applied to the analysis of routinely used ionic liquid cations and anions. For qualitative and quantitative determination of ionic liquid components the highly efficient reversed-phase liquid chromatography, ion chromatography and ion-pair chromatography methods are most recently used. An alkyl-bonded stationary phases, phenyl-bonded stationary phases or on silica packing functionalized with mixed polar and apolar functions stationary phases are applied. The usefulness of hydrophilic interaction chromatography as an interesting alternative to the classical RP-HPLC as well as IPC, especially for short-chain, polar cations separation also is discussed. Capillary electrophoresis, and recently also the isotachophoresis are also being developed.

1. Introduction

Interest in ionic liquids (ILs) is still on the increase due to the properties that this class of organic salts possesses.

---

* Corresponding author: Monika Paszkiewicz. Institute for Environmental and Human Health Protection, Faculty of Chemistry, University of Gdansk, Wita Stwosza 63, PL-80-308 Gdansk, Poland. E-mail: monikapasz@wp.pl.

As compared to classical salts with high melting points, low melting ILs generally consist of bulky, nonsymmetrical organic cations such as imidazolium, pyrrolidinium, pyridinium, ammonium or phosphonium, and numerous different inorganic or organic anions.

These compounds have unique properties, such as a negligible vapour pressure, good thermal stability, a large liquids range and tune able miscibility with water or organic solvents, depending on their ion combination, and offer numerous advantages for a vast area of applications [1-4].

They also possess a wide range of solubilising capabilities for polar and apolar molecules as well as for ions. They have interesting electrochemical properties, such as a significant dielectric constant, a high electrical conductivity and a wide range of electroactivity. In addition to the low vapour pressure, undoubtedly one of the most important advantages of ionic liquids is their ability to "tune" their properties (such as polarity, viscosity, density and solubility in water and the melting point) to the needs of a particular chemical reaction or process technology by suitable selection of the cation and anion.

Their application in analytical chemistry has been demonstrated as stationary phase components for gas chromatography [5, 6], running electrolytes for capillary electrophoresis [7-9] and mobile phase additives for high-performance liquid chromatography (HPLC) [10–12]. This intensely increasing applicability of ionic liquids will lead to their industrial production and utilisation. Consequently, simple and fast analytical techniques for the determination of ILs in various matrices will be of great significance.

The analysis of ionic liquids currently includes two main aspects:

(i) identifying and quantitatively determining the constituents of ionic liquids in a variety of matrices, and
(ii) the purity of these compounds. Determining trace amounts of the liquid ion in industrial, biological and environmental samples has been effected mainly through the analysis of cations, in most cases, to determine the biological activity of these salts on organisms used in tests of acute toxicity. The chemical analysis of ILs cations presently constitutes a variety of separation techniques able to separate the analysed quaternary cation, alkylpyridinium or alkylimidazolium, from other compounds, both charged or neutral.

Due to the fact that the presence of even trace amounts of halide ions, forming a residue after the synthesis of ionic liquids, effectively inactivates catalytic activity in the environment of these solvents as well as their effect on viscosity, density or spectral purity [13, 14], the methods analysis developed to date has focused on determining chlorides, in particular by using ion chromatography, capillary electrophoresis and inductively coupled technique plasma mass spectrometry [15-17].

The most popular technique for ILs analysis seems to be high-performance liquid chromatography (HPLC) with its various separation modes. Separation of ILs cations can be carried out by reversed-phase (RP), ion pair or cation-exchange chromatography [18-28]. However, the use of RP-HPLC for the separation of ILs cations is hampered by two important drawbacks: (1) the resolution of very polar short alkyl chain cations is difficult to achieve because these compounds are weakly retained on conventional RP columns; and (2) significant peak tailing takes place due to the electrostatic interactions between cationic solutes and ionised residual silanol groups.

Hydrophilic interaction chromatography (HILIC) is a chromatographic technique that can be used to improve retention of very polar compounds that are retained poorly by RP-HPLC [29-32]. Table 1 summarizes chromatographic methods used for separation of ionic liquids cations.

2. Chromatographic Methods

2.1. Reversed Phase and Ion-Pair Chromatography

Several papers have described the separation of alkylimidazolium RTIL cations with the reversed chromatography mode [18, 25, 33]. The reversed stationary phase has been widely used for the separation of various imidazolium cations on the alkyl-bonded stationary phase, phenyl bonded stationary phase, or on silica packings functionalised with mixed polar and apolar functions. Because the nature of the ILs is not molecular, the retention mechanism of IL salts in reversed phase mode cannot be as simple as in the case of molecular species.

Dispersive interactions between the alkyl side chains of IL cations and apolar bonded moieties of stationary phases are considered to be the main interaction of the retention mechanisms.

Furthermore, electrostatic interactions between cations and residual charged silanols also play a determinant role in retention [25, 34]. This can also involve stationary phase-anion and cation electrostatic interactions as well as charge–charge interaction and ion-pair formation between the cations and anions of the injected IL itself.

Separations of ILs cations always need a high percentage of water in the mobile phase, however, the high level of water can lead to some problems of solubility for certain ILs, such as for association with the bis(trifluoromethylsulfonyl)imide anion.

In the case of ionic liquids, wherein the alkyl chain attached to the IL cation is less than four carbon atoms, retention and selectivity in reversed mode are weak. Thus, literature data show clear limitations for the separation of short alkyl chain compounds in terms of retention and selectivity without the addition of non-volatile salts.

Better reproducibility and improvements in peak symmetry of less hydrophobic cations were observed when using mobile phases containing additive anionic species, such as ammonium acetate/acetic acid [22] and potassium phosphate/phosphoric acid mixtures [18, 23, 24]. Ion-pair chromatography can be also useful in order to improve retention of the less hydrophobic cations. The formation of ion-pairs plays a determinant role in the retention properties of imidazolium RTIL cations [25, 35].

In IP-RPLC mode, cations separation may involve the formation of neutral ion-pairs involving hydrophobic interactions with the stationary phase or a dynamic ion exchange between adsorbed counter-anions of the mobile phase and eluted species [36-38].

According to the nature of analytes that must be separated, the choice of the ion-pairing reagent appears to be decisive, as various organic phase modifiers have been reported [39-42]. However, the retention mechanism combining hydrophobic and ionic interactions often produces dissymmetric peaks and is not suitable for studying certain complex mixtures.

A routine chromatographic method for analysing ionic liquids using RPLC with electrospray ionisation mass spectrometry was firstly described by Stepnowski [22].

Table 1. Chromatographic methods used for separation and analysis of ionic liquids

Ionic liquids analyzed	Stationary phase	Mobile phase	References
[BMIM][Cl],[BMIM][Br], [BMIM][BF_4],[BMIM][Tf_2N], [BMIM][PF_6], [HMIM][Cl], [HMIM][Tf_2N],[DMIM][Cl], [OMIM][Cl],[OMIM][BF_4], [NMIM][Cl][, [NMIM][PF_6]	Phenosphere SAX (Phenomenex) LunaSCX (Phenomenex)	ACN/5 mM potassium hydrogen phthalate / 15 mM phtalic acid, pH 3.5	[19]
[EEIM][Br],[BMIM][BF_4], [BMIM][pTs],[AMIM][BF_4], [HMIM][BF_4],[OMIM][BF_4], [BzMIM][BF_4], [MBPy][BF_4]	Kromasil C-8 (Eka Chemicals)	ACN/20 mM TFA	[18]
[EMIM][Cl],[HMIM][Cl], [BMIM][Tf_2N],[OMIM][Cl] [BMMIM][Tf_2N],	MetaSil Basic C-8 (Varian)	Acetonitrile/10mM sodium 1-decanesulfonate	[20]
[EMIM][Cl],[HMIM][Cl], [BMIM][Tf_2N],[BMMIM][Tf_2N], [OMIM][Cl]	Uptisphere OH (Interchrom) YMC-Pack OH (Interchrom) Nucleosil OH (Macharey Nagel)	Acetonitrile/10mM ammonium acetate	[20]
[EMIM][Br], [EMIM][BF_4], [EMIM][PF_6], [BMIM]Cl], [BMIM][BF_4], [BMIM][PF_6], [HMIM]Cl], [HMIM][BF_4], [HMIM][PF_6], [OMIM]Cl], [OMIM][BF_4], [OMIM][PF_6]	Kromasil C-18 (Eka Azko Nobel)	Acetonitrile/water	[26]
[EMIM][Cl], [EEIM][Cl], [PMIM][Cl], [HMIM][Cl], [BMIM][Cl], [OMIM][Cl], [NMIM][Cl], [BzMIM][Cl], [EBzMIM][Cl], [MBPy][Cl]	Synergi Polar-RP (Phenomenex)	Methanol or Acetonitrile : water (5 mM KH2PO4/H3PO4)	[24]
[EEIM][Br], [BMIM][BF_4], [BMIM][pTs], [AMIM][BF_4], [HMIM][BF_4], [OMIM][BF_4], [BzMIM][BF_4], [MBPy][BF_4]	MetaSil SCX (Varian)	Acetonitrile/30 mM K_2HPO_4/H_3PO_4, pH=6,6	[18]
[BMIM][Cl], [BMIM][Br], [BMIM][BF_4], [BMIM][PF_6], [BMIM][Tf_2N] [BMIM][Cl], [BMIM][Br], [BMIM][BF_4], [BMIM][pTs], [BMIM][Tf_2N]	SphereClone SAX (Phenomenex) Metrosep A Supp 5 (Metrohm)	Acetonitrile/8 mM phthalic acid, pH=4,0 Acetonitrile/3,2 mM Na_2CO_3/1 mM $NaHCO_3$, pH 11,14	[63]
[EMIM][BF_4], [PMIM][BF_4], [BMIM] [BF_4], [AMIM] [BF_4]	Shim-pack IC-SC1, Shim-pack IC-C1 column	6.0 mM sulfuric acid/5% acetonitrile, pH 1.89	[64]

Ionic liquids analyzed	Stationary phase	Mobile phase	References
[BMIM][BF_4], [HMIM][BF_4] [OMIM][BF_4], [BMIM][PF_6] [HMIM][PF_6], [BMIM][Cl], [EMIM][PF_6] [MMIM][MS]	Zorbax SB-C18, Zorbax SB-Phenyl, Zorbax SB-CN, Zorbax NH_2	ACN/30mM phosphate buffer/50mM $NaPF_6$	[57]
[C_2OHmim][Cl], [C_3Omim][Cl], [C_2OBzmim][Cl], [C_2OAllymim][Cl], [C_4Omim][Cl], [C_5O_2mim][Cl], [C_6O_2OHmim][Cl], [C_7O_3mim][Cl]	Mixed-Mode RP-C18/Cation	ACN/10mM NaH_2PO_4/H_3PO_4	[60]

**[EEIM] 1-ethyl-, [PEIM] 1-propyl- [HEIM] 1-hexyl-3-ethylimidazolium.

[PMIM] 1-propyl-, [BMIM] 1-butyl-, [AMIM] 1-amyl-, [HMIM] 1-hexyl-, [HpMIM] 1-heptyl-, [OMIM] 1-octyl-, [NMIM] 1-nonyl-3-methylimidazolium, [MMIM] 1-Methyl-3-methyl imidazolium.

[BzMIM] 1-benzyl-, [EBzMIM] 1-(2-phenylethyl)-, [pMBzMIM] 1-p-methylbenzyl-3-methylimidazolium.

[CnOmmim]Cl of imidazolium cation-based ionic liquids (ILs), with an ether functional group on the alkyl side-chain.

[MBPy] *N*-butylmethylpyridinium, [Cl] chloride, [BF_4] tetrafluroborate, [pTs] tosylate, [Br]bromide.

[MS] methylsulfate.

On a conventional reversed phase column and a mixture of methanol and water as the mobile phase, the separation of polar components such as $[EEIM]^+$, $[PMIM]^+$ and $[PEIM]^+$ was very poor. The reason for this was mainly the continued presence of free silanols on the surface of the stationary phase, visualised as a marked tailing in the peaks of all these compounds.

As an alternative, an RP-8 column with a higher carbon load was applied. This stationary phase resulted in the satisfactory separation of short-chain ionic liquids. A mobile phase consisting of a mixture of acetonitrile and water (1% acetic acid/20 mM ammonium acetate) was used.

The separation was performed using a gradient elution from 10% to 50% of the organic modifier concentration in 25 min. The application of such conditions has enabled the separation of all analysed compounds in 23 min. (Figure 1).

A detailed study of the behaviour of alkylimidazolium cations in the reversed phase system was done by [18]. A mobile phase containing a phosphate buffer and methanol or acetonitrile as the organic modifier were applied.

The authors showed that the retention factors of the RTIL cations depended on the concentration of the organic modifier in accordance with the classical Snyder-Soczewinski model.

With an increase in the content of the organic modifier in the mobile phase, the retention time of all the cations decreased; however, differences in the retention profiles depending on the type of *N-1* functional group on the imidazolium ring were observed.

It was also shown that varying the pH values and changing the salt concentration in the mobile phase had no influence on the retention factors of the ionic liquid cations. It was also shown that varying the pH values and changing the salt concentration in the mobile phase had no influence on the retention factors of the ionic liquid cations.

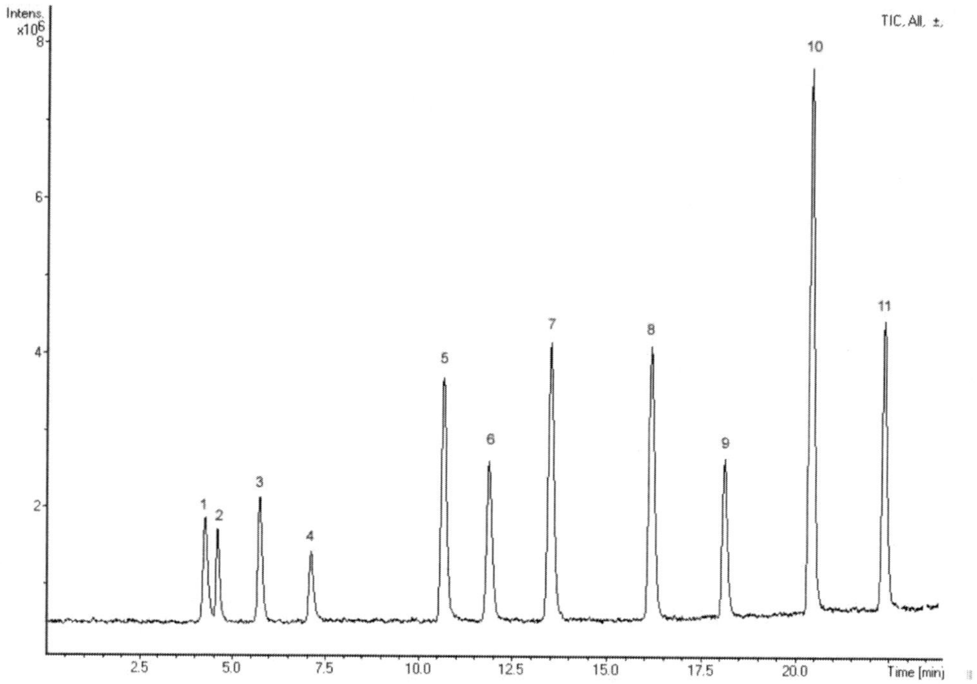

Figure 1. Gradient separation of the test mixture. Column RP C8 MetaSil Basic 250 x 4.6 mm. Mobile phase A, acetonitrile; B, water (1 % acetic acid/20 mM ammonium acetate) 0 min. 10 % B, 10 min. 20 % B, 25 min. 50 % B,. Order of elution: 1 – 1-ethyl-3-ethylimidazolium bromide; 2 - 1-n-propyl-3-methylimidazolium chloride; 3 - 1-n-propyl-3-ethylimidazolium chloride; 4 - 1-butyl-3-methylimidazolium chloride; 5 - 1-benzyl-3-methylimidazolium tetrafluoroborate; 6 - 1-amyl-3-methylimidazolium chloride; 7 - 1-(-2-phenylethyl)-3-methylimidazolium chloride; 8 - 1-p-methylbenzyl-3-methylimidazolium chloride; 9 - 1-hexyl-3-methylimidazolium chloride; 10 - 1-butyl-3-ethylimidazolium chloride; 11 - 1-heptyl-3-methylimidazolium chloride.

Changing the organic modifier from methanol to acetonitrile yielded sharper peaks and a shorter total analysis time. The authors also showed the usefulness of ion-pair chromatography in IL separation. A linear relationship between the retention factor and the content of TFA in the mobile phase was observed. With the increase in the TFA concentration (10-30 mM), the retention factors of the salts increased. By modifying the acetonitrile-water mobile phase with TFA, separation of highly hydrophilic compounds was possible.

In such a modified mobile phase (20 mM TFA), very good separation and improved peak symmetry of similar IL cations, i.e., 1-ethyl-3-ethylimidazolium and 1-butyl-3-methylimidazolium, were obtained.

The reversed phase mode with a Kromasil C18 column was used to analyse eleven 1-alkyl-3-methyl imidazolium IL salts [25]. Water-rich acetonitrile solutions (water content ≥70%, v/v) either with or without the addition of salts were applied as the mobile phases.

When salt-free mobile phases are used, imidazolium-based ILs can be differentiated by the anion and/or cation. When a buffer such as acetate or phosphate salt, or any salt, such as sodium chloride or sodium tetrafluoroborate, is added to the mobile phase, only ILs with a different alkyl chain on the imidazolium ring can be separated.

The IL retention factor depends on the concentration and nature of the added salt. Increasing the retention factors of the imidazolium cation was observed when using the mobile phase consisting of hexafluorophosphate chaotropic anions. With salt-free mobile phases, all imidazolium ILs offered mainly concave adsorption isotherms with the Kromasil C18 stationary phase producing severely fronting peaks. The addition of salts to the mobile phase completely changed the adsorption isotherms. The retention mechanism combines hydrophobic and ionic interactions producing poor peak shapes. They changed to a convex shape producing tailing peaks.

In another study, two mixtures of four 1-alkyl-3-methylimidazolium IL salts associated with the tetrafluoroborate or hexafluorophosphate anions were analysed by reversed-phase liquid chromatography with three different stationary phases: Kromasil C8, Zorbax Extend C18 and Zorbax Sb-Aq [26]. A mixture of acetonitrile–water (50/50 (v/v or 30/70 (v/v) with 0.01M of the salts NaCl, NaH_2PO_4, $NaBF_4$, $NaClO_4$ and $NaPF_6$ was used as the mobile phases (Figure 2).

The authors showed that, depending on the type of salt used as the mobile phase component, the retention of ILs increased with the increasing affinity of the inorganic anions for the apolar stationary phases (chaotropicity). The differences in the concentration of added salts causing a change in mobile phase ionic strength were also studied. Similar results were obtained for the two columns: the Kromasil C8 and the Zorbax Sb-Aq. With an increase in the ionic strength of the mobile phase, significant peak shape improvements with minimum changes in peak positions were noted.

Clearly different results were obtained using the Zorbax Extend C18 column with the increase of the mobile phase ionic strength significant increase retention time and few changes in the peak shape were observed. The authors also showed variation in the alkyl-MIM retention factor with the change of the added salt type. This effect was observed with the mobile phases containing $NaBF_4$, and especially $NaPF_6$, and reflects the affinity of the inorganic anions on the stationary phases.

IP RPHPLC was also applied for the quantitative analysis of 20 acid ionic liquids (AAILs) containing imidazolium cations and amino acid (AA) anions [43]. As a stationary phase, an Ultimate ODS column and the mobile phase containing a mixture of distilled water (A) and acetonitrile (B) were used. In order to increase the retention of low retained AAILs ions a several ion-pairing reagent were tested. It was found that HFBA gave the best performance from the tested compounds.

When 0.5 mM of HFBA was added, the imidazolium cation and the amino acid ion could be separated within 15 min for all the AAILs by gradient elution. The gradient elution programme was from 0 to 100% acetonitrile (B) content in 12 min and then 100% B for 3 min. Based on the zwitterionity of amino acid, this method was proposed to determine both the cation and the anion of AAILs simultaneously.

Satisfactory separation of 1,3-dimethylimidazolium, 1-ethyl-3-methylimidazolium, 3-methyl-1-propylimidazolium, 1-butyl-3-methylimidazolium, 1-hexyl-3-methylimidazolium and 3-methyl-1-octylimidazolium cations was achieved with using of IPC chromatography [44]. The aforementioned ionic liquid cations were separated on an Acquity UPLC BEH C18 column. To analytes separation the mobile phase containing acetonitrile and water, and sodium 1-octanesulfonate as an ion-pair reagent were applied. Sodium 1-octanesulfonate was tested at various concentrations from 0.1 to 5.0 mM in a buffered (5 mM ammonium formate, pH 3.2) ACN/water (20:80 v/v) mobile phase under isocratic conditions.

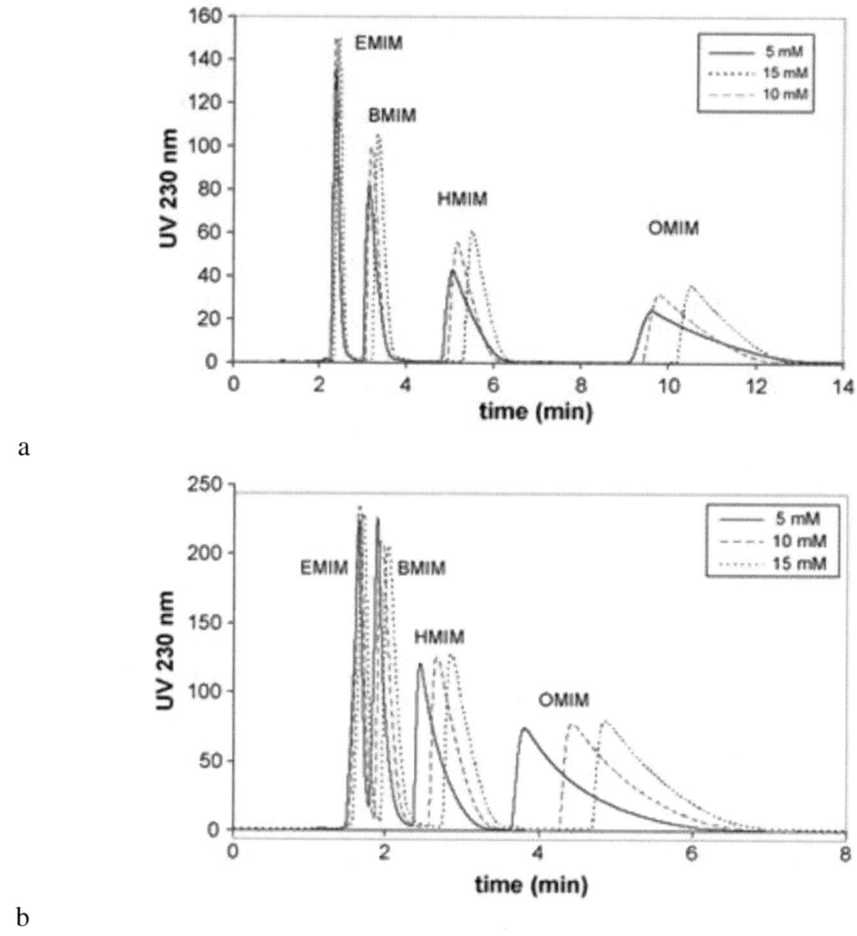

Figure 2. Superimposed chromatograms of hexafluorophosphate alkyl-3-methylimidazolium ionic liquids (each 60 nmol injected) eluted with 50/50 (v/v) acetonitrile/water mobile phases containing 5mM added NaPF6 (solid line); 10mM added NaPF6 (dashed line); 15mM added NaPF6 (dotted line); (a) Kromasil C8, 4mmI.D.; 0.7 ml/min, (b) Zorbax Extend C18, 3mmI.D.; 0.4 ml/min.

Typical IPC chromatographic behaviour of all the imidazolium cations was observed: the retention of IL cations increased with an increasing sodium 1-octanesulfonate concentration in the mobile phase. Well resolution of all the analytes was achieved by gradient elution with the acetonitrile/water mobile phase containing 1 mM sodium 1-octanesulfonate.

The effects of ion-pair reagents, acetonitrile concentration and column temperature on the retention of imidazolium ionic liquids was investigated by Gao et al., [45, 46]. A silica-based monolithic column with 1-heptanesulfonic acid sodium and acetonitrile, and the addition of citric acid as an eluent for the separations of four imidazolium ionic liquid cations by ion-pair chromatography was applied. Sodium 1-heptanesulfonate as an ion-pair reagent and acetonitrile were also applied to determine five imidazolium ionic liquid cations. Separations were performed on a ZORBAX Eclipse XDB-C18 column. The authors showed the same dependence as was described previously: an increasing retention factor was observed with an increase in the concentration of sodium 1-heptanesulfonate.

The retention time of the imidazolium cation was shortened and the peak shape was improved when the concentration of acetonitrile was increased.

Reversed-phase ion-pair chromatography with ultraviolet-visible detection was used to determination four pyridinium ionic liquid cations (*N*-ethyl-pyridinium, *N*-butyl-pyridinium, *N*-butyl-4-methyl-pyridinium, *N*-hexyl-pyridinium).

The effects of ion-pair reagent, acetonitrile concentration and column temperature on the retention and separation of the cations were elucidated [54].

Usefulness HPLC reverse phase was tested by Lamouroux et al., [47]. The separation of five imidazolium cations such as $[EMIM]^+$, $[BMIM]^+$, $[HexMIM]^+$, $[OMIM]^+$ with a different stationary phases was demonstrated. According to a classical reversed phase mechanism the retention increases with the hydrophobic character of the ILs. The authors showed the evolution of the retention factor with the percentage of acetonitrile in the mobile phase for the $[BMIM]^+$.

It was found that the retention factor increases with the increase of the ACN percentage, which is in contradiction with the classical behaviour observed in reversed phase chromate-graphy.

It suggests that there is another mechanism that governs the retention. These observations can be explain by the electrostatic (ionic) interactions between ILs and residual silanols.

2.2. Polar Reversed Phase Liquid Chromatography

Among the commercially available polar chemically-bonded stationary phases, very popular are stationary phases with ligands possessing an aromatic system in their structure. Such stationary phases are used for separations of polar compounds containing the π-electron. Examples of such stationary phases are modified by a phenyl group. The retention properties of these phases are determined by the presence of π ... π interactions [48]. Furthermore, the high polarisability of the phenyl groups makes it possible to obtain an induced dipole and, consequently, the occurrence of induced dipole-ion and dipole/dipole-induced type interactions [49]. In the case of the phenyl phase, the choice of organic modifier is important. It was found that acetonitrile, and tetrahydrofuran in contrast to methanol, have the ability to suppress the π ... π interactions between the analyte and the stationary phase ligand [50, 51]. This is a result of the presence of the π-electrons in the structure of acetonitrile and of the formation of complexes of electron donor-acceptor ligands with the analyte or stationary phase [52]. In the presence of high concentrations of acetonitrile in the mobile phase, the retention mechanism of the separated compounds with the π-electron system is analogous as for the classic reverse phase [53]. Therefore, methanol is preferred in the case of these phases [50]. Recently, fluorinated, silica-based stationary phases have become a more interesting alternative to the traditional alkyl phases. In addition to the dispersive interactions available on common C18 and C8 phases, dipole interactions of fluorosubstituted ligands, π–π interactions between solutes and phenyl-ligands as well as charge transfer and ion-exchange interactions also have a considerable impact on the selectivity of the pentafluorophenylpropyl (PFPP) columns [54].

Studies on the usefulness of the polar chemically bonded stationary phase modified with propyloxyphenyl ligands in a selective separation group of homologues of *N*-alkyl (aryl) imidazolium ionic liquids and the salts of alkyl were described by Stepnowski et al., [24].

The two main objectives of the research indicated the separation of short-chain, highly hydrophilic ILs which were difficult to separate from the reversed-phase and to determine the nature of possible interactions between the selected ionic liquids and the separating system under consideration. The chemical structure of ligands used in the stationary phase indicated that the mechanism of retention of compounds would be determined by dispersion interactions, π...π interactions and interactions arising from the possibility of hydrogen bond formation. The effect of two commonly used organic modifiers was studied – that of acetonitrile and methanol on the retention profiles of the group congeners of imidazolium ionic liquids with different alkyl chain lengths: 1-ethyl-3-methylimidazolium, 1-propyl-3-methylimidazolium, 1-butyl-3-methylimidazolium, 1-amyl-3-methylimidazolium, 1-hexyl-3-methylimidazolium, 1-octyl-3-methylimidazolium, and 1-nonyl-3-methylimidazolium chlorides.

To study the π…π interaction potential, 1-benzyl- and 1-ethylbenzyl-3-methylimidazolium entities, i.e., molecules possessing an additional aromatic system available for interactions, were included in this set.

Moreover, one *N*-butylmethylpyridinium salt, an ionic liquid cation that has already proved to be useful in many applications, was tested for comparison.

As a result of these studies, optimisation of the chromatographic conditions that allowed the separation of all the analysed cations of ILs was carried out. The application of gradient elution using a mobile phase of methanol (2-20%) and aqueous phosphate buffer solution at a concentration of 5 mM at pH 3.0 enabled the separation of all the analysed cations.

In addition, satisfactory separation was achieved for compounds that were difficult to separate using a conventional C18 and C8 stationary phase. These were chlorides, 1-ethyl-3-etyloimidazolium and 1-propyl-3-methylimidazolium.

To compare the retention mechanisms of imidazolium ionic liquid cations both A Synergi Polar-RP column packed with polar endcapped particles and a Gemini C-6 polar RP column were used as the stationary phase. The critical point in the selective separation of ionic liquids was the choice of organic modifier in the mobile phase.

To differentiate between hydrophobic and aromatic interactions, methanol and acetonitrile (concentration range 1–52% v/v) mixed with 5 mM phosphate buffer (KH_2PO_4/H_3PO_4) and adjusted to pH 3 were compared.

Separation of the ionic liquid mixture was conducted in both isocratic and gradient modes. The results of the effect of kind (acetonitrile, methanol) and concentration (5-80%) of organic solvent in the mobile phase on the retention mechanism of alkyl (aryl) imidazolium ionic liquids and alkylpyridinium indicated no linear relationship between the nature of the organic modifier content and the value of logk. The authors also pointed to a reversal elution order of two salts with a similar lipophilicity – chlorides, 1-amyl-3-methylimidazolium and 1-benzyl-3-methyloimidazolium with increasing content of acetonitrile in the mobile phase.

These studies proved that the π...π interaction determines the retention mechanism of low molecular weight, hydrophilic ionic liquids. In contrast, the significantly longer alkyl chains in 1-octyl-3-methylimidazolium and 1-nonyl-3-methylimidazolium cations led to increased hydrophobicity; in these cases the predominant mode of retention was due to hydrophobic interaction.

The π...π interaction here would be relatively weaker, and also limited because of the sterically unfavourable access of the π-system of the solute to the π-system in the stationary phase.

Contrary to expectations, the retention of ionic liquids containing an additional π-electron system, such as 1-benzyl-3-methylimidazolium and 1-ethylbenzyl-3-methylimidazolium chlorides, did not differ much from those of the alkylimidazolium entities characterised by similar lipophilicities (1-amyl-3-methylimidazolium or 1-hexyl-3-methylimidazolium cations). This suggests that only one of the aromatic moieties takes part in this particular interaction, whereas the second is involved to a much weaker extent.

The retention mechanisms of a congeneric group of imidazolium ionic liquid cations and one alkylpyridinium cation on a pentafluorophenylpropyl silica-based stationary phase were investigated [55]. Also, the effect of altering the mobile phase consisting of methanol and acetonitrile on the retention profiles of the analysed compounds was elucidated. The results were described as the relation between the concentration of organic solvent in the mobile phase and the logarithm of the k value, according to the earlier classical Soczewiński-Snyder model. It was found that for all ionic liquid cations, a combination of the reversed- and normal-phase retention mode gave rise to a "U-shaped" relationship between the retention profile and the organic modifier content. The range of the reversed-phase retention mode increased with the hydrophobicity of the compounds being analysed. This is the result of the combination of reversed- and normal-phase retention modes as well as the suppression of available π…π interactions at higher concentrations of acetonitrile. The retention of short-chain hydrophilic ionic liquid cations took place largely in the normal-phase mode; in contrast, that of the more hydrophobic cations occurred mostly in the reversed-phase mode.

Although the method of IL separation and analysis using the pentafluorophenyl phase was not very selective in the congeneric group of imidazolium IL cations, the possibility of analysing highly hydrophilic compounds at high organic modifier contents was extremely valuable.

In other studies, optimisation of the separation of alkylimidazolium and alkylpyridinium ionic liquids on a polar reverse stationary phase was performed using a variable-size simplex algorithm [56]. The authors suggested that the percentage of organic modifier (MeOH), phosphate buffer concentration and pH were expected to have a considerable effect on the chromatographic separation of the ionic liquids. Two criteria were chosen in order to evaluate the optimisation process. The first criterion was a chromatographic response function (CRF) which included the number of peaks, resolution between adjacent peaks, a specified analysis time and the individual retention times relative to a minimum retention time. The coefficient of asymmetry was calculated as the second criterion. The mobile phase at the point corresponding to the optimum consisted of 10% MeOH and 90% 15 mM KH_2PO_4/H_3PO_4 at pH 3.43. Under the optimised conditions, all critical pairs of ionic liquids were successfully separated in a single chromatographic run. By using an optimal composition of the mobile phase, good resolution (R = 0.9 – > 1.5), good peak shapes ($f_{AS} < 1$ for 8 out of 10 analytes) and an analysis time of 26.52 min were obtained.

Flieger and Czajkowska-Zelazko [57] described the optimisation of retention of several imidazolium-based ionic liquids by changing the kind of organic modifier (methanol, acetonitrile), concentration, ion-pair reagent in the mobile phase and column (Zorbax SB-Phenyl, Zorbax SB-CN, Zorbax SB-NH_2 and Supelcosil LC-F). Sodium hexafluorophosphate, perchlorate and tetrafluoroborate were tested as the ion-pair reagents.

The chaotropic salt effect on the chromatographic behaviour of ILs in the ACN/water mobile phase buffered with acidic phosphate buffer (pH: 2.9–3) was studied on Zorbax SB-C18 and Zorbax SB-Phenyl stationary phases.

According to the Hofmeister classification, five mobile phases ranging from 5 to 50 mM of sodium salts differing in anion chaotropicity were prepared. The authors suggested several important conclusions concerning the application of the chaotropic effect in the HPLC of ILs. Similar behaviour of ILs on Zorbax SB-Phenyl and Zorbax SBCN columns in chromatographic systems modified with chaotropic salt additives was observed. The presence of additional π–π interactions and increase in the retention coefficient after the addition of chaotropic salt to the mobile phase allowed analysis of ILs with a less hydrophobic cation on the Zorbax SB-Phenyl column. An increase in the retention coefficient was higher when the mobile phase contained a lower concentration of the organic solvent. It was also found that for both the improvement of peak shape and efficiency of the system, a concentration of chaotropic salt additives was crucial. To reach a plateau of k versus salt concentration dependences, about 30 mM of salt was required. As was clearly shown, the most favourable pH range to keep a satisfactory level of efficiency was between 2.9 and 5.5.

2.3. Hydrophilic Interaction Chromatography

An interesting alternative to the classical RP-HPLC as well as IPC, especially for short-chain, polar cations, is hydrophilic interaction chromatography. Historically, HILIC was considered to be a “normal-phase separation” in a “reversed-phase fashion”, in which the separation occurs on polar stationary phases with aqueous-organic mobile phases, where water is the strongly eluting solvent [58]. Under HILIC separation conditions, a water enriched layer will be established within the stationary phase and the separation is achieved by partitioning of the analytes from the eluent into this hydrophilic environment. Therefore hydrogen bonding are dominating and dipole–dipole interactions and the retention is typically an exothermic process. Due to the possibility of electrostatic interactions with the analytes stationary phases have a very significant, dimension of selectivity. However, the mechanism is not clear; several studies assume amixed mechanism of partitioning and adsorption. Different stationary phases have been reported for HILIC separations of a wide range of charged or polar solutes: silica or bounded silica with various aminopropyl, amide, poly (succinimide), diol silica, cyanopropyl or sulfoalkyl betaine functionalities which have been recently reviewed [35]

The first study utilizing HILIC for the separation of imidazolium IL cations on diol stationary phase was carried out by Le Rouzo et al., [20]. The retention mechanisms of imidazolium ionic liquids using HILIC diol stationary phases were determined by studying the influence of organic modifier content and salt concentration in the mobile phase. The effect of acetonitrile content in the mobile phase was studied over the range from 30 to 95%. For acetonitrile compositions varying from 60 to 95%, typical HILIC behaviour was observed, as reported in the literature [58, 59]. With increasing organic modifier content in the mobile phase, both an increase in the retention of the analysed compounds and improvement of the resolution of chromatographic signals were shown. This dependence was due to the presence of dominant hydrophilic interactions between the analyte and the stationary phase in the presence of high concentrations of the organic solvent.

As explained previously, additive salts also played a determinant role in HILIC retention. Capacity factors decreased when the ammonium acetate concentration was raised from 5 to 20 mM.

At low salt concentrations the retention of compounds was strong. With increasing salt concentrations the excess of acetate counter anions promoted the formation of neutral ion-pairs, with the charged solutes being more soluble in the mobile phase, which resulted in shorter retention times. Optimal conditions for separation of all the investigated ILs were obtained using the following mobile phase composition: 90% acetonitrile/10% 10 mM ammonium acetate. The HILIC method was validated and applied to the quantification of imidazolium RTIL cations in irradiated samples.

The studies conducted by Lamouroux et al., [47] have attempted to answer the following question: "What were the most important interactions for the separation of ionic liquids?" In this paper the mechanisms that govern the retention of ILs in HILIC have been described.

In order to do this, two types of phases have been tested: (i) phases grafted with neutral groups, such as cyanopropyl, aminopropyl, diol and carbamate, and (ii) phases grafted with ionic groups. HILIC chromatography was then tested on classical phases (amide, amino, cyano, carbamate and l) and on mixed-mode phases. The usefulness of traditional diol columns for the separation of five imidazolium salts was also described. The imidazolium cations were eluted as a function of the decrease in their hydrophobic character, which is consistent with the HILIC mechanism.

The observed low retention could be explained by the fact that the amides functions and cyanopropyl are essentially donors-acceptors of the electron and are not engaged in the hydrogen interactions. With the increase of the salt concentration, a non-linear decrease of the retention factor was shown.

This behaviour is classically observed for charged compounds with HILIC and we suggest that the formation of neutral ion pairing increases with the salt concentration that contributes to a weaker retention. The authors concluded that only the neutral phases grafted with diol are convenient for the separation of RTILs under HILIC mode. The presence of both partitioning and ion exchange with the silanol surface group interactions were observed. Only diol-grafted phases allow both a good retention and selectivity, which were explained by the greater capability of diol to create a thick aqueous layer.

Furthermore, the stationary phase, based on a mixed interaction associating ion exchange and hydrophobic properties, was studied. The results showed that the essential interactions needed to assure a good retention of ILs were the ionic interactions, and that the hydrophobic interactions played a role in the selectivity of the separation.

The potential of the ethylene bridged hybrid silica phase for the analysis of twelve imidazolium cations was investigated by Orentienė et al. While studying the chromatographic behaviour of twelve analytes, a number of parameters such as buffer concentration, pH, acid additive, organic modifier and column temperature were taken into account [60]. The authors pointed out that the ethylene bridged hybrid silica column offered a dual retention mechanism enabling a choice between two retention modes (hydrophilic interaction chromatography (HILIC) and per aqueous liquid chromatography (PALC) modes) with opposite separation selectivity just by changing the composition of the mobile phase.

In the HILIC mode both the partitioning and ion-exchange interactions as mechanisms responsible for retention were mentioned. It was possible, under HILIC conditions, to almost complete separation of the twelve substances in less than 4 min by using an isocratic elution.

In contrast, under PALC conditions selectivity was not sufficient to distinguish between analytes with very similar hydrophobicity.

2.4. Chromatography on Mixed-Mode Stationary Phases

Different types of columns with specific structural properties for the separation of mixtures of ionic liquid cations are applied [33]. The retention mechanism of ionic liquid cations on stationary phases with cholesterol ligands chemically bonded to silica (SG-CHOL) and mixed stationary phases (SG-MIX) containing cyanopropyl, aminopropyl, phenyl and octadecyl ligands is one of the widely studied problems. RP-18e Innovation ChromolithTM Performance and Macrosphere 300 C4 packings were also used. The authors concluded that stationary phases are unsuitable for the separation of ionic liquid cation mixtures under the given analysis conditions (KH_2PO_4; pH 4; methanol as the organic modifier). The best results were obtained for butyl and silica monolith with bonded octadecyl groups. This was probably due to the lowest heterogeneity of these ligands in comparison to the other stationary phases. By using a chemically bonded butyl stationary phase, determination of the analysed cations in 22 min was possible, except for 1-amyl-3-methylimidazolium and *N*-butyl-4-methylpyridinium, whose separation was not satisfactory. The complete resolution of the six ionic liquid cations was done by applying monolithic silica modified with an octadecyl chain column. A long analysis time (50 min) was observed when using an isocratic elution, but this problem was solved by applying a gradient elution.

A comprehensive study of interactions between imidazolium ILs and stationary phases was first described by Studzińska et al. [27]. The retention mechanism of a homologous group of 1,3-alkylimidazolium ILs on eight home-made stationary phases (Octadecyl (SG-C18), Octyl (XDB-C8), Octyl endcapping (SG-C8END), aminopropyl (SG-NH_2), alkylamide (SG-AP), cholesterolic (SG-CHOL), alkylamide/cholesterolic (SG-CHOL/AP) and mixed (SG-MIX)) was widely studied. The chromatographic behaviour and strength of interactions of the IL cations on the stationary phase surface was investigated by designation of the retention factors. The authors pointed out that the main interactions responsible for the retention of analytes appear to be electrostatic, π–π and hydrophobic. The results showed comparable values of retention factors for the SG-AP and SG-C8END phases, especially in the case of the most hydrophobic hexyl-3-methylimidazolium tetrafluoroborate. Lower *k*-values for 1-ethyl-3-methylimidazolium, 1-propyl-3-methylimidazolium, 1-butyl-3-methylimidazolium and 1-amyl-3-methylimidazolium cations on the octyl stationary phase were noted. Higher *k* values were obtained for SGC18 due to a considerable part of hydrophobic interactions between the analytes and stationary phase surface. The occurrence of a $\pi \ldots \pi$ interaction was characteristic of the SG-MIX, SGCHOL/AP, SG-CHOL and SG-Ph stationary phases.

A high value of the retention factor on the SG-Ph stationary phase indicates a greater share of $\pi \ldots \pi$ interactions of the studied imidazolium cations than the hydrophobic interactions.

With the use of SG-AP, the contribution of alkyl groups in the retention mechanism was analysed. It was found that the retention factors were several times higher than in the case of SG-CHOL, SG-CHOL/AP and SG-MIX due to the hydrophobic interactions between the IL cations and stationary phase surface.

Participation of the cholesterol molecules in the retention mechanism of IL cations was also performed. The influence of the mobile phase buffer pH on the interactions between imidazolium cations and stationary phases was also widely discussed. With the exception of $SGNH_2$, which is the most polar stationary phase, the following tendency was shown: with an increase in the mobile phase pH, the retention of imidazolium cations also increased.

It was found how essential electrostatic interactions are in the retention of IL cations, connected either with protonisation or dissociation of the surface groups due to a change of the pH of the mobile phase.

The possibility of analyzing imidazolium cation-based ionic liquids (ILs), with an ether functional group on the alkyl side-chain by high performance liquid chromatography (HPLC) was investigated [61]. The mixed-mode reversed/cation exchange stationary phase with the aqueous-acetonitrile mobile phase was used. This stationary phase offers multimode retention mechanisms, including RP, cation exchange, and hydrophilic interaction. These interactions can be controlled by changing pH of the mobile phase. In acidic medium carboxylic groups are not ionized and mainly RP retention mechanism is observed. In basic medium carboxylic groups are ionized, and both cation exchange and hydrophobic interaction contribute to retention. In highly organic mobile phase, the mixed mode column can serve as a hydrophilic interaction liquid chromatography (HILIC) column. The authors found that if the contents of acetonitrile in the eluent was higher than 30% all studied ILs eluted in order of increasing lipophilicity factor. Under isocratic conditions satisfactory separation was not achieved. By using acetonitrile as the organic modifier on mixed-mode stationary phase, partial separation of all analyzed IL was achieved by gradient elution.

2.5. Ion Chromatography

Studies on the behaviour of ionic liquid cations to the strong cation exchanger were conducted by Stepnowski et al. [18]. The influence of mobile phase composition, pH and buffer concentration on the selectivity of the separation of a group of 7 homologues of 1-alkyl-3-methylimidazolium and 4-methyl-*N*-butylpyridinium ionic liquids was tested. Ion chromatography separation was performed on MetaSil ion exchange columns. Using a SCX stationary phase, several interactions depending on the mobile phase composition were possible. Cation exchange, nonspecific hydrophobic interactions and adsorption chromatography behaviour were observed. An increase in the concentration of the organic modifier resulted in a significant decrease in retention times and reversal of the elution order.

The influence of the buffer concentration on the retention factor of all the analysed compounds was also studied. As expected, an increase in the buffer concentration reduced the retention time of all the ionic liquids. Additionally, it was found that above 30 mM the buffer concentrations in the mobile phase selectivity of the separation of analysed compounds did not change.

Validation was done using optimised chromatographic conditions with acetonitrile and 30 mM K_2HPO_4 (40:60, v/v) as the mobile phase.

The usefulness of a cation exchange system with conductometric detection for alkylimidazolium cation analysis was first described by Stepnowski and Markowska [62].

The usefulness of a phosphate buffer (KH_2PO_4/H_3PO_4), adjusted to different pHs and modified with acetonitrile in different proportions, was tested. The optimal mobile phase for alkylimidazolium cation separation consisted of 18 mM potassium hydrogen phthalate and 2 mM phthalic acid (pH = 4.46) mixed with 60% (v/v) of acetonitrile.

Studies on the simultaneous separation of cations (1-butyl-3-methylimidazolium, 1-hexyl-3-methylimidazolium, 1-octyl-3-methylimidazolium, 1-nonyl-3-methylimidazolium, 1-decyl-3-methylimidazolium) of ionic liquids in one chromatographic run were also done.

For this purpose, a tandem chromatographic column containing silica-based strong anion exchange and strong cation exchange packings was applied [63]. The use of the following mobile phase composition: 5 mM potassium hydrogen phthalate and 15 mM phthalic acid, and 45% (v/v) acetonitrile at pH 3.05 allowed the separation of all analytes in one chromatographic analysis. The method can be used successfully for a routine simultaneous analysis of the cations and anions of ionic liquids.

The chromatographic behaviour of imidazolium and pyridinium cations on a cation exchange column employing crown ether and carboxylate as well as phosphonate cation exchange sites was investigated by Molíková et al. [64]. The effect of the concentration and pH of methanesulfonic acid (MSA) as well as volume percentage of acetonitrile on the retention of analytes was defined. The authors pointed to the typical order of elution as in the reversed-phase if the mobile phase containing 2.5, 5, 7.5, and 9 mM MSA and up to 60% of acetonitrile was used. When using a higher percentage of acetonitrile (in excess of 65%, v/v) and lower concentrations of MSA (1 mM), normal-phase retention behaviour was observed.

Unfortunately, cations such as 1-ethyl-3-methylimidazolium, 1-propyl-3-methylimidazolium, and 1-butyl-3-ethylimidazolium, N-butyl-4-methylpyridinium and 1-ethylbenzyl-3-methylimidazolium, 1-hexyl-3-methylimidazolium cations were not separated properly.

For the separation of imidazolium ionic liquid cations, a simple, selective and accurate chromatographic method using a carboxyl acid cation exchange column and direct conductivity detection was developed by Gao and Yu [65].

Some validation parameters were determined. The described method was applied to determine two synthesised ionic liquids.

Conductometric detection was also used to analyse a broad structural variety of IL cations [66]. The selection of cationic entities covered the chemical space of IL cations as comprehensively as possible and contained the most common head groups substituted with different alkyl- and functionalised side chains. The influence of the mobile phase composition (concentration of nitric acid and organic modifier) on the separation of imidazolium-based ILs with alkyl substituents, imidazolium-based ILs with functionalized side chains and different cyclic (aromatic and aliphatic) and non-cyclic ILs cations on short Metrosep C4 column was discussed.

3. Electrophoretic Methods

3.1. Capillary Electrophoresis (CE)

Currently, capillary electrophoresis (CE) is becoming an alternative to chromatographic techniques and is applicable to the analysis of various groups of compounds, including the separation of ionic liquids. The first attempts to separate groups of 1,3-dialkylimidazolium salts and isomers using capillary electrophoresis was performed by Qin et al. [67]. By selecting the composition of the buffer, it was possible to control the selectivity of separations of the analysed compounds. The presence of an acidic buffer solution (5 mM triethylamine) allowed to obtain a good symmetry of the signals, while the addition of the buffer 2 mM α-cyclodextrin allowed the separation of two 1-butyl-3-methylimidazolium structural isomers.

All compounds, both of ionic and neutral molecules, could be separated within 8 min, thereby achieving low limits of detection of analytes in the range from 0.4 to 1.4 ppm. This was done in a running buffer consisting of 5.0 mM triethylamine and 2.0 mM α-cyclodextrin adjusted to pH 4.5 with acetic acid. CE was equipped with a UV detector. Bare fused-silica capillaries with an external polyimide coating (50 μm i.d.; 360 μm o.d.) were used.

CE was used to separate seven compounds: imidazole, 1-methylimidazole, 1,2-dimethylimidazole, 1-ethyl-3-methylimidazolium, 2-ethylimidazole and two 1-butyl-3-methylimidazolium structural isomers.

In other cases, CE was applied to separate the cationic constituents of imidazolium ionic liquids [68]. Separation of the cationic compounds of these ten imidazolium ILs was achieved with a citric buffer as the background electrolyte and a bare fused-silica capillary.

The electropherogram of the standard mixture of ionic liquid imidazolium cations obtained under selected optimum conditions is shown in Figure 3.

3.2. Isotachophoresis (ITP)

Capillary isotachophoresis with conductivity detection was used for the analysis of the homologues of imidazolium ionic liquids [69]. Optimisation of separation conditions for 5 ionic liquid cations and two anions was described. Analyses were performed using a polymeric fluoroethylenpropyl column FEP. The leading electrolyte mixture was 10 mM ammonium acetate/0.1% hydroxyethyl cellulose, pH 5.20, and terminating electrolyte was 5 mM tetrabutylammonium perchlorate.

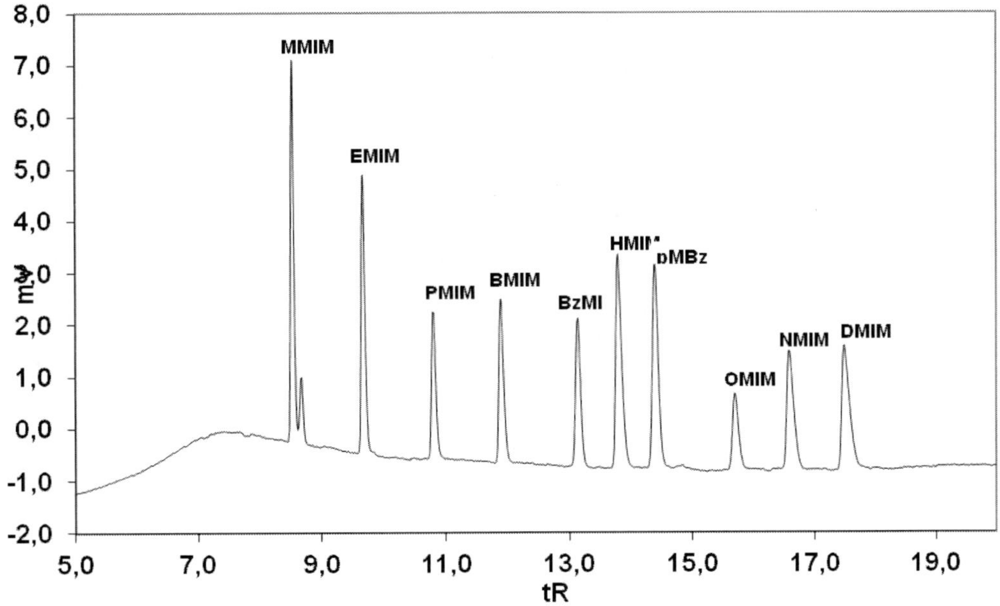

Figure 3. Electropherogram of the standard mixture of ionic liquids imidazolium cations obtained under selected optimum conditions (X, unidentified peak). Running BGE, 200 mM citric buffer, pH 4.0; temperature, 20°C; separation voltage, 12 kV; injection, 50 s.

The optimal conditions allowed the separation of all the analysed ions: 1-propyl-3-methylimidazolium, 1-butyl-3-methylimidazolium, 1-amyl-3-methylimidazolium, 1-hexyl-3-methylimidazolium and 1-octyl-3-methylimidazolium in 20 min.

Isotachophoretic methods for separating and analysing IL cations and anions were applied to analyse the selected ILs in natural matrices, such as in freshwater or soil [70]. An ITA Chrom EA 101 isotachophoretic analyser was used. Both columns (the first capillary column and the analytical column) were made of a fluorinated ethylene-propylene polymer. The detection current was 250 μA in the first column and 50 μA in the analytical column. The instrument was equipped with two conductivity detectors. A mixture of seven cations (1-ethyl-3-methylimidazolium, 1-propyl-3-methylimidazolium, 1-butyl-1-methylpyrrolidinium, 4-(dimethylamine)-1-hexylpyridinium, triisobutyl(methyl)phosphonium, tetraethyl-ammonium, tetrabutylphosphonium) was used for the separation experiment.

Seven short-chain imidazolium, pyridinium and non-chromophoric ammonium and phosphonium IL cations were separated with 10 mM potassium hydroxide + 10 mM acetic acid as LE, and 5 mM β-alanine + 5 mM acetate as TE. An isotachopherogram of the mixture of seven cations is shown in Figure 4.

Conclusion

Ionic liquids (ILs) are a new class of non-molecular solvents with specific properties. They possess a wide range of solubilising capability for polar and apolar molecules as well as for ions.

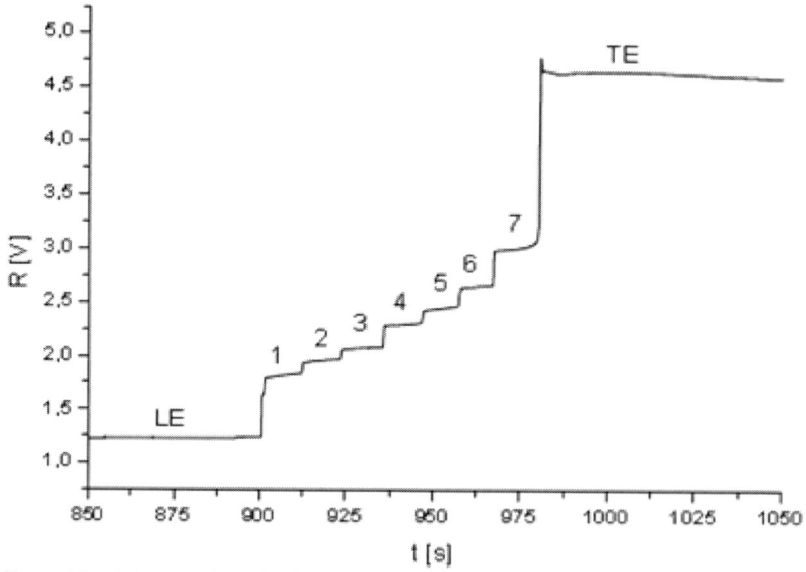

Running buffers: 10 mM potassium hydroxide + 10 mM acetic acid as LE, and 5 mM β-alanine + 5 mM acetic acetate as TE.

Figure 4. Isotachopherogram of a mixture of 7 cations. Order of migration: 1 – 1-ethyl-3-methylimidazolium bromide; 2 - 1-propyl-3-methylimidazolium chloride, 3 - tetraethylammonium, 4 - 1-butyl-1-methylpyrrolidinium 5 - 4-(dimethylamine)-1-hexylpyridinium, 6 - triisobutyl(methyl)phosphonium, 7 - tetrabutylphosphonium).

They also have good thermal and chemical stabilities associated with an extremely low volatility and flammability. The growing interest in ionic liquids (ILs) has resulted in an exponentially increasing production of analytical applications. Reversed-phase liquid chromategraphy seems to be the most common technique involving C18, C8 or short-chain bonded silica stationary phases. Nevertheless, due to the non-molecular nature of ILs, the RP-HPLC retention mechanism of ILs is not as simple. Mixed mechanisms involving ion-pairing, ion-exchange and hydrophobic interaction combine to produce anion retention, and maybe, in a different way, cation retention. The retention time of a basic component can be effectively adjusted by pH, although stronger anionic and hydrophobic counter-ions may require a higher buffer concentration. The spectrophotometer, conductometer and mass spectrometer can be used as detection systems.

REFERENCES

[1] N. V. Plechkova, K. R. Seddon, In: A. Perosa, F. Zecchini (Eds.), *Methods and Reagents for Green Chemistry*, Wiley, New York, 2007, p. 105.

[2] P. Wasserscheid, T. Welton (Eds.), *Ionic Liquids in Synthesis*, Wiley–VCH, Weinheim, 2003.

[3] R. D. Rogers, K. R. Seddon, S. Volkov (Eds.), Green Industrial Applications of Ionic Liquids, NATO Science Series. II. *Mathematics Physics and Chemistry*, vol. 92, Kluwer, Dordrecht, 2002.

[4] J. G. Huddleston, A. E. Visser, W. M. Reichert, H. D. Willauer, G. A. Broker, R. D. Rogers, Characterization and comparison of hydrophilic and hydrophobic room temperature ionic liquids incorporating the imidazolium cation, *Green Chem.* 3 (2001) 156-164.

[5] S. Pandey, Analytical applications of room-temperature ionic liquids: A review of recent efforts, *Anal. Chim. Acta* 556 (2006) 38 - 45.

[6] Armstrong, D. W., He, L., Liu, Y. S. Examination of Ionic Liquids and Their Interaction with Molecules, When Used as Stationary Phases in Gas Chromatography. *Anal. Chem.,* 1999, 71(17), 3873–3876.

[7] Jiang, T. F., Gu, Y. L., Liang, B., Li, J. B., Shi, Y. P., Ou, Q. Y. Dynamically coating the capillary with 1-alkyl-3-methylimidazolium-based ionic liquids for separation of basic proteins by capillary electrophoresis. *Anal. Chim. Acta*, 2003 479(2), 249–254.

[8] Qin, W., Li, S. F. Y. Electrophoresis of DNA in ionic liquid coated capillary. *Analyst,* 2003, 128(1), 37–41.

[9] M. Vaher, M. Koel, M. Kaljurand, Application of 1-alkyl-3-methylimidazolium-based ionic liquids in non-aqueous capillary electrophoresis, *J. Chromatogr.* A 979 (2002) 27.

[10] R. Kaliszan, M. P. Marszal, M. J. Markuszewski, T. Baczek, J. Pernak, Suppression of deleterious effects of free silanols in liquid chromatography by imidazolium tetrafluoroborate ionic liquids, *J. Chromatogr.* A 1030 (2004) 263-271.

[11] L. He, W. Zhang, L. Zhao, X. Liu, S. Jiang, Effect of 1-alkyl-3-methylimidazolium-based ionic liquids as the eluent on the separation of ephedrines by liquid chromatography, *J. Chromatogr.* A 1007 (2003) 39-45.

[12] X. Xiaohua, Z. Liang, L. Xia, J. Shengxiang, Ionic liquids as additives in high performance liquid chromatography: Analysis of amines and the interaction mechanism of ionic liquids, *J. Anal. Chim.* 519 (2004) 207-211.

[13] I. Billard, G. Moutiers, A. Labet, A. El Azzi, C. Gaillard, C. Mariet, K. Lutzenkirchen, Stability of divalent europium in an ionic liquid: spectroscopic investigations in 1-methyl-3-butylimidazolium hexafluorophosphate, *Inorg. Chem.* 42, 2003, 1726-1733.

[14] S. Park, R. J. Kazlauskas, Improved preparation and use of room-temperature ionic liquids in lipase-catalyzed enantio- and regioselective acylations, *J. Org. Chem.* 66, 2001, 8395-8401.

[15] C. Villagran, M. Deetlefs, W. R. Pitner, C. Hardacre, Quantification of halide in ionic liquids using ion chromatography, *Anal. Chem.* 76, 2004, 2118-2123.

[16] D. Berthier, A. Varenne, P. Gareil, M. Digne, C. P. Lienemann, L. Magna, H. Olivier-Bourbigou, Capillary electrophoresis monitoring of halide impurities in ionic liquids, *Analyst*, 129, 2004, 1257-1261.

[17] K. McCamley, N. A. Warner, M. M. Lamoureux, P. J. Scammells, R. D. Singer, Quantification of chloride ion impurities in ionic liquids using ICP-MS analysis, *Green Chem.* 7, 2004, 341-344.

[18] Stepnowski, P., Mrozik, W. Analysis of selected ionic liquid cations by ion exchange chromatography and reversed-phase high performance liquid chromatography. *J. Sep. Sci.*, 2005, 28(2), 149-154.

[19] Markowska, A., Stepnowski, P. Simultaneous determination of ionic liquid cations and anions using ion chromatography with tandem ion exchange columns: a preliminary assessment. *Anal. Sci.*, 2008, 24(10), 1359-1361.

[20] Le Rouzo, G., Lamouroux, Ch., Bresson, C., Guichard, A., Moisy, P., Moutiers, G. Hydrophilic interaction liquid chromatography for separation and quantification of selected room-temperature ionic liquids. *J. Chromatogr.* A, 2007, 1164(1-2), 139-144.

[21] Gao, W., Yu, H., Zhou, S. Determination of Imidazolium Ionic Liquid Cations by Ion-Pair Chromatography Using a Monolithic Column and Direct Conductivity Detection. *Chromatographia,* 2010, 71(5-6), 475.

[22] Stepnowski, P., Müller, A., Behrend, P., Ranke, J., Hoffmann, J., Jastorff, B. Reversed-phase liquid chromatographic method for the determination of selected room-temperature ionic liquid cations. *J. Chromatogr. A,* 2003, 993(1-2), 173-178.

[23] Buszewski, B., Kowalska, S., Stepnowski, P. Influence of stationary phase properties on the separation of ionic liquid cations by RP-HPLC. *J. Sep. Sci.,* 2006, 29(8), 1116-1125.

[24] Stepnowski, P., Nichthauser, J., Mrozik, W., Buszewski, B. Usefulness of π...π aromatic interactions in the selective separation and analysis of imidazolium and pyridinium ionic liquid cations. *Anal. Bioanal. Chem.,* 2006, 385(8), 1483-1491.

[25] Ruiz-Angel, M. J., Berthod, A. Reversed phase liquid chromatography of alkyl-imidazolium ionic liquids, *J. Chromatogr.* A, 2006, 1113(1-2), 101-108.

[26] Ruiz-Angel, M. J., Berthod, A. Reversed-phase liquid chromatography analysis of alkyl-imidazolium ionic liquids: II. Effects of different added salts and stationary phase influence. *J. Chromatogr.* A, 2008, 1189(1-2), 476-482.

[27] S. Studzińska, B. Buszewski: Study of retention mechanism of imidazolium-based ionic liquids in HPLC, *J. Sep. Sci.* 33 (2010), 1264-1273.

[28] Molíková, M., Markuszewski, M. J., Kaliszan, R., Jandera, P.: Chromatographic behaviour of ionic liquid cations in view of quantitative structure-retention relationship. *J. Chromatogr.* A. 2010, 1217(8), 1305-1312.
[29] Hemström, P., Irgum, K., "Hydrophilic interaction chromatography"; *Journal of Separation Science* 29 (12), 2006, 1784-1821.
[30] Alpert, A. J. Hydrophilic-interaction chromatography for the separation of peptides, nucleic acids and other polar compounds. *J. Chromatogr.*, 1990, 499, 177-196.
[32] Y. Guo, S. Gaiki, J. Retention Behavior of Small Polar Compounds on Polar Stationary Phases in Hydrophilic Interaction, *Chromatography Chromatogr.* A 2005, 1074 71-80.
[33] Stepnowski, P. Application of Chromatographic and Electrophoretic Methods for the Analysis of Imidazolium and Pyridinium Cations as Used in Ionic Liquids. *Int. J. Mol. Sci.*, 2006, 7(11), 479-509.
[34] Kowalska, S., Buszewski, B. Effect of stationary phase polarity on the retention of ionic liquid cations in reversed phase liquid chromatography. *J. Sep. Sci.*, 2006, 29(8), 2625-2634.
[35] A. Berthod, M. J. Ruiz-Angel, S. Huguet, Non molecular Solvents in Separation Methods: Dual Nature of Room Temperature Ionic Liquids, *Anal. Chem.* 2005, 77, 4071-4080.
[36] F. Gritti, G. Guiochon, Effect of the ionic strength of salts on retention and overloading behavior of ionizable compounds in reversed-phase liquid chromatography: I. XTerra-C_{18}, *J. Chromatogr.* A 2004, 1033, 43-55.
[37] F. Gritti, G. Guiochon, Effect of the ionic strength of salts on retention and overloading behavior of ionizable compounds in reversed-phase liquid chromatography: II. Symmetry-C_{18}, *J. Chromatogr.* A 2004,1033, 57-69.
[38] F. Gritti, G. Guiochon, Role of the buffer in retention and adsorption mechanism of ionic species in reversed-phase liquid chromatography: I. Analytical and overloaded band profiles on Kromasil-C_{18}, *J. Chromatogr.* A 2004,1038, 53-66.
[39] J. Dai, P. W. Carr, Role of ion pairing in anionic additive effects on the separation of cationic drugs in reversed-phase liquid chromatography *J. Chromatogr.* A 2005, 1072, 169-184.
[40] J. Dai, S. D. Mendonsa, M. T. Bowser, C. A. Lucy, P. W. Carr, Effect of anionic additive type on ion pair formation constants of basic pharmaceuticals, *J. Chromatogr.* A 2005, 1069, 225-234.
[41] Y. Kazakevich, R. LoBrutto, R. Vivilecchia, Reversed-phase high-performance liquid chromatography behavior of chaotropic counteranions, *J. Chromatogr.* A 2005, 1064, 9-18.
[42] L. Pan, R. LoBrutto, Y. V. Kazakevich, R. Thompson, Influence of inorganic mobile phase additives on the retention, efficiency and peak symmetry of protonated basic compounds in reversed-phase liquid chromatography, *J. Chromatogr.* A 1049 (2004) 63-73.
[43] Tang, F., Wu, K., Nie, Z., Ding, L., Liu, Q., Yuan, J., Guo, M., Yao, S. Quantification of amino acid ionic liquids using liquid chromatography–mass spectrometry. *J. Chromatogr.* A, 2008, 1208(1-2), 175–181.
[44] Orentiené, A., Olšauskaité, V., Vičkačkaité, V., Padarauskas, A. UPLC a Powerful Tool for the Separation of Imidazolium Ionic Liquid Cations. *Chromatographia,* 2011, 73(1-2), 17–24.

[45] Gao, W., Yu, H., Ma, Y. Determination of imidazolium ionic liquid cations by ion-pair chromatography using gradient elution. *Chin. J. Chromatogr.*, 2010, 28(6), 556-60.
[46] Gao, W., Yu, H., Zhou, S. Determination of Imidazolium Ionic Liquid Cations by Ion-Pair Chromatography Using a Monolithic Column and Direct Conductivity Detection. *Chromatographia,* 2010, 71(5-6), 475.
[47] Lamouroux, C., Foglia, G., Le Rouzo, G. How to separate ionic liquids: use of hydrophilic interaction liquid chromatography and mixed mode phases. *J. Chromatogr.* A, 2011, 1218(20), 3022-3028.
[48] R. Brindle, K. Albert, Stationary phases with chemically bonded fluorene ligands: A new approach for environmental analysis of π-electron containing solutes, *J. Chromatogr. A* 757, 1997, 3-20.
[49] S. Kayillo, G. R. Dennis, R. A. Shalliker, An assessment of the retention behaviour of polycyclic aromatic hydrocarbons on reversed phase stationary phases: Selectivity and retention on C18 and phenyl-type surfaces, *J. Chromatogr. A* 1126, 2006, 283-297.
[50] D. H. Marchand, K. Croes, J. W. Dolan, L. R. Snyder, Column selectivity in reversed-phase liquid chromatography: VII. Cyanopropyl columns, *J. Chromatogr. A* 1062, 2005, 57-64.
[51] U. D. Neue, K. VanTran, P. C. Iraneta, B. A. Alden, Characterization of HPLC packings, *J. Sep. Sci.* 26, 2003, 174 -186.
[52] J. Horak, N. M. Maier, W. Lindner, Investigations on the chromatographic behavior of hybrid reversed-phase materials containing electron donor–acceptor systems: II. Contribution of π–π aromatic interactions, *J. Chromatogr. A* 1045, 2004, 43-58.
[53] M. Yang, S. Fazio, D. Munch, P. Drumm, Impact of methanol and acetonitrile on separations based on π–π interactions with a reversed-phase phenyl column, *J. Chromatogr. A* 1097, 2005, 124-129.
[54] M. R. Euerby, A. P. McKeown, P. Peterson, Chromatographic classification and comparison of commercially available perfluorinated stationary phases for reversed-phase liquid chromatography using Principal Component Analysis, *J. Sep. Sci.* 26, 2003, 295-306.
[55] Nichthauser, J., Stepnowski, P. Retention Mechanism of Selected Ionic Liquids On a Pentafluorophenylpropyl Polar Phase: Investigation using RP-HPLC. *J. Chromatogr. Sci.*, 2009, 47(3), 247-253.
[56] Nichthauser, J., Paszkiewicz, M., Składanowski, A. C., Stepnowski, P. Simplex-optimized chromatographic resolution of selected ionic liquid cations utilizing a polar reversed-phase system. *Anal. Sci.*, 2008, 24(10), 1355-1358.
[57] Flieger, J., Czajkowska-Zelazko, A. Identification of ionic liquid components by RP-HPLC with diode array detector using chaotropic effect and perturbation technique, *J. Sep. Sci.*, 2012, 35(2), 248–255.
[58] Wang, X. D., Li, W. Y., Rasmussen, H. T. Orthogonal method development using hydrophilic interaction chromatography and reversed-phase high-performance liquid chromatography for the determination of pharmaceuticals and impurities. *J. Chromatogr.* A, 2005, 1083(1-2), 58-62.
[59] Yanagida, A., Murao, H., Ohnishi-Kameyama, M., Yamakawa, Y., Shoji, A., Tagashira, M., Kanda, T., Shindo, H., Shibusawa, Y. Retention behavior of oligomeric proanthocyanidins in hydrophilic interaction chromatography. *J. Chromatogr.* A, 2007, 1143(1-2), 153- 161.

[60] Orentiene, A., Olšauskaite, V., Vičkačkaite, V., Padarauskas, A. Retention behaviour of imidazolium ionic liquid cations on 1.7 μm ethylene bridged hybrid silica column using acetonitrile-rich and water-rich mobile phases. *J. Chromatogr.* A, 2011,1218(39) 6884-6891.

[61] Leicunaite, J., Klimenkovs, I., Kviesis, J., Zacs, D., Kreismanis, J. P. Liquid chromatography and characterization of ether-functionalized imidazolium ionic liquids on mixed-mode reversed-phase/cation exchange stationary phase. *Com. Ren. Chim.* 2010, 13, 1335–1340.

[62] Markowska, A. Applying of ion chromatography, ion-pair, and capillary isotachophoresis to separations and analysis of ionic liquids components, *PhD thesis*, 2010.

[63] Stepnowski, P., Markowska, A. How to analyze ionic liquid anions? Investigation into the application of suppressed and non-suppressed ion chromatography. *Austr. J. Chem.*, 2008, 61, 409.

[64] Molíková, M., Studzińska, S., Kosobucki, P., Jandera, P., Buszewski, B. Determination of imidazolium and pyridinium ionic liquid cations by ion chromatography. *J. Liquid Chromatogr. and Related Technol.* 2009, 33(2), 225-238.

[65] Gao, W., Yu, H. Determination of Homologue Imidazolium Ionic Liquid Cations by Ion Chromatography Using a Carboxyl Acid Cation-Exchange Column with Direct Conductivity Detection. *Anal. Lett.,* 2011, 44(5), 922-931.

[66] Stolte, S., Steudte, S., Markowska, A., Arning, J., Neumann, J., Stepnowski, P. Ion chromatographic determination of structurally varied ionic liquid cations and anions—a reliable analytical methodology applicable to technical and natural matrices. *Anal. Methods*, 2011, 3(4), 919-926.

[67] Qin, W., Wei, H., Li, S. F. Y. Separation of ionic liquid cations and related imidazole derivatives by α-cyclodextrin modified capillary zone electrophoresis. *Analyst*, 2002, 127(4), 490–493.

[68] Markuszewski, M. J., Stepnowski, P., Marszałł, M. P. Capillary electrophoretic separation of cationic constituents of imidazolium ionic liquids. *Electrophoresis*, 2004, 25(20), 3450–3454.

[69] Kosobucki, P., Buszewski, B. Isotachophoretic separation of selected imidazolium ionic liquids. *Talanta,* 2008, 74(5), 1670-1674.

[70] Markowska, A., Stepnowski, P. *Capillary isotachophoresis for the analysis of ionic liquid entities*, unpublished data.

In: Ionic Liquids: Synthesis, Characterization and Applications ISBN: 978-1-63117-716-3
Editor: Anson Brooks

Chapter 11

PHASE BEHAVIOR MODELING IN AROMATICS EXTRACTION FROM GASOLINE BY IONIC LIQUIDS

***Mohammad Ilbeigi*[1]*, Farnoosh Naziri*[2]*, Alireza Fazlali*[2]*, Seyed Foad Aghamiri*[1]*, Mohammad R. Talaee*[1] *and Amir H. Mohammadi*[3,4*]**

[1]Chemical Engineering Department, Faculty of Engineering, Isfahan University, Isfahan, Iran

[2]Department of Chemical Engineering, Arak University, Arak, Iran

[3]Institut de Recherche en Génie Chimique et Pétrolier (IRGCP), Paris Cedex, France

[4]Thermodynamics Research Unit, School of Chemical Engineering, University of KwaZulu-Natal, Howard College Campus, Durban, South Africa

ABSTRACT

In this communication, separation of aliphatic and aromatic hydrocarbons from artificial gasoline by ionic liquids, [EMpy][ESO_4] or [EMim][ESO_4], has been investigated theoretically using the Wilson, NRTL, UNIQUAC, UNIFAC and ASOG activity models. For this purpose, phase equilibria of ternary systems containing pure hexane, heptane, octane, cyclohexane, cyclooctane, or methylcyclohexane as artificial gasoline and pure toluene, benzene, xylenes or ethyl benzene [BTEX] as aromatic compounds has been predicted. In each system, binary interaction parameters between components or subgroups have been obtained and reported. Model predictions were compared with the experimental phase equilibrium data. The results show that NRTL activity model is more capable in prediction of the phase equilibrium data with the root mean square deviation (rmsd) of 0.26 % for the studied systems

Keywords: Liquid-liquid extraction, Gasoline, Aromatic compounds, Ionic liquid, Thermodynamic modeling, Activity coefficient.

* Corresponding Authors: M. Illbeigi: E-mail: ms.2006.illbeigi@gmail.com & A.H. Mohammadi, E-mail: a.h.m@irgcp.fr.

List of Symbols

a_{ij} interaction parameter in UNIQUAC model
a_{mn} group energy interaction parameter in UNIFAC model
A group interaction parameter in ASOG model
G NRTL parameter
l bulk factor
M number of tie-lines
NC number of component
NG number of groups in mixture
q pure component molecular-structure constants
Q_k surface area parameters
r pure component molecular-structure constants
R universal gas constant
R_k volume parameter
u interaction energy
v molar volume ($m^3 mole^{-1}$)
x mole fraction
X group mole fraction
z coordination number

Greek Letters

α non-randomness parameter in NRTL model
γ activity coefficient
Γ group activity coefficient
θ area fraction
Λ adjustable parameter in Wilson model
μ chemical potential
ν number of groups
τ energy interaction parameter in NRTL model
φ segment (volume) fraction in UNIQUAC model
Φ volume fraction in UNIFAC model

Superscripts/Subscripts

C combinatorial
FH Flory–Huggins (combinatorial part)
G groups (residual part)
R residual
calc calculated value
exp experimental value

INTRODUCTION

Separation of aromatic compounds from some petroleum streams, especially gasoline, is an important issue since these compounds are considered as the most toxic substances which cause deadly disease and permanent damages to organism and the environment. Several industrial processes are using these aromatic compounds as raw materials for producing a number of important petrochemicals. Aromatic compounds are normally mixed with aliphatic hydrocarbons such as hexane, heptane, octane, etc.

Aromatic hydrocarbons, due to their close boiling points, and their azeotrope points cannot be separated in an efficient and economic way by distillation (González et al., 2010). Liquid-liquid extraction is a conventional process for separation of these aromatic and aliphatic hydrocarbon mixtures (Meindersma et al., 2006). There are several solvents that can be used for extraction of aromatics compounds such as N-ethylimidazo and N-formylmorpholine, propylene carbonate or combinations of these solvents (González et al., 2009). In the liquid-liquid extraction after formation of immiscible phases in order to recover the solvent, additional distillation is required The high energy consumption by the solvent purification processes (González et al., 2009) and other major problems of the conventional solvents (e.g., toxicity, flammability and volatility) have encouraged the researchers for searching for other possible solvents (Kamankesh et al., 2012).

One solution to overcome these problems is using the ionic liquid (IL) (Seoane et al., 2012). ILs are organic salts which exist in the liquid state below 100 ^{0}C or even at the room temperature (Seoane et al., 2012). The latter are called room temperature ILs.

Among of the numerous applications of ILs, one of their significant applications is in the separation process especially in aromatics extraction (González et al., 2011). Ionic liquids have very low vapor pressure at the temperature and pressure conditions of a typical extraction process.

Hence, their concentration in the vapor phase is very low during the environmental concerns (González et al., 2011). Other advantages of ILs are: no flammability, easy to recover, high thermal stability (González et al., 2011). The liquid extraction process using ILs is the same as other solvents.

The only difference is that one of the two immiscible phases is containing a slight amount of the solvent (González et al., 2011). This can be contributed to the low solubility of aliphatic hydrocarbons in IL (González et al., 2011). Consequently, the recovery of the ionic solvents would be economically improved comparing to the conventional organic solvents (González et al., 2011).

In order to utilize ionic liquids in the separation process, liquid-liquid equilibrium (LLE) data are essential for designing of extraction processes and for calculation of the thermodynamic restrictions and parameters of extraction systems (Meindersma et al., 2006). Mirkhani et al., (Mirkhani et al., 2011) measured LLE data for the ternary systems containing the ionic liquid [EMpy][ESO_4], heptane and benzene or toluene at T = 298.15 K and atmospheric pressure.

Domanska et al., (Domanska et al., 2007) studied the separation of aromatic hydrocarbons from alkanes using ammonium ionic liquid. Wang et al., (Wang et al., 2008) analyzed the (liquid + liquid) equilibria for the (benzene + cyclohexane + 1-methyl-3-

methylimidazolium dimethylphosphate or 1- ethyl-3-methylimidazolium diethylphosphate) system.

In this work, thermodynamic modeling and parameters estimation of various activity models for aromatics extraction from gasoline have been studied. For this purpose, the Wilson, NRTL, UNIQUAC, UNIFAC and SAOG models are used and binary interaction and model parameters are calculated for each model by minimization of the objective function.

2. Thermodynamic Modeling

For thermodynamic equilibrium in multi-phases systems, chemical potential of the components, μ_i, in existing phases should be equal. Chemical potential is defined as follows:

$$\mu_i = \mu_i^{o} + RT\, ln\left(x_i \gamma_i\right) \tag{1}$$

where, μ_i^0 is defined as the chemical potential of the component i in the reference pressure and temperature, x_i is the mole fraction of the component i in the mixture, and γ stands for the activity coefficient.

Equality of chemical potential of the species i in two existing liquid phases (thermodynamic equilibrium) leads to the following equation:.

$$(x_i \gamma_i)^I = (x_i \gamma_i)^{II} \tag{2}$$

In this work, the Wilson, NRTL, UNIQUAC and SAOG models have been used for calculation of the activity as follows:

2.1. Wilson Model

Wilson activity model (Wilson, 1964) contains only two adjustable parameters and it is mathematically simple which can be expressed as follows:

$$\ln \gamma_i = -\ln\left(\sum_j^n x_j \Lambda ij\right) + 1 - \sum_k^n \frac{x_k \Lambda_{ki}}{\sum_j^n x_j \Lambda_{kj}} \tag{3}$$

In which that Λ_{ij} is adjustable parameter and related to pure component molar volume, v_i, and characteristic energy difference, λ_{ij}- λ_{ij} as follows:

$$\Lambda_{ij} = \frac{v_j}{v_i} \exp\left(-\frac{\lambda_{ij} - \lambda_{ii}}{RT}\right) \tag{4}$$

2.2. NRTL Model

In the NRTL (Renon and Prausnitz, 1969) activity model, the activity coefficient of the component i is calculated as follows:

$$\ln\gamma_i = \frac{\sum_j^n \tau_{ji} G_{ji} x_j}{\sum_k^n G_{ki} x_k} + \sum_j^n \frac{x_j G_{ij}}{\sum_k^n G_{kj} x_k}\left(\tau_{ij} - \frac{\sum_k^n x_k \tau_{kj} G_{kj}}{\sum_k^n G_{kj} x_k}\right) \tag{5}$$

where

$$\tau_{ij} = \frac{g_{ij} - g_{jj}}{RT} = \frac{\Delta g_{ij}}{RT} \tag{6a}$$

$$G_{ij} = \exp(-\alpha_{ij}\tau_{ij}) \tag{6b}$$

$$\alpha_{ij} = \alpha_{ji} = \alpha \tag{6c}$$

α_{ij} is related to the non randomness in the mixture; τ_{ij} and G_{ij} are the NRTL parameters which are calculated using adjustable parameters α and Δg_{ij}.

Although the NRTL equation was originally developed for the non-electrolytes systems, and it has been widely used in modeling both liquid-liquid and vapor-liquid equilibria of various systems, but it has been found that it can successfully predict LLE data of ternary systems containing ionic liquids.

2.3. UNIQUAC Model

UNIQUAC activity model (Abrams and Prausnitz, 1975) is mathematically more complex than the NRTL with the advantages: (1) (2) UNIQUAC's parameters often have a smaller dependence on temperature, and (3) UNIQUAC activity model is applicable to solutions containing small or large molecules, including polymers (Poling et al., 1977). UNIQUAC activity model is comprised of two combinatorial and residual parts for activity coefficients, which are calculated as follows:

$$\ln\gamma_i = \ln\gamma_i^C + \ln\gamma_i^R = \ln\frac{\varphi_i}{x_i} + \frac{Z}{2}q_i \ln\frac{\theta_i}{\varphi_i} + l_i - \frac{\varphi_i}{x_i}\sum_j^n x_j l_j - q_i \ln\left(\sum_j^n \theta_j \tau_{ji}\right) + q_i - q_i \sum_j^n \frac{\theta_j \tau_{ij}}{\sum_k^n \theta_k \tau_{kj}} \tag{7}$$

where,

$$\varphi_i = \frac{r_i x_i}{\sum_i^n r_i x_i} \tag{8a}$$

$$\theta_i = \frac{q_i x_i}{\sum_i^n q_i x_i} \tag{8b}$$

$$\tau_{ij} = \exp(\frac{-\Delta u_{ij}}{RT}) \quad or \quad \tau_{ij} = \exp(\frac{-a_{ij}}{T}) \tag{8c}$$

$$l_i = (\frac{z}{2})(r_i - q_i) - (r_i - 1) \tag{8d}$$

In this model, r_i and q_i are pure component molecular-structure constants and φ_i and θ_i are the segment (volume) fractions and area fraction, respectively, and z is coordination number equals to 10. In Table 1, the structural parameters have been depicted. In equation (8c), a_{ij} is interaction parameter between component i and j.

In three aforementioned models (Wilson, NRTL and UNIQUAC) the interaction parameters are only considered between the molecules in solution nevertheless in UNIFAC and ASOG models, the interactions are considered between all the subgroups of the molecules resulting in the additional number of adjustable parameters in these models. These models are also known as Group-contribution models. The activity coefficient in the Group-contribution models is divided into two parts, namely a combinatorial (molecular size contribution) and a residual (intermolecular forces). In this study, two group contribution based models of UNIFAC and ASOG have been used.

2.4. UNIFAC Model

The UNIFAC model (Fredenslund et al., 1975) is an extension of the quasi-chemical theory of liquid mixtures (UNIQUAC).

$$\ln\gamma_i = \ln\gamma_i^C + \ln\gamma_i^R \tag{9}$$

The combinatorial part of activity coefficient is expressed by the following expression:

$$\ln\gamma_i^C = \ln\frac{\Phi_i}{x_i} + \frac{z}{2} q_i \ln\frac{\theta_i}{\Phi_i} + l_i - \frac{\Phi_i}{x_i}\sum_{j=1}^{NC} x_j l_j \tag{10}$$

where θ_i, Φ_i and l_i are the area fraction, volume fraction and bulk factor, respectively and defined as follows:

$$\theta_i = \frac{q_i x_i}{\sum_{j=1}^{NC} q_j x_j} \tag{11a}$$

$$\Phi_i = \frac{r_i x_i}{\sum_{j=1}^{NC} r_j x_j} \tag{11b}$$

$$l_i = \frac{z}{2}(r_i - q_i) - (r_i - 1) \tag{11c}$$

The lattice coordination number z is set equal to 10. The structural group volume parameter (R_k) and surface area parameters (Q_k) are used for calculation of r_i and q_i as follows:

$$r_i = \sum_{k=1}^{NG} v_{k,i} R_k \tag{12a}$$

$$q_i = \sum_{k=1}^{NG} v_{k,i} Q_k \tag{12b}$$

where $v_{k,i}$ is the number of groups of type k in molecule i.

Table 1. Volume and surface area parameters for component or subgroup

Components or Subgroups	R_i or r_i	Q_i or q_i	Ref
CH	0.4469	0.2280	(Poling et al., 1977)
CH_2	0.6744	0.5400	(Poling et al., 1977)
CH_3	0.9011	0.8480	(Poling et al., 1977)
ACH	0.5313	0.4000	(Poling et al., 1977)
$ACCH_3$	1.2663	0.9680	(Poling et al., 1977)
$ACCH_2$	1.0396	0.6600	(Poling et al., 1977)
C_5H_5N	2.9993	2.1130	(Poling et al., 1977)
n-Octane	5.8486	4.9360	(Poling et al., 1977)
Xylene	4.6578	3.5360	(Poling et al., 1977)
n-Heptane	5.1742	4.3960	(Poling et al., 1977)
Toluene	3.9228	2.9680	(Poling et al., 1977)
Benzene	3.1878	2.4000	(Poling et al., 1977)
Hexane	4.4998	3.8560	(Poling et al., 1977)
Cyclooctane	5.3952	4.3200	(Poling et al., 1977)
Ethylbenzene	4.5972	3.5080	(Poling et al., 1977)
Cyclohexane	4.0464	3.2400	(Poling et al., 1977)
Methylcyclohexane	4.7200	3.7760	(Poling et al., 1977)
$[EMpy]^+$	5.0359	3.7330	This work
$[EMim]^+$	4.2753	4.2480	This work
$[ESO4]^-$	3.6647	2.9620	(Vatani et al., 2012)
[EMpy] [ESO4]	8.7006	6.6950	This work
[EMim] [ESO4]	7.9400	7.2100	(Alonso et al., 2008)

Table 2. Binary interaction parameters for the Wilson, UNIQUAC and NRTL models

Triple system T (K)	Activity model	Δu_{ij}, Δg_{ij} and Λ_{ij} for the UNIQUAC, NRTL and Wilson equations						α for NRTL model
		1-2	2-1	1-3	3-1	2-3	3-2	
$[EMpy][ESO_4]$ (1) + octane (2) + p-xylene(3) 313.15 K	Wilson	48.6	8924.5	-5892.1	8928.4	-98.2	1978.4	0.28
	UNIQUAC	458.5	9247.1	-78.3	989.5	-0.5	1.4	
	NRTL	13899.1	7859.4	847.4	-98.3	-278.4	39.6	
$[EMpy][ESO_4]$ (1) + heptane (2) + p-xylene (3) 313.15 K	Wilson	189.5	938.3	7859.6	9.4	19.3	0.9	0.25
	UNIQUAC	858.9	78.4	9237.2	171.2	0.5	857.1	
	NRTL	9876.3	8593.2	89.1	-678.5	348.1	11.6	
$[EMpy][ESO_4]$ (1) + heptane (2) + toluene (3) 298.15 K	Wilson	123.6	1876.3	-8769.3	0.2	11.5	-568.4	0.26
	UNIQUAC	443.4	4125.6	-1689.2	42.1	-1398.3	-1765.8	
	NRTL	11058.3	467.3	-3548.9	45.6	2349.1	0.4	
$[EMpy][ESO_4]$ (1) + heptane (2) + benzene (3) 298.15 K	Wilson	-89.3	3492.3	23.5	-568.2	0.9	9478.2	0.30
	UNIQUAC	2630.2	3653.2	2234.5	510.5	550.3	14.6	
	NRTL	1009.4	-8589.5	98.4	1.4	-34.9	2989.2	
$[EMpy][ESO_4]$ (1) + hexane (2) + benzene (3) 283.15 K	Wilson	784.2	84.1	9647.2	6.5	0.4	-748.5	0.19
	UNIQUAC	674.4	8627.2	2332.9	-1117.8	737.8	27.8	
	NRTL	-3716.5	51002.1	9276.4	28561.7	1845.3	-562.4	
$[EMpy][ESO_4]$ (1) + cyclooctane (2) + benzene (3) 298.15 K	Wilson	178	73.4	987.4	8844.3	11754.5	947.7	0.2
	UNIQUAC	7754.4	-847.4	0.6	8617.1	-1137.5	7.9	
	NRTL	7849.4	17978.5	-4387.5	15885.4	-3460.4	5832.1	
$[EMpy][ESO_4]$ (1) + cyclooctane (2) + toluene (3) 298.15 K	Wilson	456.4	9487.5	68.4	-11.6	3478.5	0.8	0.1
	UNIQUAC	656.6	5792.3	-8573.6	1.8	34.9	23574.6	
	NRTL	235.9	23876.4	-7935.5	52654.6	-7844.5	6400.3	
$[EMpy][ESO_4]$ (1) + cyclooctane (2) + ethylbenzene (3) 298.15 K	Wilson	74.6	837.6	1975.6	-5748.6	65.9	1.9	0.09
	UNIQUAC	785.9	8373.6	17.6	34773.6	-768.4	11746.6	
	NRTL	11324.6	-2568.5	-6986.4	33768.6	-7854.3	12354.5	
$[EMim][ESO_4]$ (1) + o-xylene (2) + hexane (3) 298.15 K	Wilson	73.6	9874.3	19485.4	657.4	385.9	0.1	0.1
	UNIQUAC	-380.5	2850.4	530.1	171678.4	1190.3	-578.2	
	NRTL	-3478.5	17579.3	58847.5	53478.6	5895.3	-2768.5	

Triple system T (K)	Activity model	Δu_{ij}, Δg_{ij} and Λ_{ij} for the UNIQUAC, NRTL and Wilson equations						α for NRTL model
		1-2	2-1	1-3	3-1	2-3	3-2	
$[EMim][ESO_4]$ (1) + p-xylene (2) + hexane (3) 298.15 K	Wilson	91.4	7865.3	-67.3	477.3	17845.3	6305.	0.15
	UNIQUAC	-1009.4	6705.2	580.4	79748.3	-1568.3	2438.5	
	NRTL	-2789.8	40857.4	2289.5	33675.5	-1700.5	2945.4	
$[EMim][ESO_4]$ (1) + m-xylene (2) + hexane (3) 298.15 K	Wilson	2585.3	-7689.2	39.5	6584.2	0.4	8478.9	0.1
	UNIQUAC	670.2	4200.5	590.3	154839.3	-1499.8	2467.3	
	NRTL	-1955.3	46857.4	2567.5	30757.4	-5387.5	7858.4	
$[EMim][ESO_4]$ (1) + cyclohexane (2) + ethylbenzene (3) 298.15 K	Wilson	101.3	-423.1	2159.7	6849.9	-3005.2	28190	0.3
	UNIQUAC	1011.8	2770.7	-650.9	3450.2	-131.3	105.2	
	NRTL	8150.3	18753.2	2430.5	17703.1	-363.4	1025.8	
$[EMim][ESO_4]$ (1) + cyclooctane (2) + ethylbenzene (3) 298.15 K	Wilson	19.7	-543.4	3450.9	4170.3	-1090.5	1804.1	0.2
	UNIQUAC	1790.2	2050.8	-553.4	2950.3	-1125.1	1223.3	
	NRTL	10705.6	20115.5	4102.3	18450.9	-663.9	1650.3	
$[EMim][ESO_4]$ (1) + methylcyclohexane (2) + ethylbenzene (3) 298.15 K	Wilson	0.9	1150.1	443.2	-112.3	1704.2	32.9	0.25
	UNIQUAC	453.2	129.9	-1899.3	34.7	-790.1	6649.2	
	NRTL	753.9	-1897.1	-0.4	18.3	1905.9	11812.7	

Table 3. rmsd for liquid-liquid equilibrium calculation using the UNIQUAC, NRTL and Wilson models

Sys. no.	Triple extraction system	No. of tie-line	rmsd (%)			T (K)	ref.
			NRTL	UNIQUAC	Wilson		
1	[EMpy][ESO_4] (1) + octane (2) + p-xylene(3)	9	0.29	0.33	4.32	313.15	[1]
2	[EMpy][ESO_4] (1) + heptane (2) + p-xylene (3)	9	0.27	0.51	3.98	313.15	[1]
3	[EMpy][ESO_4] (1) + heptane (2) + toluene (3)	7	0.25	0.54	3.32	298.15	[2]
4	[EMpy][ESO_4] (1) + heptane (2) + benzene (3)	8	0.31	0.48	4.91	298.15	[2]
5	[EMpy][ESO_4] (1) + hexane (2) + benzene (3)	10	0.25	0.38	4.11	283.15	[3]
6	[EMpy][ESO_4] (1) + cyclooctane (2) + benzene (3)	10	0.25	0.26	4.43	298.15	[4]
7	[EMpy][ESO_4] (1) + cyclooctane (2) + toluene (3)	10	0.21	0.48	5.54	298.15	[4]
8	[EMpy][ESO_4] (1) + cyclooctane (2) + ethylbenzene (3)	7	0.32	0.75	4.19	298.15	[4]
9	[EMim][ESO_4] (1) + o-xylene (2) + hexane (3)	10	0.28	0.26	5.78	298.15	[5]
10	[EMim][ESO_4] (1) + p-xylene (2) + hexane (3)	8	0.34	0.57	3.93	298.15	[5]
11	[EMim][ESO_4] (1) + m-xylene (2) + hexane (3)	10	0.26	0.49	4.99	298.15	[5]
12	[EMim][ESO_4] (1) + cyclohexane (2) + ethylbenzene (3)	10	0.30	0.35	5.65	298.15	[5]
13	[EMim][ESO_4] (1) + cyclooctane (2) + ethylbenzene (3)	8	0.22	0.34	4.58	298.15	[5]
14	[EMim][ESO_4] (1) + methylcyclohexane (2) + ethylbenzene (3)	9	0.17	0.32	4.45	298.15	[5]
	Total		0.26	0.43	4.60		

[1], (Kamankesh et al., 2012); [2], (Mirkhani et al. 2012); [3], (González et al., 2010); [4] (González et al., 2011).

Table 4. UNIFAC group interaction parameters

Sys no.		CH_n	ACH	$ACCH_m$	Empy	ESO_4
1, 2, 3, 4, 5, 6, 7, 8	CH_n	0	61.13*	76.5*	12.49[a]	33.28[a]
	ACH	-11.12*	0	167*	56.78[a]	-351.43[a]
	$ACCH_m$	-69.7*	-146.8*	0	-588.45[a]	-265.89[a]
	Empy	324.78[a]	-497.23[a]	-98.27[a]	0	-409.67[a]
	ESO_4	193.11[a]	91.35[a]	23.76[a]	351.33[a]	0
9, 10, 11, 12, 13, 14		CH_n	ACH	$ACCH_m$	Emim	ESO_4
	CH_n	0	61.13*	76.5*	220.34[a]	33.28[a]
	ACH	-11.12*	0	167*	332.25[a]	-351.43[a]
	$ACCH_m$	-69.7*	-146.8*	0	193.45[a]	-265.89[a]
	EMim	265.18[a]	360.97[a]	469.23[a]	0	-151.27[a]
	ESO_4	193.11[a]	91.35[a]	23.76[a]	-219.23[a]	0

* Reference (Poling et al., 1977), [a] estimated in this work, n= 0, 1, 2, 3, m= 1, 2, 3.

The residual part of the activity coefficient is given by the following equations:

$$\ln \gamma_i^R = \sum_{k=1}^{NG} v_{k,i} (\ln \Gamma_k - \ln \Gamma_k^{(i)}) \tag{13a}$$

$$\ln \Gamma_k = Q_k (1 - \ln(\sum_{m=1}^{NG} \theta_m \psi_{mk}) - \sum_{m=1}^{NG} \frac{\theta_m \psi_{km}}{\sum_{n=1}^{NG} \theta_n \psi_{nm}}) \tag{13b}$$

where θ_m in Eq. (13b) is the area fraction of group *m*.

$$\theta_m = \frac{Q_m X_m}{\sum_{n=1}^{NG} Q_n X_n} \tag{14}$$

and X_k is the mole fraction of group *k* in the mixture.

$$X_k = \frac{\sum_{j=1}^{NC} x_j v_{k,j}}{\sum_{j=1}^{NC} (x_j \sum_{1}^{NG} v_{k,j})} \tag{15}$$

Interaction parameters ψ_{mn} are evaluated by the following equation

$$\psi_{mn} = \exp(\frac{-a_{mn}}{T}) \tag{16}$$

In which, a_{mn} are the group energy interaction parameter between main group *m* and *n* that are given in Table 4.

2.5. ASOG Model

ASOG (Wilson and Deal, 1962) and UNIFAC models are similar in principle with some differences in details.

In ASOG, the Flory–Huggins, FH, (Flory, 1942; Huggins, 1942) equation is used for the combinatorial part of the activity coefficient and the Wilson equation (Wilson, 1964) is used for determination of the group residual activity coefficients.

The activity coefficient of component *i*, γ_i, can be calculated by the following equations, where the superscripts FH and G stand for "Flory–Huggins" (combinatorial part) and "groups" (residual part), respectively:

$$\ln\gamma_i = \ln\gamma_i^{FH} + \ln\gamma_i^G \tag{17}$$

The combinatorial and residual terms are given by eq. 18 and 19, respectively

$$\ln\gamma_i^{FH} = 1 + \ln\left[\frac{v_i^{FH}}{\sum_{j=1}^{NC} x_j v_j^{FH}}\right] - \frac{v_i^{FH}}{\sum_{j=1}^{NC} x_j v_j^{FH}} \quad (18)$$

$$\ln\gamma_i^{G} = \sum_{k=1}^{NG} v_{k,i}(\ln\Gamma_k) - \sum_{k=1}^{NG} v_{k,i}(\ln\Gamma_k^{(i)}) \quad (19)$$

where $v_{k,i}$ and v_i^{FH} are the total number of atoms (except hydrogen atoms) in group k of molecule i and number of atoms (except hydrogen atoms) in molecule i, respectively. Γ_k stands for activity coefficient of group k in the mixture of groups and $\Gamma_k^{(i)}$ is activity coefficient of group k in the standard state The standard state depends on molecule i (standard state containing only molecules of type i):

$$\ln\Gamma_k = -\ln\left(\sum_{i=1}^{NG} X_i A_{ki}\right) + \left(1 - \sum_{i=1}^{NG} \frac{X_i A_{ik}}{\sum_{m=1}^{NG} X_i A_{im}}\right) \quad (20)$$

where, the binary interaction between main groups i and k, A_{ki} and A_{ik} ($A_{ki} \neq A_{ik}$), can be obtained by the following temperature dependent equation:

$$A_{ki} = \exp\left(m_{ki} + \frac{n_{ki}}{T}\right) \quad (21)$$

where, X_i can be evaluated by equation 15.

3. Parameters Estimation

The interaction parameters for the components (for Wilson, NRTL and UNIQUAC methods) or subgroups (for UNIFAC and ASOG methods), have been evaluated using the minimization of the following objective function (O.F):

$$O.F = \sum_{k=1}^{n}\sum_{j=1}^{2}\sum_{i=1}^{3}\left(x_{ijk}^{exp} - x_{ijk}^{cal}\right)^2 \quad (22)$$

in which that k, i and j are the number of tie lines, components and liquid phase in equilibrium respectively, while superscripts '*exp*' and '*calc*' refer to the experimental and calculated values of the liquid phase concentration.

There are numerous algorithms or methods for minimizing of the objective function. In this work, the objective function was minimized using a code with nested loops using Maple program. The method is similar to the Genetic algorithm (GA) (Vatani et al., 2012) but with some differences in details.

Deviations between results of the models and experimental data have been determined using the root mean square deviation (rmsd) as follows:

$$rmsd = 100\sqrt{\frac{\sum_{i}^{M}\sum_{j}^{N}(x_{ij}^{\mathrm{I,exp}} - x_{ij}^{\mathrm{I},calc})^2 + (x_{ij}^{\mathrm{II,exp}} - x_{ij}^{\mathrm{II},calc})^2}{6M}} \tag{23}$$

where M and N are number of tie-lines and components of the mixture, respectively.

4. Results and Discussion

In this work, 15 systems containing pure hexane, heptane, octane, cyclohexane, cyclooctane, and methylcyclohexane as artificial gasoline and pure toluene, benzene, xylene and ethyl benzene (BTEX) which are common aromatic compounds in gasoline produced in refineries, have been studied. The aromatic compounds in mixtures of these aliphatic and aromatics were extracted by pure [EMpy][ESO_4] (1-ethyl-3-methylpyridinium ethylsulfate) and [EMim][ESO_4] (1-ethyl-3-methylimidazolium ethylsulfate) ionic liquids. Table 1 shows the structural group volume parameter (R_k) and surface area parameters (Q_k) which has been used in the models, as mentioned earlier. As can be observed, new parameters have been reported in this table in addition to those reported already in the literature.

Table 2 reports the Δu_{ij}, Δg_{ij} and Λ_{ij} for the UNIQUAC, NRTL and Wilson equations, respectively, calculated by minimizing of O.F. in equation 22. The first column of Table 2 is the triple extraction systems (components).

Table 3 shows the result of the thermodynamic model based on molecular approach. For each system, rmsd has been calculated for three models mentioned in Table 2. As can be seen, using the NRTL model yields better results for the investigated systems in comparison with the UNIQUAC and Wilson models. The UNIQUAC model yields better results comparing with the Wilson model.

Three models mentioned earlier consider the system in molecular state but in some models such as the UNIFAC and ASOG, the system is considered as set of subgroups, as pointed out earlier. Therefore, interactions between the subgroups must be calculated instead of molecules. Tables 4 and 5 report the subgroups interactions for the UNIFAC and ASOG models, respectively. As can be seen, new parameters are reported in these tables in addition to those reported in the literature.

Rmsd's for extraction systems that mentioned in Table 3 have been depicted in Table 6. It is clear that the UNIFAC model produces better results for the investigated systems in comparison with the ASOG model. In overall, the accuracy of the models for the studied systems can be summarized as: NRTL > UNIQUAC > UNIFAC > ASOG > Wilson. As an example, Figure 1 *A-E* summarizes the experimental data (Mirkhan et al., 2011) and results of five mentioned activity models for the system of toluene-heptane-[EMpy][ESO_4].

Table 5. ASOG group interaction parameters

Sys no.		CH_2		ArCH		CyCH		Empy		ESO_4	
		m	n	m	n	m	n	M	n	m	N
1,2,3,4,5,6,7,8	CH_2	0	0	-0.7457[*]	146[*]	0.1530[*]	2.1[*]	2.45[a]	23.98[a]	0.38[a]	399.1[a]
	ArCH	0.7297[*]	-176.8[*]	0	0	-0.3288[*]	156.3[*]	-0.59[a]	-347.99[a]	9.5[a]	487.3[a]
	CyCH	-0.1842[*]	0.3[*]	0.5301[*]	-251[*]	0	0	7.45[a]	467.23[a]	3.87[a]	-328.4[a]
	Empy	0.78[a]	423.23[a]	-0.79[a]	155.45[a]	5.55[a]	-4.98[a]	0	0	5.89[a]	1.45[a]
	ESO_4	-3.9[a]	349.5[a]	4.05[a]	178.43[a]	0.49[a]	12.49[a]	-2.8[a]	347.5[a]	0	0
9,10,11, 12,13,14		CH_2		ArCH		CyCH		EMim		ESO_4	
		m	n	m	n	m	n	M	n	m	N
	CH_2	0	0	-0.7457[*]	146[*]	0.1530[*]	2.1[*]	1.17[a]	-32.76[a]	0.38[a]	399.1[a]
	ArCH	0.7297[*]	-176.8[*]	0	0	-0.3288[*]	156.3[*]	-8.11[a]	231.89[a]	9.5[a]	487.3[a]
	CyCH	-0.1842[*]	0.3[*]	0.5301[*]	-251[*]	0	0	-9.3[a]	177.11[a]	3.87[a]	-328.4[a]
	EMim	2.4	125.7	3.6700	56.98	-4.76	491.19	0	0	3.7[a]	-171.4[a]
	ESO_4	-3.9[a]	349.5[a]	4.0500[a]	178.43[a]	0.49[a]	12.49[a]	2.51[a]	-86.45[a]	0	0

[*] Reference (Tochigi et al., 1990), [a] estimated in this work;

Table 6. rmsd for liquid-liquid equilibrium calculation using the UNIFAC and ASOG models

Sys no.	Triple extraction system	No. of tie-line	rmsd (%)		T (K)	References
			UNIFAC	ASOG		
1	[EMpy][ESO_4] (1) + octane (2) + p-xylene(3)	9	1.08	1.34	313.15	(Kamankesh et al., 2012)
2	[EMpy][ESO_4] (1) + heptane (2) + p-xylene (3)	9	1.05	1.17	313.15	(Kamankesh et al., 2012)
3	[EMpy][ESO_4] (1) + heptane (2) + toluene (3)	7	1.11	1.26	298.15	(Mirkhani et al., 2011)
4	[EMpy][ESO_4] (1) + heptane (2) + benzene (3)	8	1.27	1.45	298.15	(Mirkhani et al., 2011)
5	[EMpy][ESO_4] (1) + hexane (2) + benzene (3)	10	0.89	1.37	283.15	(González et al., 2009)
6	[EMpy][ESO_4] (1) + cyclooctane (2) + benzene (3)	10	0.71	1.54	298.15	(Gonzalez et al., 2010)
7	[EMpy][ESO_4] (1) + cyclooctane (2) + toluene (3)	10	1.26	1.29	298.15	(Gonzalez et al., 2010)
8	[EMpy][ESO_4] (1) + cyclooctane (2) + ethylbenzene (3)	7	0.61	1.14	298.15	(Gonzalez et al., 2010)
9	[EMim][ESO_4] (1) + o-xylene (2) + hexane (3)	10	1.24	1.32	298.15	(Gonzalez et al., 2011)
10	[EMim][ESO_4] (1) + p-xylene (2) + hexane (3)	8	0.87	1.08	298.15	(Gonzalez et al., 2011)
11	[EMim][ESO_4] (1) + m-xylene (2) + hexane (3)	10	1.13	1.28	298.15	(Gonzalez et al., 2011)
12	[EMim][ESO_4] (1) + cyclohexane (2) + ethylbenzene (3)	10	0.85	1.18	298.15	(González et al., 2011)
13	[EMim][ESO_4] (1) + cyclooctane (2) + ethylbenzene (3)	8	0.95	1.11	298.15	(González et al., 2011)
14	[EMim][ESO_4] (1) + methylcyclohexane (2) + ethylbenzene (3)	9	0.83	1.26	298.15	(González et al., 2011)
	total		0.98	1.27		

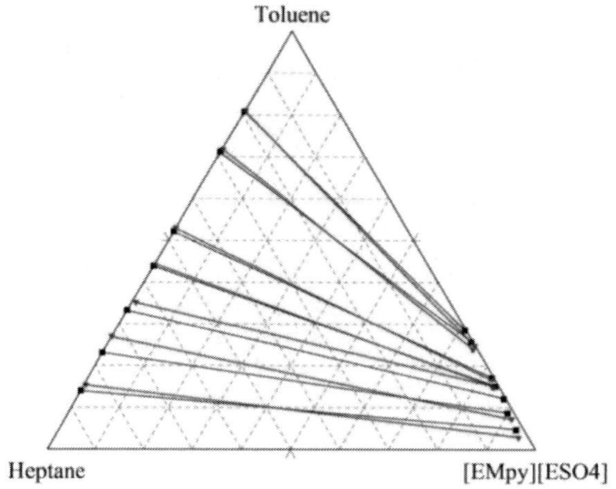

A

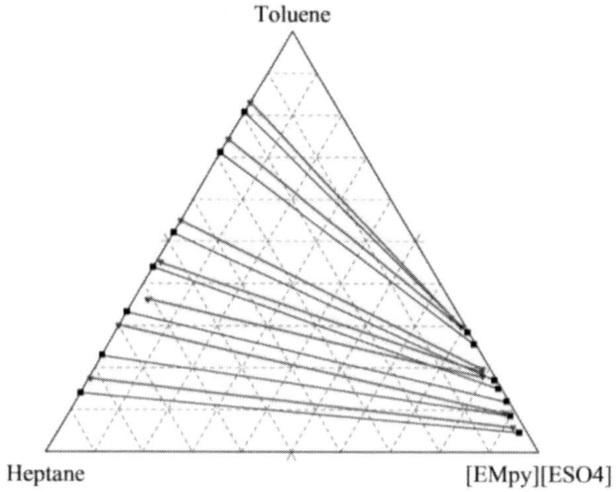

B

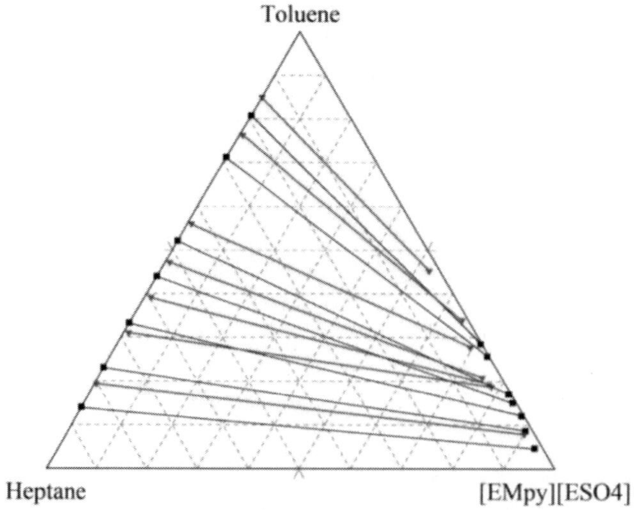

C

Figure 1. Continued on next page.

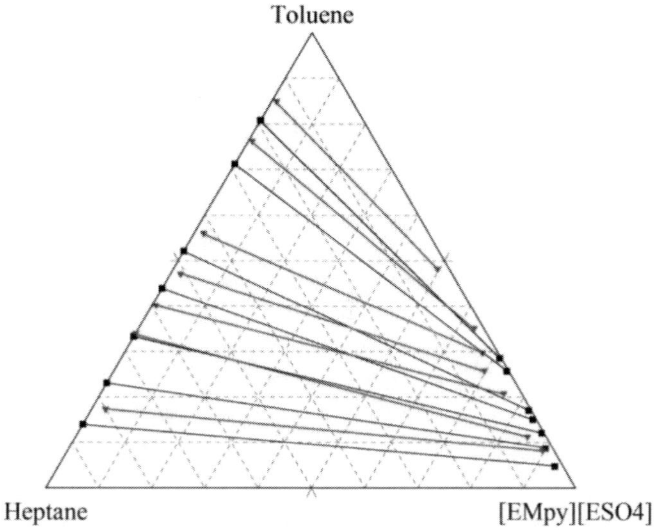

D

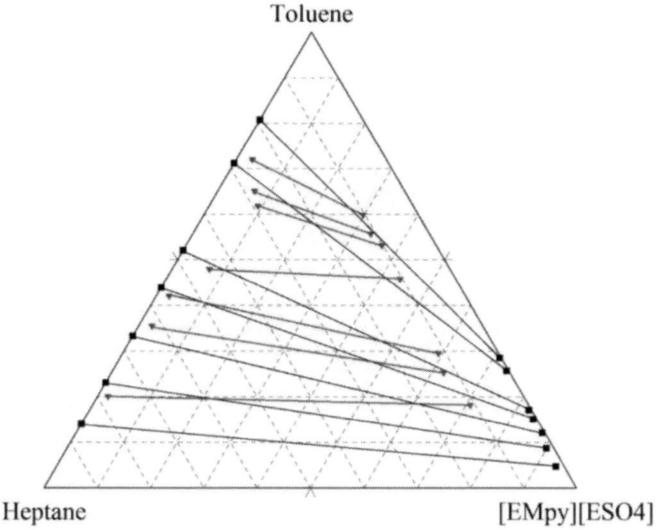

E

Figure 1. Experimental [7] (-■-) and calculated (-▲-) Ternary diagram of LLE system of Heptane-Toluene-[EMpy][ESO_4], (A) NRTL, (B) UNIQUAC, (C) UNIFAC, (D) ASOG and (E) Wilson model at T=298.15 K.

CONCLUSION

In this work, LLE modeling for the ternary systems containing artificial gasoline and common aromatic compounds (BTEX) that exist in gasoline were investigated by various activity models including the NRTL, UNIQUAC and Wilson molecular models and the UNIFAC and ASOG group contribution models. The models performances for the studied systems are summarized as follows: NRTL > UNIQUAC > UNIFAC > ASOG > Wilson.

REFERENCES

Abrams, D. S., and Prausnitz, J. M., *AIChE J.*, 21(1975):116-128.

Alonso, L., Arce, A., Francisco, M., Soto, A., *Fluid Phase Equilib.* 270 (2008) 97–102.

Domanska, U., Pobudkowska, A., Królikowski, M., *Fluid Phase Equilib.* 259 (2007) 173–179.

Flory, P. J., *J. Chem. Phys.* 10 (1942) 51–61.

Fredenslund, A., Jones, R. L., Prausnitz, J. M., AIChE J. 21 (1975) 1086-1099.

González, E. J., Calvar, N., González, B., Domínguez, Á., *J. Chem. Thermodynamics* 41 (2009) 1215–1221.

Gonzalez, E. J., Calvar, N., Gonzalez, B., Dominguez, A., *Fluid Phase Equilib.* 291 (2010) 59–65.

González, E. J., Calvar, N., González, B., Domínguez, Á., *J. Chem. Thermodynamics* 42 (2010) 752–757.

Gonzalez, E. J., Gonzlez, B., Calvar, N., Angeles Dominguez, *Fluid Phase Equilib.* 305 (2011) 227–232.

González, E. J., Calvar, N., Dominguez, I., Domínguez, Á., *J. Chem. Thermodynamics* 43 (2011) 725–730.

Huggins, M.L., *Ann. N. Y. Acad. Sci.* 43 (1942) 1–32.

Kamankesh, A., Vossoughi, M., Shamloo, A., Mirkhani, S.A., Akbari, J., *Fluid Phase Equilib. 3*32 (2012) 48–54.

Meindersma, G. W., Podt, Anita J. G., Haan, A. B. de., *Fluid Phase Equilibria* 247 (2006) 158–168.

Mirkhani, S.A., Vossoughi, M., Pazuki, G.R., Safekordi, A.A., Heydari, A., Akbari, J., Yavari, M., *J. Chem. Thermodynamics* 43 (2011) 1530–1534.

Poling, B.E., Prausnitz, J.M., O'Connell, J.P., The Properties of Gases and Liquids, Fifth ed., McGraw- Hill Companies, Inc., *USA*, 1977.

Renon, H., and Prausnitz, J. M., *Ind. Eng. Chem. Process Design Develop.,* 8 (1969) 413-419.

Seoane, R. G., González, E. J., González, B., *J. Chem. Thermodynamics* 53 (2012) 152–157.

Tochigi, K., Tiegs, D., Gmehling, J., Kojima, K., *J. Chem. Eng. Japan* 23 (1990) 453–463.

Vatani, M., Asghari, M., Vakili-Nezhaad, G., *J. Ind. Eng. Chem.* 18, (2012) 1715-1720.

Wang, R., Wang, J., Meng, H., Li, C., Wang, Z., *J. Chem. Eng. Data* 53 (2008) 1159– 1162.

Wilson, G. M., and Deal, C. H., *Ind. Eng. Chem. Fundam.,* 1 (1962) 20-23.

Wilson, G. M., *J. Am. Chem. Soc.* 86 (1964)127-130.

Index

A

E

F

G

H

I

J

K

L

M

N

O

P

Q

R

S

T

U

V

W

X

Y

Z